# The Political Ecology of Agrofuels

This book explores the political ecology of agrofuels as an encompassing socio-spatial transformation process consisting of a series of changing contexts, political reconfigurations, and the restructuring of social and labour relations. It includes conceptual chapters as well as case studies from different world regions (North America, Europe, Latin America, sub-Saharan Africa, Asia) and levels (local, national, transnational).

*The Political Ecology of Agrofuels* advances a conceptualization of agrofuels that helps to fill existing research gaps. It covers global food regimes and agrarian politics as well as political arenas such as energy, climate, transport and trade. It reflects on the biophysical materiality of agrofuels, new forms of nature appropriation, struggles, discursive framings, the building of hegemony, shifting geopolitical constellations, socio-spatial configurations of power, the construction of territory, the agency of social movements and the different ways in which agrofuels are politicized at different scales.

This book asks how patterns of mobility, emissions regulation, food and energy production and consumption, and social relations (e.g. labour, class and gender relations) are shaped and re-shaped by the materiality and representations of agrofuels in both the Global South and North. The book provides tools for thinking about the diversity of the conflicts, struggles and spatial, socio-ecological and politico-economic reconfigurations and perpetuations engendered by current production and consumption patterns in the agrofuel sector.

**Kristina Dietz** is a post doctoral Research Fellow at the International Research Network on Interdependent Inequalities in Latin America, Berlin, Germany.

**Bettina Engels** is Deputy Assistant Professor for Development Politics and Politics of Africa at the University of Bayreuth, Germany.

**Oliver Pye** teaches Southeast Asian studies at Bonn University, Germany.

**Achim Brunnengräber** is Associate Professor at the Department of Political and Social Sciences, Freie Universität Berlin, Germany.

**Routledge ISS studies in rural livelihoods**



Routledge and the Institute of Social Studies (ISS) in The Hague, the Netherlands have come together to publish a new book series in rural livelihoods. The series will include themes such as land policies and land rights, water issues, food policy and politics, rural poverty, agrarian transformation, migration, rural-oriented social movements, rural conflict and violence, among others. All books in the series will offer rigorous, empirically grounded, cross-national comparative and inter-regional analyses. The books will be theoretically stimulating, but will also be accessible to policy practitioners and civil society activists.

1 **Land, Poverty and Livelihoods in an Era of Globalization**
Perspectives from developing and transition countries
*Edited by A. Haroon Akram-Lodhi, Saturnino M. Borras Jr. and Cristóbal Kay*

2 **Peasants and Globalization**
Political economy, agrarian transformation and development
*Edited by A. Haroon Akram-Lodhi and Cristóbal Kay*

3 **The Political Economy of Rural Livelihoods in Transition Economies**
Land, peasants and rural poverty in transition
*Edited by Max Spoor*

4 **Agrarian Angst and Rural Resistance in Contemporary Southeast Asia**
*Edited by Dominique Caouette and Sarah Turner*

5 **Water, Environmental Security and Sustainable Rural Development**
Conflict and cooperation in Central Eurasia
*Edited by Murat Arsel and Max Spoor*

6 **Reforming Land and Resource Use in South Africa**
Impact on livelihoods
*Edited by Paul Hebinck and Charlie Shackleton*

7 **Risk and Social Change in an African Rural Economy**
Livelihoods in pastoralist communities
*John G. McPeak, Peter D. Little and Cheryl R. Doss*

8 **Public Policy and Agricultural Development**
*Edited by Ha-Joon Chang*

9 **Social Conflict, Economic Development and the Extractive Industry**
Evidence from South America
*Edited by Anthony Bebbington*

10 **The Ecotourism-Extraction Nexus**
Political economies and rural realities of (un)comfortable bedfellows
*Edited by Bram Büscher and Veronica Davidov*

11 **Rural Livelihoods, Regional Economies and Processes of Change**
*Edited by Deborah Sick*

12 **Rural Development and the Construction of New Markets**
*Edited by Paul Hebinck, Jan Douwe van der Ploeg and Sergio Schneider*

13 **The Political Ecology of Agrofuels**
*Edited by Kristina Dietz, Bettina Engels, Oliver Pye and Achim Brunnengräber*

# The Political Ecology of Agrofuels

Edited by Kristina Dietz,
Bettina Engels, Oliver Pye and
Achim Brunnengräber

Routledge
Taylor & Francis Group
LONDON AND NEW YORK

First published 2015
by Routledge
2 Park Square, Milton Park, Abingdon, Oxon OX14 4RN

and by Routledge
711 Third Avenue, New York, NY 10017

*Routledge is an imprint of the Taylor & Francis Group, an informa business*



*British Library Cataloguing in Publication Data*
A catalogue record for this book is available from the British Library

*Library of Congress Cataloging in Publication Data*
The political ecology of agrofuels / edited by Kristina Dietz, Bettina Engels, Oliver Pye and Achim Brunnengräber.
pages cm
Includes bibliographical references and index.
1. Biomass energy–Political aspects. 2. Political ecology. I. Dietz, Kristina, 1972–
HD9502.5.B542P65 2014
333.95'39–dc23 2014021425

ISBN: 978-1-138-01315-5 (hbk)
ISBN: 978-1-315-79540-9 (ebk)

Typeset in Times New Roman
by Wearset Ltd, Boldon, Tyne and Wear

# Contents

# Contributors

**Maria Backhouse** is a sociologist and works as a research assistant at the Institute for Latin American Studies at Freie Universität Berlin. She recently finished her PhD thesis on green grabbing in the Brazilian Amazon in the context of the state-supported expansion of palm oil plantations. Her regional focus is on Brazil and her disciplinary interests are political ecology, critical development and postcolonial studies.

**Achim Brunnengräber** is an Associate Professor at the Department of Political and Social Sciences, Freie Universität Berlin, and coordinator of the research project 'Multi Level Governance-Perspective on Management of Nuclear Waste Disposal' at the Environmental Policy Research Centre (FFU). Key teaching and research areas are: global and multi-level governance, international political economy (IPE), international environmental energy and climate policy.

**Kristina Dietz** is currently a postdoctoral Research Fellow at the International Research Network on Interdependent Inequalities in Latin America, desiguALdades.net, at the Ibero-Amerikanisches Institut, Berlin. Her key teaching and research areas are: political ecology, conflicts over land and resources in Latin America, climate and energy policy and spatial and democracy theory.

**Bettina Engels** is currently Deputy Assistant Professor of Development Politics and Politics of Africa at the University of Bayreuth. Her key teaching and research areas are: conflict over land and resources, spatial and action theory and resistance, urban protest and social movements in Africa.

**Ruth Hall** is Associate Professor at the Institute for Poverty, Land and Agrarian Studies (PLAAS) at the University of the Western Cape, South Africa. She is a co-founder and convenor of the Land Deal Politics Initiative, and coordinator of the Future Agriculture Consortium's research on land in Africa. Her work addresses land rights, land reform, farm workers, agrarian reform, land grabs and corporate expansion in African agriculture and food value chains.

**Carol Hunsberger** is Assistant Professor in the Department of Geography at the University of Western Ontario in London, Canada. Her key teaching and research areas are: political ecology, agrofuels, energy justice and environmental discourses.

**Julia*** (this is her full name) is a Presidium member of the Kalimantan Women Alliance for Peace and Gender Justice (AlPeKaJe) based in West Kalimantan, Indonesia. Her present work focuses on research and advocacy of women and environment issues in West Kalimantan. She is presently involved in research on the political ecology of the Kapuas River in collaboration with Bonn University and Bremen University. She is co-author (with Ben White) of 'Gendered experiences of dispossession: oil palm expansion in a Dayak Hibun community in West Kalimantan' (*The Journal of Peasant Studies*, 2012).

**Aaron Leopold** is the Global Energy Advocate at Practical Action, where he represents the organization globally on energy issues and is responsible for designing and coordinating Practical Action's advocacy work on energy across its country offices. Prior to this he was Director for Environment and Sustainable Development at the Global Governance Institute, which he co-founded. He also served as a Team Leader and Editor for Sustainable Energy at the International Institute for Sustainable Development, and as a Research Associate for the 'Fair Fuels?' research project at the Freie Universität, Berlin.

**Philip McMichael** is a Professor of Development Sociology at Cornell University. In his research he examines capitalist modernity through the lens of agrarian questions, food regimes, agrarian/food sovereignty movements, and, most recently, the implications for food systems of agrofuels and land grabbing.

**Victoria Marin-Burgos** is a researcher from Colombia. She holds a Master's degree in International and European Law and in International Relations. In 2014 she completed her PhD at the University of Twente, in the Netherlands. In her dissertation she combines concepts and theories from different disciplines – political ecology, political economy, ecological economics, critical sociology, peasant studies, environmental justice and human rights – to investigate the accelerated expansion of oil palm cultivation in Colombia since 2000 and the socio-environmental conflicts connected with this expansion.

**Shishusri Padran** is a Lecturer in International Development at the University of Edinburgh. Her work focuses on the role of science and innovation in South Asian development. Her research centres on energy policy, politics of the energy economy, environmental sustainability, energy technologies, rural development, governance and gender studies. She researches on energy and related development policies and their socio-economic impact in developing regions like India and Africa.

**Clara M. Park** is a Gender Officer with the Food and Agriculture Organization of the United Nations (FAO), and a PhD candidate at the International Institute of Social Studies in The Hague. Her research has focused mainly on agricultural and land policy, tenure and property rights, land grabbing, gender and social inequalities and women and minorities' rights in Africa and Asia. Her current research explores the intersection of land grabbing, climate change mitigation strategies and conflicts and gender in Cambodia and Myanmar.

**Oliver Pye** teaches Southeast Asian studies at Bonn University. Key teaching and research areas include political ecology, globalization and social movements. Current research projects include labour in the palm oil industry and the political ecology of the Kapuas river in West Kalimantan.

**Emmanuel Sulle** is a researcher and PhD candidate at the Institute for Poverty, Land and Agrarian Studies at the University of the Western Cape, South Africa. He has experience in both policy and biofuels research in East and Southern Africa. His research focuses on understanding inclusive business models of land-based investments, land tenure and livelihoods in sub-Saharan Africa.

**Thomas Vogelpohl** is a political scientist and has worked as a researcher at the Institute for Ecological Economy Research (IÖW) in Berlin since 2009. His key research area is the analysis of environmental, energy and climate policies, especially their instruments and effects, with a special focus on German and European agrofuel policy.

**Ben White** is Emeritus Professor of Rural Sociology at the International Institute of Social Studies (ISS), The Hague. His research interests focus on agrarian change and the anthropology of childhood and youth – particularly in Indonesia, where he has been engaged in research since the early 1970s. He is a founder member of the Land Deal Politics Initiative.

**Markus Wissen** is Professor of Social Sciences, with a focus on socio-ecological transformation, at the Berlin School of Economics and Law. His work focuses on theories of society–nature relations, energy politics and the transformation of dominant patterns of production and consumption.

# Preface and acknowledgements

The conceptualizations and analyses of agrofuels in political ecological terms presented in this volume are the result of vivid and sustained debates and interactions among the contributors. In this sense, the book represents more than just a compilation of diverse articles. It is, rather, a collective endeavour to deepen our understanding of how (hidden) power relations and the potential for political movements are inscribed in the 'agrofuels project' in complex and sometimes unexpected ways. This undertaking grew out of the fruitful encounter of different research projects in workshop discussions. The start to compiling a book on the political ecology of agrofuels was made at the international workshop 'The Political Ecology of Agrofuels' that took place in Berlin, 11–12 October 2012. The workshop was jointly organized by the Junior Research Group 'Fair Fuels? Between Dead End and Energy Transition: A Socio-ecological Multilevel Analysis of Transnational Biofuel Policy' and the Institute of Latin American Studies of the Freie Universität Berlin. It was funded by the German Federal Ministry of Education and Research (BMBF). This connected to ideas evolving from another research project with palm oil workers in Southeast Asia and to contributors and discussions leading back to the 2009 global workshop on 'Biofuels, Land and Agrarian Change', organized by the *Journal of Peasant Studies, Initiatives in Critical Agrarian Studies* (ICAS) and International Development Studies (IDS) at Saint Mary's University in Halifax, Canada. The book in hand is the product of this encounter and the subsequent exchanges.

There are many people we need to thank for making this book possible; more than can be mentioned here. First of all we would like to express particular gratitude to all the contributors to this book, as well as to Bernd Belina and two anonymous reviewers whose comments on some of the chapters and on the outline of the book were of great help. The Latin American Institute and the Otto Suhr Institute at Freie Universität Berlin, as well as the Institute of Southeast Asian Studies at Bonn University, provided institutional support for the realization of the project. Irene Wilson did a splendid job of translating some of the contributions and preparing the text for publication; Lisa Thomson, Emily Kindleysides and Natalie Tomlinson supported the project at Routledge and were always reassuring. Saturnino M. Borras, together with Haroon Akram-Lodhi, Cristobál Kay and Max Spoor, by accepting the volume for their series, encouraged us to push the project of the book forward. Finally, we wish to thank the German Federal Ministry of Education and Research and, especially, its Social-Ecological Research Programme and the DFG for funding the research that made this book possible.

The Editors

Berlin and Bonn, May 2014

# 1 An introduction to the political ecology of agrofuels

*Kristina Dietz, Oliver Pye, Bettina Engels and Achim Brunnengräber*

In recent years, agrofuels – i.e. substitutes for petrol and diesel made from plants – have been put forward as a partial technical solution to the crisis of global warming and peak oil. In particular, the USA and the European Union have enacted legislation that sets mandatory blending targets and subsidies for agrofuels. Other countries have followed suit or, as in the case of Brazil, had already passed legislation long before. By the end of 2013, 62 countries worldwide had approved agrofuels mandates (Biofuels Digest 2013). The publications and policy documents that prepare and support these policies discuss agrofuels in a technical and managerial way, focusing on issues such as the amount of carbon that each different kind of agrofuel could save, how 'safeguards' can be put in place to ensure that agrofuels are 'sustainable', or the best land practices that could guarantee that they would contribute to rural development and curtail deforestation (cf. Langeveld and Van Keulen 2014). A political ecology analysis of agrofuels rejects this 'technical rendering' and sees agrofuels as a political issue. It is defined by unequal power relations, competing coalitions of actors at different spatial scales, conflicts between these different groups, their competing discourses, the transformation or perpetuation of social power relations (e.g. gender and class, or ethnic relations) and politicization.

For example, a look at the main actors who prepared and lobbied for government policies supporting agrofuels is illuminating in order to understand what the agrofuel project is all about. In early 2005, the EU Commission set up an advisory body called the Biofuels Research Advisory Council (BIOFRAC) to help them develop their agrofuel policy and prepare a corresponding research agenda. In addition to representatives from the agricultural and forestry sectors, and some (technically inclined) academics, this council was made up of people from agribusiness, three automotive companies (Peugot, Volkswagen and Volvo) and three oil companies (Neste Oil, Shell and Total) (CEO 2007). In their recommendation report to the Commission, this group of representatives of large corporations – who are interested in making profits from selling as many cars and as much petrol as possible – propose substantial government subsidies in order to achieve their 'vision' that agrofuels should account for 25 per cent of the fuel to be used in the transport sector by 2030 (BIOFRAC 2006).[1] Similarly, in the USA, the agribusiness corporation ADM heavily influenced the passing of legislation that provides billions of dollars' worth of subsidies to the ethanol sector (see Leopold, Chapter 13, in this volume).

As various contributions in this book show, actor coalitions supporting the agrofuels project, in addition to collaboration between governments and oil, automotive and agribusiness corporations in the North, also include transnational corporations from the South and state actors that perceive agrofuels as part of a developmentalist project (for Brazil, see Backhouse, Chapter 10 and for Colombia, see Marin-Burgos, Chapter 9 in this volume). In India, national agrofuel policies were developed and implemented by coalitions made up of state agencies, private and state-owned oil and energy companies and academics (see Pradhan, Chapter 12 in this volume). Just as illuminating is the fact that environmental organizations such as Greenpeace or Friends of the Earth have not become part of the agrofuels project. Rather, they are increasingly involved in transnational campaign networks that link place-based actors in countries in the Global South affected by agrofuel crop expansion with activist networks opposing pro-agrofuel legislation (Pye 2010).

In this introductory chapter, we chart the contours of the political ecology of agrofuels. In the first part we explore critical themes and theoretical starting-points that constitute such a perspective and frame the contributions that form the core of this collection. We draw on recent theoretical debates in the field of political ecology such as discourse theory, which demands a more serious consideration of the construction of knowledge, and the politics of meaning and representation. We furthermore take up the idea formulated by Michael Watts and Richard Peet (2004: 6) of 'a broader and more sophisticated sense of the *forms* of political contention and deeper conception of *what* is contended', and who is engaged in these contentions, not only to identify underlying power relations but also to mark the emancipatory potential of processes of politicization around agrofuels. We also take into consideration critical and non-essentialist accounts on the materiality of nature and on the spatiality of agrofuels. This involves taking the biophysical materiality of agrofuel crops seriously in order to get a deeper understanding of how specific natures, understood as 'historical products of material, representational and symbolic practices' (Bakker and Bridge 2006: 18), can be productive in the generation of social configurations and the way social relations of power unfold. The latter stresses the need to reflect on how social inequalities and (geo-)political and transnational configurations of power are produced, perpetuated or altered through space. In the second and final part of this introductory chapter, the structure of the book and the main lines of the contributions are briefly outlined.

## Politics of representation and competing discourses

The 'agrofuels project' (McMichael 2008: 14) is characterized by a confluence of transnational public-private actor coalitions, national legislations, political incentives and specific discursive frames through which it is promoted but also contested. The competing representations of agrofuels can be seen first and foremost in the choice of words to describe the project itself. Proponents use the word 'biofuel' to suggest that this is a 'green' project and part of a 'sustainable' solution to climate change, peak oil and rural poverty. The discursive power of 'biofuels' appears in the accelerated increase in crop production applicable to agrofuels at a global scale. As Fairhead *et al.* (2012: 241) suggest with regard to carbon trading, we claim that there would not be so many flexible possibilities for valuating palm oil or sugar cane without the global discourses that have discerned agrofuels as a 'flex solution' to multiple crises, with wide-ranging structural and material implications. So one important argument for a deeper involvement with agrofuels, from a political ecological perspective, relates to the mutuality between the apparently immense power of discourses through which agrofuels inevitably have come into being and their social and biophysical materiality.

As Vogelpohl (Chapter 14, this volume) suggests, in Europe the sustainable biofuel narrative forms part of a 'neo-liberal "green transformation" meta-discourse'. Part of this is the ongoing hype around the so-called '2nd and 3rd generation of biofuels',[2] which are represented as promising to overcome the social, ecological and economic 'limitations' of the so-called 'first generation of biofuels', based on food crops like corn, wheat, rape seed, palm oil or soy, and which have been widely criticized for their adverse impacts on global climate and food security. The main argument in favour of these 'advanced biofuels' (Krauss 2014), based on residues, cellulose or algae, is that they would not be in direct competition with food production.

In contrast, critics purposely introduced the term 'agrofuels', because it 'not only reminds us of crop land competition and fuel displacing food' (McMichael 2010: 621), but analytically it also reflects the dominance of capitalist societal nature relations, i.e. a dominance of strategies to appropriate and represent nature that are guided mainly by capitalism's social logic of accumulation (Görg 2011). Because the term 'biofuels' suggests a cautious and reflexive treatment of nature, promoted as mitigating climate change, the capitalist logic behind them remains invisible. Instead, the term agrofuels alludes to agro-industrial agriculture and its inherent commodification of nature, and to the modernization and industrialization of rural areas with all impacts on social relations and ecological conditions. To talk about agrofuels instead of biofuels is thus both an outcome of empirical observations and an analytical intention.

The discursive power of 'biofuels' is not limited to specific spatio-political contexts (the EU) or scales (global scales). Carol Hunsberger (Chapter 8, this volume) argues with reference to Kenya that the crop jatropha has a 'high discursive flexibility' that combines a 'large-scale discourse' of climate change mitigation and national development with a 'small-scale discourse about jatropha as a rural development strategy and a source of 'clean energy for development'. A similar discourse can be seen in India (see Pradhan, Chapter 12 in this volume). Related to the debates that surround the 'agrofuels project' in Africa, and more generally, Festus Boamah (2011) points in the same direction, emphasizing that these debates are underpinned by two competing discourses: a powerful sustainable managerial discourse which sees the 'agrofuels project' as a way out of climate change and rural poverty and hunger; and a more populist and critical discourse which sees agrofuel investments as a motor for climate change and a threat to local livelihoods and food security. Both discourses are expressed through contextualized narratives (stories) that help to simplify complex issues and express interests.

Being sensitive to the discourses and underlying narratives, knowledge constructions, cultural and social practices and representations through which agrofuels come into being – particularly through discourses of sustainable development, ecological modernization and green economy – helps on the one hand to further understand how the changing relationship between nature and capital, framed as 'green capitalism' (Brand 2012), the '"ecological phase" of capital' (Escobar 1996: 326) or 'post-Fordist societal nature relations' (Brand *et al.* 2008) is articulated by this discourse (see Backhouse in Chapter 10 of this volume). On the other hand, a consideration of the competing discursive practices and related social micro-practices opens new paths to conceptualizing the 'agrofuels project' as a highly contested locus of constructing hegemony. The assumption that 'biofuels' are a stable end result of a process of socio-ecological transformation of the fuel sector, or a 'technological fix' of the climate crisis, can be questioned. It directs our view to the social and discursive micro-practices through which opposition and discontent, but also consent, are articulated and produced. These are embedded in everyday practices and struggles that do not always catch the attention of the media (see Pye in Chapter 11 of this volume). Nevertheless, in order to understand how the normalization of agrofuels works, we need to take these 'other' discursive and social practices into account as they can lead to both the emergence of new antagonisms 'that might trigger a crisis of the dominant rationalities justifying the process of policymaking' (Gottweis 1998: 264) or the stabilization of existing tendencies. What is contested, fought over and negotiated is thereby an empirical question and depends on the positions and interests of the actors involved (see Marin-Burgos, Chapter 9 in this volume).

## Politicization, contention and transnational activism

The discursive power of the win-win narrative (see Vogelpohl in Chapter 14 of this volume) and the economic power embedded within the global networks of agrofuel production are connected to political, regulative power. Powerful political economy networks like the oil and automotive industry formation in the European Union (see Brunnengräber, Chapter 5 in this volume), the transnational agribusiness companies in the United States (see Leopold in Chapter 13 of this volume), the state-corporate alliance in India and Brazil (Pradhan, Chapter 12 and Backhouse, Chapter 10 in this volume) or the 'palm oil industrial complex' in Malaysia (Pye 2008), influence government bodies at the national or supra-national scale to enact pro-agrofuel legislation without which the agrofuel industry would hardly be economically viable.

But the hegemony of the agrofuels project is increasingly being challenged by different sets of actor alliances. Land conflicts connected to agrofuel expansion have seen the emergence of place-based actors – small-scale farmers and indigenous groups – who oppose the expropriation of their land and forests in the name of 'sustainable biofuels'. Activists quickly developed transnational campaign links. In June 2007, European NGOs (non-governmental organizations) launched a 'Call for an immediate moratorium on EU incentives for agrofuels, EU imports of agrofuels and EU agroenergy monocultures' (EcoNexus 2007) linking 250 organisations working on issues as varied as global issues of climate justice and neoliberalism, food sovereignty and land rights, conservation or transport policies. Transnational campaigning was particularly strong around the production of agrodiesel from palm oil. European media coverage (including several television programmes) could debunk the ecological credentials of agrofuels by showing that they contributed to rainforest destruction in Southeast Asia and to (violent) land conflicts in Indonesia and Colombia (Pye 2010; Coronado *et al.* 2013; Grajales 2011; Marin-Burgos, Chapter 9 in this volume). Most notably, the confederation of small-scale farmers, La Vía Campesina, has been quite vocal in its rejection of agrofuels as 'a false solution to climate change.' (La Vía Campesina 2009). In the global climate justice protests that took place during the Climate Summit in Copenhagen in 2009, Vía Campesina activists connected opposition to agrofuels with their alternative of food sovereignty and a small-scale eco-agriculture that they proposed could 'cool the planet'. Linking local struggles against the agro-industrial expansion of 'flex crops', with climate justice movements calling for genuine action on climate change and the transformation of fossil-fuel-based capitalism (Klimaforum09 2009) created new alliances between urban and rural activists in the North and the South. As linkages between food prices and agrofuels became clear (Leopold, Chapter 13 in this volume), activists started to include other spatial dynamics of agrofuels by targeting speculative commodity trading and the involvment of banks in agrofuel land grabs (Clapp 2014).

Agrofuel hegemony (Leopold, Chapter 13 in this volume), the complex and contradictory international political economy of agrofuels and the very vocal public campaigning politicized the agrofuels project in different ways. Particularly in Europe, agrofuel regulation became a contested arena of different national, private and public interests. This was already the case during the debates leading up to the passing of the Renewable Energy Directive (European Union 2009; Brunnengräber, Chapter 5 and Vogelpohl, Chapter 14 in this volume). The contested nature of agrofuels in Europe appeared again in October 2012, when the European Commission published a proposal to put its agrofuel policy on an ecologically sound basis. With the intention of minimizing climate impacts and meeting its 10 per cent renewable energy target by 2020, a 5 per cent cap of first-generation biofuels was suggested.[3] Although until the beginning of May 2014 no final decision had been taken on the issue, the proposal itself fuelled new contentions around the pros and cons of agrofuels in Europe. Representatives of the European agrofuel industry have rejected the proposed cap, arguing that it would destroy the industry, cut off European farmers from important world markets and jeopardize desperately needed jobs in rural areas.[4] On the other hand, environmental NGOs such as WWF welcome the Commission's plans. Others, like the former UN Special Rapporteur on the Right to Food, Jean Ziegler, continue to demand a ban on agrofuels altogether, calling them a 'crime against humanity'.[5]

## Materiality, socio-spatial transformations and conflicts

Markus Wissen (Chapter 2, this volume) suggests that a political ecology of agrofuels should explore the dimensions of power, space and materiality. Starting with the latter, in political ecology terms, agrofuels represent a new round in the 'commodification of nature', in which – through government regulation and discursive framings – a new resource is created, leading to new lines of conflict and socio-spatial transformation. This is most evident in the case of jatropha, a crop that is being promoted as an agrofuel marvel crop because it cannot be eaten or used for much else. Only when the agrofuels project started to develop did jatropha become a profitable commodity, creating a new political ecology in agrarian development in parts of Africa and Asia in the process (Hunsberger, Chapter 8 and Pradhan, Chapter 12 in this volume; Ariza-Montobbio *et al.* 2010; McCarthy *et al.* 2012). But jatropha is the exception rather than the rule for agrofuels. Major agrofuel feedstocks such as corn, sugarcane, oil palm, rape seed and soy are by contrast 'flex crops', i.e. 'crops that have multiple uses (food, feed, fuel, industrial material) that can be easily and flexibly inter-changed' (Borras *et al.* 2012, 851). When analysing the social effects of the territorial expansion of oil palm plantations, e.g. on access rights or on the emergence of new subject positions, we do not know with certainty whether these effects can be attributed to the agrofuels project. The 'material flexibility' (Hunsberger, Chapter 8 in this volume) of these crops allows investors to diversify investments in one single crop according to market signals: when sugar prices are high, they sell sugar, and when they are low, they can still sell ethanol or speculate and wait for more profitable ethanol market conditions to emerge. Thus, in the final instance, we cannot predict whether the palm oil produced in a certain location will be used for cooking or to make hygiene products or agrodiesel.

The 'flex-crop materiality' of agrofuels feedstocks is a key defining feature of the political ecology of agrofuels. Additional, large-volume production of the five globalized flex crops: sugarcane, corn, rape seed, palm oil and soy, excacerbates the existing production trends and structures, reinforcing large-scale monoculture production, industrial processing, precaritized wage labour, and transnational production and trading networks (see Dietz *et al.*, Chapter 3 in this volume). The most direct consequence of this materiality is the expansion of large-scale monocultures into areas that for different reasons have hitherto not been attractive for major capital investments, a situation that has been changing in many parts of the world due to the emergence of agrofuels and the discovery of 'land' as a valued capital investment in times of crises (Borras and Franco 2012).

In some cases, the agrofuel expansion is complemented by repressive power and forced through by new rounds of 'internal territorialization', marginalizations and (violent) displacements, for example in Colombia (Marin-Burgos, Chapter 9 and Dietz *et al.*, Chapter 3 in this volume). Against this background critical scholars from different disciplines and fields of research have so far mainly focused on agrofuels' feedstock expansion related to the impacts on land use, land tenure, access to land, land-related conflicts and the disputed construction of new territorialities in the Global South. Mançano Fernandes *et al.* (2010), for example, have been analysing territorial disputes between private companies, state agents and agrarian reform settlers (often once landless workers) who worked on a massive expansion of sugar cane plantations for ethanol production in the Pontar region, state of São Paulo (Brazil), at the beginning of the twenty-first century. In this case, the challenge for the peasants and their movements was to maintain their autonomy over the territories they had gained in their struggle for land reform. But it was not only unproductive and illegally titled land surrounding the rural settlements that was converted into sugar cane plantations, but land for settlement as well, so companies enticed settlers with extraordinary gains that could be made if they would lease their lands to the company or allow them to plant sugar cane on their lots (ibid. 801f.) Both led to a fragmentation of the settler movement and most often a loss of independence and control over the land, since many of the farmers and families, shortly after they had signed contracts and allowed sugar cane plantations on their lots, found themselves forced to work as badly-paid sugar cane cutters in the companies' fields. Similar processes of exclusion, de-peasantization and marginalization of certain land users as opposed to others can be found in many other places in the South, for example in India (Ariza-Montobbio and Lele 2010), Mozambique and many other countries in sub-Saharan Africa (Nhantumbo and Salomão 2010; Borras *et al.* 2011; Sulle and Hall, Chapter 7 in this volume). The expansion of monoculture crop plantations for agrofuels has also been conceptualized as a form of territorialization of a global corporate-driven agro-industrial production system and the emergence of a new 'food-for-fuel regime' (see McMichael, Chapter 6 in this volume).

Because of the growing number of land conflicts in the South that can (though often not exclusively, see Marin-Burgos, Chapter 9 in this volume) be traced back to the expansion of flex crop production, agrofuels are often seen as reinforcing existing patterns of imperialist power relations and economic dependencies between the North and the South (Holt-Giménez and Shattuck 2009). But at the same time, new sets of either South–South or North-South–South formations are emerging through and beyond the nation state. South–South relations have emerged, for instance, through the activities of the Brazilian state and private investors in the expansion of sugar cane in different parts of southern Africa (Richardson 2010). North-South–South formations can be seen, among others, in new triangular alliances between transnational capital and actors based in the North with South–South partnerships, e.g. the agreement between the EU, Brazil and Mozambique from 2010 on investments and technology transfer related to sugar cane-based ethanol production in Mozambique (Dauvergne and Neville 2009; Danker *et al.* 2013). Another expression of changing spatio-political power relations that cannot easily be captured by one-dimensional concepts of North-South relations, are newly emerging transnational networks consisting of private companies, state agents, academics, research institutions and local producers. Building on concepts such as technoscience, (actor) networks (Latour 1987: 174; Latour 2005; Pradhan, Chapter 12 in this volume) and (global) assemblages (Sassen 2008; Collier and Ong 2005), James Smith 'trace[s] the global assemblages of expert knowledge, technologies, political and economic domains that shape and constitute the context in which [agro]fuels are evolving' (Smith 2010: 11).

Land conflicts are one expression of agrofuels-related agrarian change, but not the only one. As McCarthy (2010) shows for Jambi, Indonesia, additional rural investment and smallholder inclusion lead to processes of social differentiation and to 'adverse incorporation' into global markets. Agrarian transition is accompanied by the transformation of gender relations, as women's control over common land resources and customary tenure arrangements give way to private land ownership in smallholder schemes and women are given inferior work in the proletarization processes accompanying large-scale plantations (White *et al.*, Chapter 4 in this volume).

The 'other end' of the materiality of agrofuels is also crucial, e.g. the materiality of greenhouse gas production and climate change, that forces political responses of which agrofuels are a part. The entrenched interests of 'fossilism' (Brunnengräber, Chapter 5 in this volume) and the continued reliance on oil and coal are the material context in which agrofuels are developing. A crucial point here is that agrofuels feature material characteristics that are compatible with fossil fuels. Agrofuels are liquid, they can be easily transported from one place to another, they can be produced in many places in the world and they can be easily blended with fossil fuels. Agrofuels could thus be considered as the perfect cure for a system ('fossilism'), the most important energy source of which, fossil fuel energy, sooner or later will come to an end. But 'capitalist social life is [not only] profoundly dependent on the abundant provision of fossil fuel energy' (Huber 2008: 105) or appropriate substitutions, it is also transformed by fossil fuel. In this vein, this book also aims at exploring the dialectics of the production and consumption of agrofuels, and the socio-spatial transformations that accompany it.

The political ecology of agrofuels is therefore connected not only to agrarian change but to the debates around energy production, transportation systems, the climate debt and the transformation of global capitalism itself. Related to the latter, in this volume agrofuels are considered as part of the current endeavour of greening capitalism. In this sense, some of the contributions to this volume conceptualize agrofuels in terms of green grabbing (see Backhouse Chapter 10), neoliberalizing nature (Brunnengräber Chapter 5; Vogelpohl Chapter 14) and financializing nature (Dietz *et al.* Chapter 3). Green grabbing is defined as the valuation of nature justified by environmental or climatic ends (Fairhead *et al.*

2012). Neoliberalizing nature is understood by Noel Castree (in reference to David Harvey's concept of a spatial fix) as 'environmental fixes' constituted by 'conservation and its two antitheses of destroying existing and creating new biophysical resources' (Castree 2008: 150). And financialization can be conceptualized as a process by means of which global financial markets, institutions, instruments and actors gain ever more influence over nature, e.g. through derivates, futures, mortgages etc. (Tricarico and Löschmann 2012).

## Doing political ecology of agrofuels

Since its boom as a technical solution to the climate and energy crises, and as a possible way out of rural poverty and unemployment at the beginning of the twenty-first century, agrofuels have also been critically scrutinized through political and ecological lenses (cf. Borras *et al.* 2010). So why a further book on this topic? Our answer is twofold. First, there is currently no single volume that seeks to bring together the diverse research across world regions on the particularities of the political ecology of agrofuels. In saying this, we do not pretend that this book captures the political ecology of agrofuels in an exhaustive way. Rather, the 14 chapters collected in this volume explore, both theoretically and empirically, some of the major themes, perspectives and politics that are central to a political ecological analysis of agrofuels, and that until now have rarely been discussed under one umbrella. Besides global food regimes and agrarian politics, the chapters encompass political arenas such as energy, climate, transport and trade. They reflect on the biophysical materiality of agrofuels, new forms of nature appropriation, struggles, discursive framings, the building of hegemony, shifting geopolitical constellations, socio-spatial configurations of power, the construction of territory, the agency of social movements and the different ways of politicization of agrofuels on different scales. They ask how patterns of mobility, emissions regulation, food and energy production and consumption, and social relations (e.g. labour, class and gender relations) are shaped and re-shaped by the materiality and representations of agrofuels, in both the Global South and North.

Having said this, the book builds upon existing endeavours to critically understand the 'agrofuels project', e.g. as a topic of agrarian change (McMichael, Chapter 6 in this volume) or of global political economic re-configurations (Borras *et al.* 2010). At the same time, it goes beyond these insights by linking them to other powerful political arenas and aspects of political ecology, newly emerging scales of regulation and divergent processes of the transformation of social relations. Second, with this volume, we aim to offer bundled reading material to teachers, postgraduate students, doctoral students and activists alike concerned with agrofuels as a specific topic of contentious politics, such as the financialization of nature, social struggles, gendered inequalities, precarious labour regimes, transnational political, social and economic spaces, forms of inclusion and exclusion through space and the re-definition of social power relations in the realm of current climate, land and energy policies.

The book is organized into four sections. In the first three chapters, conceptual and theoretical contours of a political ecology of agrofuels are outlined. In Chapter 2, Markus Wissen investigates the contribution to a critical understanding of the social and political content of agrofuels. Starting from the origins and basic assumptions of political ecology and demonstrating how it transcends other approaches to the environmental crisis, he discusses core theoretical concepts that are crucial for a political ecological analysis of agrofuels: power, materiality and space. In Chapter 3, Kristina Dietz, Bettina Engels and Oliver Pye build on these insights and explore the spatial dynamics of the political ecology of agrofuels. They show how spatial categories enable a comprehensive analysis that succeeds in linking the macro-structures of the global political economy to concrete, place-based struggles related to agrofuels. Three core socio-spatial dynamics of agrofuel politics are highlighted and applied to empirical findings: territorialization, the financial sector as a new scale of regulation and transnational spaces of resource and capital flows. In Chapter 4, Ben White, Clara Park and Julia shed light on a hitherto marginalized topic within the field of agrofuels: gender dimensions. Starting from a feminist political ecological and agrarian studies perspective, they excavate the intricate and shifting relations between the expansion of the large-scale, corporate driven production of agrofuels destined for commercial use and consumption in distant places and the gender relations involved. They focus on the gendered access to and control of land, the gendered division of labour and participation in production, the gendered access to food and the gender dimensions of voice and participation in decision-making processes at the household and community level. These introductory chapters provide a conceptual background to the field of the political ecology of agrofuels against which the other chapters of the book can be situated.

The three chapters that follow contextualize agrofuels in those wider political arenas to which they are intrinsically linked: climate and energy, the global food regime and land rights. In Chapter 5, Achim Brunnengräber undertakes a critical revision of the European Union's endeavours to tackle climate change and achieve energy security via agrofuels. He argues that agrofuels have become a decisive element for the consolidation and prolongation of 'fossilism', a fossilist energy system that suits the capitalist mode of production and reflects existing power constellations in the energy, transport, automotive and agriculture sector within the EU. Insisting on the powerful possibilities of a historical-materialist perspective on food production and consumption, Philip McMichael, in Chapter 6, approaches the agrofuels project through the lens of food regime analysis. In this reprint from his 2010 article 'Agrofuels in the food regime', published in *The Journal of Peasant Studies*, and based on Marx's concept of the 'metabolic rift', the author considers the 'rush to agrofuels' as the ultimate demystification of capitalism's subjection of food to the commodity form at the expense of human habitats, non-capitalist societal nature relations and ecologies. The relationship between agrofuels and changing land rights are at the centre of Emmanuel Sulle and Ruth Hall's text in Chapter 7. Focusing on both recent processes of agrofuel expansion and apparent failures of such expansion efforts in Africa, they examine the impacts of these projects on land rights and livelihoods of the rural poor. To sustain their arguments, they draw on their own empirical research on agrofuels and land rights in Mozambique and Tanzania.

In the remaining chapters, we move from concepts and contexts to case studies, struggles, discourses and conflicts in different countries and world regions. In Chapter 8 Carol Hunsberger enters the political ecology of agrofuels through the lens of discourse. While major agrofuel feedstocks can be considered to be 'flex crops', the oilseed plant *Jatropha curcas* stands as an exception to this pattern. Jatropha is unsuitable for human or animal consumption and can be grown in dry conditions. This low *material* flexibilty is accompanied by a high *discursive* flexibility. With reference to fieldwork from Kenya, the author demonstrates how a large-scale discourse of jatropha – as a climate change mitigation strategy and a source of national economic development – has co-existed with a small-scale discourse about jatropha as a rural development strategy and a source of 'clean energy for development'. In her Chapter she explores these intertwined strands – low *material* flexibility and high *discursive* flexibility – and how each has influenced its trajectory as an agrofuel crop. She argues that paying close attention to these two dimensions and their interactions can contribute to a more nuanced understanding of the political ecology of jatropha in particular as well as 'flex crops' more generally. Victoria Marin-Burgos, in Chapter 9, examines the complex relations among the material, socio-economic and political factors that shape land-related conflicts in Colombia connected with the expansion of agrofuel crops, particularly palm oil. Focusing on both government policies that foster agrofuel production and consumption and a specific case in the province of Cesar, where a local association of peasants resists planting oil palms on their lots, she shows how conflicts over land and land use, especially when located at the sites of expanding commodity frontiers, are at the same time struggles over territory, inclusion and exclusion. Commodity frontiers, crop booms, socio-environmental conflicts and territorialization are the central figures through which Marin-Burgos explores the conflictual political ecology of agrofuels in Colombia. Chapter 10 continues with experiences from Latin America. Maria Backhouse, in her case study on the expansion of oil palm monocultures in the eastern Brazilian Amazon, points in the same direction, although she argues that this expansion represents a form of 'green grabbing'. On the basis of a re-interpretation of the Marxist concept of primitive accumulation, she differentiates three interrelated analytical dimensions that together serve as a framework for a more nuanced analysis of processes of green grabbing: a material, political and discursive dimension. The latter, in particular, becomes highly important in order to understand how the agrofuels project is legitimized as a green solution to climate change and poverty. In Backhouse's case, it is especially the strong narrative of 'degraded areas' where agro-industrial landscapes and resources come into being, societal agreement is organized and resistance is made difficult. From Latin America we go straight on to Asia. In Chapter 11, Oliver Pye delves into a topic that has until now remained marginal to debates around agrofuels, i.e. the issue of labour. He considers how the experiences of migrant workers from Indonesia who work in the palm oil industry, plantations and mills in Malaysia contribute to a new transnational 'labour geography'. He looks at the life experiences of these workers in the palm oil estates and mills and shows how the temporary and precarious nature of their existence also precludes the emergence of political organization to articulate discontent. The production of a 'mill-estate-scale' and a 'reserve army' of illegalized and outsourced workers serves to isolate and discipline the workers. On the national scale, the Malaysian state disenfranchises the workers by territorializing the plantations as national space, turning them into 'denizens' and attempting to outsource their reproduction across the border. However, based on a longitudinal study with ten groups of workers consisting of biographical interviews, subsequent in-depth interviews and finally group discussions, Pye shows – with a specific focus on everyday micro-practices – how migrant workers in the Malaysian plantations develop agency as they create a multitude of networks linking specific places to each other and producing transnational social spaces. He argues that, across time and space, the linkages between specific sites in Indonesia to places and networks in Malaysia can empower workers, who develop various means of everyday resistance that reassert their own humanity over their alienation as labour power on the plantations. In the following Chapter 12, Shishusri Pradhan explores agrofuel networks in India, looking particularly at the Indian state of Chhattisgarh. Drawing on network theory, Pradhan shows that a state-led alliance between government agencies, Indian corporations and academia developed an ambitious programme to plant 160 million jatropha seedlings, by incorporating a secondary network of local institutions and farmer groups. In the Chhattisgarh context, the promise of rural development played a key discursive role in generating enthusiasm to attain the status of a 'biofuel-reliant state' by 2015. However, the initial enthusiasm waned as the programme targeted 'wasteland' on which to plant the crop, leading to the appropriation of common property resources on which the poorest rural citizens relied. The politicization that this created undermined the grandiose agrofuel plans of the Chhattisgarh state. From India, we turn our attention back to the Global North. In Chapter 13 Aaron Leopold applies a neo-Gramscian analytical perspective to agrofuel policies in the United States and asks whose interests are being served by these policies. He explores the dominant role of the agriculture industry, especially of agricultural multinational Archer Daniels Midland (ADM), in shaping and controlling political economic discourses on agrofuels. While the hegemonic control of these discourses has waned over time due to factors such as the global financial crisis, questions from civil society about their social and ecological credentials and the commodity price crises, these shifts have barely affected the agrofuels industry, despite calls for reform. The author thus concludes that political support for agrofuels in the United States has first and foremost benefited deeply entrenched, hegemonic business interests, and not produced the broader economic, environmental and social benefits claimed by proponents of agrofuels in general. In the final chapter Thomas Vogelpohl takes us back to Europe and explores the strong and strange persistence of the win-win narrative that has dominated the agrofuel policies in Europe, despite ongoing criticism. From a discourse analytical approach, he shows how this narrative developed a neoliberal twist that led to both its reproduction and manifestation in policy decisions, which in turn gave rise to specific material effects.

## Notes

1 With the publication of their final report in June 2006, BIOCRAF was dissolved and EBTP (European Biofuels Technology Platform) was launched, consisting of much the same corporations and research institutions that formally had made up BIOCRAF (EBTP 2013).

2 In the case of the 'second generation', these are synthetic or BTL (Biomass-to-Liquid) fuels, produced from any organic materials (food crop waste, agricultural residues) and fuels based on cellulose, produced from wood or straw. Biofuels made from algae are often labelled 'third generation'.

3 See http://europa.eu/rapid/press-release_IP-12–1112_en.htm (accessed: 11 March 2014).

4 See http://ethanolproducer.com/articles/9220/eu-releases-proposal-to-alter-biofuel-policy (accessed: 11 March 2014).

5 See www.theguardian.com/global-development/poverty-matters/2013/nov/26/burning-food-crops-biofuels-crime-humanity (accessed: 12 March 2014).

## References

Ariza-Montobbio, Pere and Lele Sharachchandra (2010) 'Jatropha plantations for biodiesel in Tamil Nadu, India: viability, livelihood trade-offs, and latent conflict', *Ecological Economics* 70(2), 189–195.

Ariza-Montobbio, Pere, Lele Sharachchandra, Giorgos Kalli and Joan Martinez-Alier (2010) 'The political ecology of *Jatropha* plantations for biodiesel in Tamil Nadu, India', *The Journal of Peasant Studies* 397(4), 875–898.

Bakker, Karen and Gavin Bridge (2006) 'Material worlds? Resource geographies and the "matter of nature"', *Progress in Human Geography* 30(1), 5–27.

BIOFRAC (2006) 'Biofuels in the European Union. A vision for 2030 and beyond. Final report of the Biofuels Research Advisory Council', Brussels: Commission of the European Communities. Available at: http://ec.europa.eu/research/energy/pdf/biofuels_vision_2030_en.pdf (accessed: 11 February 2014).

BiofuelsDigest (2013) 'Biofuels mandates around the world 2014'. Available at: www.biofuelsdigest.com/bdigest/2013/12/31/biofuels-mandates-around-the-world-2014/ (accessed: 29 April 2014).

Boamah, Festus (2011) 'Competition between biofuel and food? Evidence from a jatropha biodiesel project in Norhern Ghana', in Matondi, Prosper B., Kjell Havnevik and Atakilte Beyene (eds) *Biofuels, Land Grabbing and Food Security in Africa*, London, New York: Zed Books, 159–175.

Borras, Saturnino M., David Fig and Sofía Monsalve Suárez (2011) 'The politics of agrofuels and mega-land and water deals: insights from the ProCana case, Mozambique', *Review of African Political Economy* 38(128), 215–234.

Borras, Saturnino M. and Jennifer C. Franco (2012) 'Global Land Grabbing and Trajectories of Agrarian Change: A Preliminary Analysis', *Journal of Agrarian Change* 12(1), 34–59.

Borras, Saturnino M., Jennifer C. Franco, Sergio Gómez, Cristóbal Kay and Max Spoor (2012) 'Land grabbing in Latin America and the Caribbean', *The Journal of Peasant Studies* 39(3–4), 845–872.

Borras, Saturnino M., Philip McMichael and Ian Scoones (2010) 'The politics of biofuels, land and agrarian change: editors' introduction', *The Journal of Peasant Studies* 37(4), 575–592.

Brand, Ulrich (2012) 'Green economy and green capitalism: some theoretical considerations', *Journal für Entwicklungspolitik (JEP)* 28(3), 118–137.

Brand, Ulrich, Christoph Görg, Joachim Hirsch and Markus Wissen (2008) *Conflicts in Global Environmental Regulation and the Internationalization of the State. Contested Terrains*, London: Routledge.

Castree, Noel (2008) 'Neoliberalising nature: the logics of deregulation and reregulation', *Environment and Planning A* 40, 131–152.

CEO (2007) 'The EU's agrofuel folly: policy capture by corporate interests'. Available at: http://archive.corporateeurope.org/agrofuelfolly.html (accessed: 29 April 2014).

Clapp, Jennifer (2014) 'Financialization, distance and global food politics', *The Journal of Peasant Studies*, DOI: 10.1080/03066150.2013.875536.

Collier, Stephen J. and Aihwa Ong (2005) 'Global assemblages, anthropological problems', in Ong, Aihwa and Stephen J. Collier (eds) *Global Assemblages: Technology, Politics, and Ethics as Anthropological Problems*, Malden: Blackwell, 3–21.

Coronado Delgado, Sergio and Kristina Dietz (2013) 'Controlando territorios, reestructurando relaciones socio-ecológicas: la globalización de agrocombustibles y sus efectos locales, el caso de Montes de María en Colombia', *IberoAmericana* 49, 93–116.

Danker, Hans-Christian, Kristina Dietz, Nicola Jaeger and Wiebke Thomas (2013) 'Die Globalisierung der Agrarkraftstoffe. Produktion, Handel und Akteure', Berlin: Fair Fuels? Working Paper 7.

Dauvergne, Peter and Kate J. Neville (2009) 'The changing North–South and South–South political economy of biofuels', *Third World Quarterly* 30(6), 1087–1102.

EBTP (2013) 'EBTP Steering Committee'. Available at: www.biofuelstp.eu/steering.html#mems (accessed: 27 May 2013).

EcoNexus (2007) 'Call for an immediate moratorium on EU incentives for agrofuels, EU imports of agrofuels and EU agroenergy monocultures'. Available at: www.econexus.info/call-immediate-moratorium-eu-incentives-agrofuels-eu-imports-agrofuels-and-eu-agroenergy-monocultur-0 (accessed: 24 April 14).

Escobar, Arturo (1996) 'Construction Nature. Elements for a post-structuralist political ecology', *Futures* 28(4), 325–343.

European Union (2009) 'Directive 2009/28/EC of the European Parliament and of the council of 23 April 2009 on the promotion of the use of energy from renewable sources and amending and subsequently repealing Directives 2001/77/EC and 2003/30/EC', Brussels.

Fairhead, James, Melissa Leach and Ian Scoones (2012) 'Green grabbing: a new appropriation of nature?', *The Journal of Peasant Studies* 39(2), 237–261.

Görg, Christoph (2011) 'The societal relationships with nature: a dialectical approach to environmental politics', in Biro, Andrew (ed.) *Critical Ecologies. The Frankfurt School and Contemporary Environmental Crises*, Toronto: University of Toronto Press, 43–72.

Gottweis, Herbert (1998) *Governing Molecules: The Discursive Politics of Genetic Engineering in Europe and the United States*, Cambridge, MA: MIT Press.

Grajales, Jacobo (2011) 'The rifle and the title: paramilitary violence, land grab and land control in Colombia', *The Journal of Peasant Studies* 38(4), 771–792.

Holt-Giménez, Eric and Annie Shattuck (2009) 'The agrofuels transition. restructuring places and spaces in the global food system', *Bulletin of Science, Technology & Society* 29(3), 180–188.

Huber, Matthew T. (2008) 'Energizing historical materialism: fossil fuels, space and the capitalist mode of production', *Geoforum* 40, 105–115.

Klimaforum09 (2009) 'System Change not Climate Change. A People's Declaration from Klimaforum09'. Available at: http://declaration.klimaforum.org/files/declaration_english_screen.pdf (accessed: 6 May 2014).

Krauss, Clifford (2014) 'Biofuel technology continues to advace, but demand is stalling', *The New York Times International Weekly in collaboration with Süddeutsche Zeitung* (2 May 2014).

Langeveld, John Dixon and Herman Van Keulen (eds) (2014) *Biofuel Cropping Systems. Carbon, land and food*, Abingdon: Routledge.

Latour, Bruno (1987) *Science in Action*, Cambridge: Harvard University Press.

Latour, Bruno (2005) *Reassembling the Social: An Introduction to Actor-Network-Theory*, Oxford: Oxford University Press.

La Vía Campesina (2009) *Small Scale Sustainable Farmers are Cooling down the Earth*, Jakarta: LVC.

McCarthy, John F. (2010) 'Processes of inclusion and adverse incorporation: oil palm and agrarian change in Sumatra, Indonesia', *The Journal of Peasant Studies* 37(4), 821–850.

McCarthy, John F., Jacqueline A.C. Vel and Suraya Afiff (2012) 'Trajectories of land acquisition and enclosure: development schemes, virtual land grabs, and green acquisitions in Indonesia's Outer Islands', *The Journal of Peasant Studies* 39(2), 521–549.

McMichael, Philip (2008) 'Agrofuels, food security, and the metabolic rift', *Kurswechsel* 3, 14–22.

McMichael, Philip (2010) 'Agrofuels in the food regime', *The Journal of Peasant Studies* 37(4), 609–629.

Mançano Fernandes, Bernardo, Clifford Andrew Welch and Elienaí Constantino Gonçalves (2010) 'Agrofuel policies in Brazil: paradigmatic and territorial disputes', *The Journal of Peasant Studies* 37(4), 793–819.

Nhantumbo, Isilda and Alda Salomão (2010) *Biofuels, Land Access and Rural Livelihoods in Mozambique*, London: IIED.

Pye, Oliver (2008) 'Nachhaltige Profitmaximierung. Der Palmöl-Industrielle Komplex und die Debatte um "nachhaltige Biotreibstoffe"', *Peripherie* 112, 429–455.

Pye, Oliver (2010) 'The biofuel connection – transnational activism and the palm oil boom', *The Journal of Peasant Studies* 37(4), 851–874.

Richardson, Ben (2010) 'Big Sugar in southern Africa: rural development and the perverted potential of sugar/ethanol exports', *The Journal of Peasant Studies* 37(4), 917–938.

Sassen, Saskia (2008) 'Neither global nor national: novel assemblages of territory, authority and rights', *Ethics & Global Politics* 1(1–2), 61–79.

Smith, James (2010) *Biofuels and the Globalization of Risk*, London: Zed Books.

Tricarico, Antonio and Heike Löschmann (2012) 'Finanzialisierung – ein Hebel zur Einhegung der Commons', in Helfrich, Silke and Heinrich-Böll-Stiftung (eds) *Commons. Für eine neue Politik jenseits von Markt und Staat*, Bielefeld: Transcript, 184–195.

Watts, Michael and Richard Peet (2004) 'Liberating political ecology', in Peet, Richard and Michael Watts (eds) *Liberation Ecologies, Second Edition. Environment, Development, Social Movements*, London, New York: Routledge, 3–47.

# 2 The political ecology of agrofuels

## Conceptual remarks

*Markus Wissen*

### Introduction

In recent years, the social and political dimensions of environmental issues have moved to the centre of political and scientific agendas: climate change is framed in terms of justice by NGOs and social movements; struggles around the democratization and ecological transformation of energy provision have become more visible; in some places, there is an increasing discussion on food sovereignty; the acknowledgement of the fact that climate change is taking place has given rise to research on, and politics of, adaptation, which cannot but tackle the social dimensions of climate change, e.g. in terms of vulnerability; finally, the renewed scientific and political interest in the social dimensions of the ecological crisis has been driven by contested strategies for mitigating climate change, such as the expansion of agrofuel production.

These developments have created a need for conceptual approaches that unveil and explain the social and political dimensions of the ecological crisis. One such approach is political ecology, which has been used since the 1970s to investigate environmental problems, particularly in the global South. Political ecology represents a *perspective* on the ecological crisis rather than a theory, or to put it more precisely 'a "frame of research" consisting of a more or less diverse set of questions, modes of explanation and methods for analysis' (Martín 2013: 4), which nevertheless has been nourished by various critical theories (Marxist, Gramscian, feminist and Foucauldian). It emerged from the critique of approaches to the environmental crisis, which focused on issues of population growth and resource scarcity rather than on the ways in which environmental destruction was mediated by various social relations of power and domination.

Nowadays, political ecology provides a conceptual framework for understanding societal nature relations beyond approaches which emphasize the threat stemming from climate change and other environmental crisis phenomena, but neglect the social relations underlying these phenomena. This also implies that political ecology makes it possible to criticize ostensibly technical fixes of the environmental crisis and to reveal the social forces and the relations of power and domination which are inscribed in them. Since the production and utilization of agrofuels can be understood as such a technical fix, approaching it from a political ecology perspective appears promising. This has indeed already been undertaken by several authors.[1]

This chapter aims to contribute to the corresponding discussions on a conceptual level. Starting with the origins and basic assumptions of political ecology and demonstrating how it transcends other approaches to the environmental crisis, I will discuss core theoretical concepts which are crucial for an analysis of the 'agrofuels project' (McMichael 2008) from a political ecology perspective: power, materiality and space. In the final section, these concepts are applied to current energy- and agrofuel-related issues and their use-value for a *political ecology of agrofuels* is discussed.

## Origins and basic assumptions of political ecology

Political ecology had its origins in the 1970s and 1980s when (mainly) geographers and anthropologists began to question the dominant way of dealing with environmental problems, which until then had been conceptualized as 'limits to growth' (Meadows *et al.* 1972), as a result of poverty and 'overpopulation' (Ehrlich 1968) or as a 'tragedy of the commons' (Hardin 1968), the latter meaning the tendency to overuse natural resources that are not protected by property rights or state regulation. According to these approaches, *humankind* was threatened because of the destructive ways in which *humans* treated nature. The *political* dimensions of the relationship between humans and nature were hardly addressed; basic social relations – class, gender, ethnicity, unequal North–South relations – were not taken into account as a factor of environmental degradation. As Budds (2008: 62) has put it, '[m]any scientific assessments exclude social processes or make generalized assumptions about the human causes of environmental degradation, especially by failing to disaggregate the actions of different social groups'.

In contrast to such an 'apolitical ecology', which takes basic social relations for granted and denies its own implicit normativity by claiming 'the objectivity of disinterest' (Robbins 2004: 5), political ecologists started to insist on the *inherently political character of environmental problems*: it is not simply the fact that humans overuse natural resources, neither is it the sheer number of humans inhabiting the Earth which produces environmental problems; nor are we 'all in the same boat'. Instead, it is the socially and geographically uneven distribution of the consequences of environmental degradation which should be tackled; and it is the social relations through which the access to, and the use of, resources and sinks are mediated that count (Paulson *et al.* 2003).

Environmental problems in political ecology are thus seen through the lens of power and domination. Social relations of power and domination cause environmental problems. At the same time, they are constituted by the way in which nature is appropriated and transformed: social power and domination depend on who controls the access to natural resources; they are mediated through control over nature. That means that environmental problems are inseparably linked to social inequality: while some suffer from them, others may benefit, or – as

Blaikie and Brookfield (1987: 14) put it – '[o]ne person's degradation is another's accumulation'.

This focus on the political dimension of environmental degradation has provoked critique from within the discipline of political ecology itself. Vayda and Walters have complained that the 'overreaction to the "ecology without politics" of three decades ago is resulting now in a "politics without ecology".' According to them, 'what are actually studied are political controls or political contests over natural resources and not, or at least not to any significant extent, how the resources are affected by those controls or contests' (Vayda and Walters 1999: 168–169). This critique is important to the extent that it stresses the role of biophysical processes, or nature's materiality, as a constitutive dimension of societal nature relations, a point which has to be taken into account when studying the political ecology of agrofuels. However, it should not lead us to abandon theoretically informed research or to give up making theoretical assumptions explicit, since the need to overcome seemingly 'apolitical' and under-theorized approaches is precisely what gave rise to political ecology.

More recently, and against the background of climate change and biodiversity loss, this need has taken the form of a new debate on natural limits. In contrast to the debate of the 1970s, it focuses less on the exhaustion of natural *resources* than on the environmental impacts of resource use: the destruction of *ecosystems* and the overstretching of the capacity of *sinks* to absorb the emissions produced by human activity. A prominent example here is the notion of 'planetary boundaries' – a quantified measure for the 'limits outside of which the Earth system cannot continue to function in a stable [...] state' (Rockström *et al.* 2009b: 474). According to Rockström and his colleagues, these boundaries have already been transgressed in three so-called subsystems of the Earth: biodiversity loss, climate change and the nitrogen cycle. These findings are indeed alarming, since they indicate 'the risk of crossing thresholds that will trigger non-linear, abrupt environmental change within continental- to planetary-scale systems' (Rockström *et al.* 2009a). As such, they may also have a politicizing effect, urging governments, or the more environmentally sensitive state apparatuses, to strengthen environmental policy efforts. However, like in the 1970s, the social relations mediating both the production of the environmental crisis and the unequal distribution of its consequences remain hidden in the scientific representation. There is hardly any reference to the *political content* of the crisis from the resilience research side (Folke 2006) to which Rockström and his colleagues belong.[2]

This is not just a scientific shortcoming. In contrast, since dominant representations of environmental problems also shape the corridor within which solutions are sought, notions like that of 'planetary boundaries' or, to give another example, 'ecosystem services' (Millennium Ecosystem Assessment 2005) may encourage technical and market-oriented strategies of ecosystem protection which are blind to complex and diverse local societal nature relations and thus run the risk of subordinating the latter to the logic of commodification, thereby threatening the very social preconditions which have sustained the ecosystems to be protected.

It is the merit of recent political ecology debates on 'green grabbing', i.e. 'the appropriation of land and resources for environmental ends' (Fairhead *et al.* 2012: 238), to have problematized these dangers and come up with first empirical findings on them (McAfee 2012). In doing so, political ecology scholars have stressed the essentially political character of environmental limits: where the prevailing scientific contributions spread the notion of *planetary boundaries*, political ecology has introduced the critical territorial concept of *new frontiers* of land control (Peluso and Lund 2011), which has already been applied to the study of the agrofuels issue (Backhouse 2013b; see Backhouse, Chapter 10 in this volume).

It is important to note that the current debate on limits or boundaries shows parallels to that of the 1970s but, nevertheless, it is not the same. Neither the findings by Rockström and his colleagues nor the concept of ecosystem services draw on neo-Malthusian arguments. Furthermore, they are very reflexive concerning uncertainties and knowledge gaps. Finally, there is no causal relationship between prevailing definitions of the environmental crisis on the one hand and technical and market-oriented solution strategies on the other. In political ecology, it has been pointed out that in order to understand the step from the scientific concept to the commodification of ecosystem services, one has to take into account the neoliberal societal and institutional environment in which the scientific debate is taking place (Gómez-Baggethun and Ruiz-Pérez 2011). Furthermore, there is a fundamental ambivalence within prevailing definitions of the environmental crisis. As Mansfield has put it regarding ecosystem services, these 'might be both a tool of dispossession and a tool for challenging dispossession' (Mansfield, cited in Dempsey and Robertson 2012: 16). The question is therefore how the concept 'can escape its own foundational commodity logic' (ibid.: 16).

By addressing the political content of prevailing forms of representing and producing nature, political ecology constitutes an important part of a critical political economy and theory of society. The principal intention of critical theory is to *de-naturalize social relations of power*, i.e. to elucidate their historical, human-made character and thus the possibility of transforming them in an emancipatory manner (Sayer 1992: 42). The specific contribution of political ecology in this context is to *de-naturalize the relations between society and nature*, i.e. to show that the predominating, destructive forms of dealing with nature are by no means simply given, 'natural' and without alternatives, but that they are only one of many possible societal nature relations and that they can and should be overcome. Thus, political ecology is part of a critical political economy and theory of society because it shares the same general intention. At the same time it goes beyond them by explicitly addressing the role of societal nature relations which in other critical approaches often remain implicit.

It is for this reason that political ecology constitutes an appropriate framework for understanding the politics and ecology of agrofuels. In particular the concepts of *power*, *materiality* and *space* and their respective theoretical underpinnings can shape a perspective from which agrofuel-related research questions can be fruitfully addressed. I will consider these concepts in the following section.

## Central concepts of political ecology

### *Power*

Political ecology draws upon different theoretical approaches, the common feature of which is nevertheless the centrality of power and domination. The concept of power may vary depending on the theoretical basis used. One widespread approach is to conceptualize power in political economy terms, i.e. as rooted in the social relations of production and distribution. To exert power means to control access to, and the distribution of, natural resources. The ability to do so depends on a person's, an organization's or a group's position in the social relations of class, gender and ethnicity, as well as with respect to the state and the capitalist world market. This position decides over the possibility to satisfy fundamental needs as well as over the exposure to environmental crisis phenomena like land degradation, water scarcity or biodiversity loss (which themselves are understood as a result of social power relations) and the opportunity to cope with them. This structural concept of power was prevalent in earlier contributions which, like the one by Blaikie and Brookfield, define political ecology as combining 'the concerns of ecology and a broadly defined political economy' (Blaikie and Brookfield 1987: 17; see also Bryant and Bailey 1997: 38–47). It is also characteristic of feminist contributions, which have pointed out the close connection between environmental degradation on the one hand and the asymmetric distribution between men and women of responsibilities, political and economic means and resource rights, on the other (e.g. Rocheleau *et al.* 1996; Agarwal 1998). Finally, the political economy concept of power is prevalent in more recent vulnerability studies which, rather than understanding vulnerability as the problem of being exposed to an external risk, conceptualize it as a distributional issue depending on access to physical infrastructure and natural resources, on class, gender and ethnicity, and on the chances of political participation (Dietz 2011).

A second concept of power in political ecology debates is influenced by Gramsci and Foucault. Of course, there are considerable differences between these two thinkers. Gramsci, as a Marxist, focused on the relations of social forces, which are ultimately based in the capitalist mode of production, whereas Foucault was interested in the 'micro-dynamics of everyday life' (Jessop and Sum 2006: 164). However, both stress the cultural rather than (Foucault), or in addition to (Gramsci), the political economy dimensions of power. Thus, practising political ecology with a Gramscian concept of hegemony would mean analysing the constitution of certain definitions of nature and the environmental crisis, the struggles over their generalization, i.e. their becoming common sense, their institutionalization as well as the marginalization of other definitions and the respective social forces (cf. Mann 2009). Similarly to the Gramscian approach, political ecology in a Foucauldian tradition focuses on the normalization of certain forms of knowledge about nature and the environmental crisis and, going beyond Gramscian contributions, on the constitution of subjectivities.

Both Foucauldian and Gramscian approaches can thus help us to understand how powerful social interests emerge and become effective, not in the form of an external constraint which simply subordinates competing interests, but as they are strengthened through the 'active consensus' of the subaltern (Gramsci) or are internalized by the latter and become a constitutive part of individual identities (Foucault). In this sense, Goldman (2004) has analysed the notion of nature as a service provider and of local populations as ecosystem managers. Similarly, Agrawal has shown the transformation of identities in rural India through new forms of environmental policy participation, with 'national and state governments [...] striving to make rural populations accomplices in environmental and their own control' (Agrawal 2005: 14).

Focusing on the cultural dimensions of power in such a way is an important extension of the political economy concept insofar as environmental conflicts are always also struggles about the concrete definition of the environmental problem which has to be treated. Depending on which definition becomes dominant or even hegemonic, the 'solution' will benefit some actors and disadvantage others. Thus, it makes a difference if climate change or the erosion of biodiversity are defined as global problems of humankind, to be solved by 'global resource managers' (Goldman 1998) and within offset schemes, or if they are considered as problems of resource control, which should be treated by strengthening the territorial rights of local communities (cf. Fogel 2004; Newell and Paterson 2010).

### *Materiality*

Materiality as a concept is central not only in political ecology but also in critical political economy and social theory in the tradition of Marx. Materiality in this tradition refers to *social forms* like the value form and the state form, i.e. social relations which, up to a certain degree, have become independent of the multiple actions of individuals and groups and in turn orientate these actions in a way that enables the reproduction of capitalist societies in spite of their inherent contradictoriness. As Sonja Buckel puts it, social forms are 'the curdled social relations which guide action in a not immediately obvious way and make fundamental social contradictions processable' (Buckel 2008: 118–119, my translation). If materiality is connected with *nature*, the latter is used more in a metaphoric sense, in order to describe the fact that human relations in capitalist societies take a commodity form and as such strike the human beings as an external, 'natural' force.[3]

There is, however, an important exception from understanding materiality only in a social sense: the elder Critical Theory of the Frankfurt School, particularly Horkheimer and Adorno. Although also using the metaphoric concept of a 'second nature' for grasping the materiality of reified social relations, their interest was directed to the dialectical relationship between physical nature and capitalist societies. In the 'Dialectic of Enlightenment', Horkheimer and Adorno identified the pursuit of an ever more perfect domination of nature as a constitutive feature of capitalist modernity. Driven by the idea of reducing the dependency of society on nature, the pursuit of domination has even enhanced this dependency since it has neglected, or abstracted from, the peculiarities of nature (Horkheimer and Adorno 1990 [1944]).[4] Today's most visible expression of this dialectic is the various phenomena of the environmental crisis.

Although in international political ecology debates there is little explicit reference to the Frankfurt School (for exceptions see Görg 2003 and Biro 2011), political ecology has a strong concept of physical materiality which resembles that of Critical Theory. This applies particularly to those contributions which have shown how the production and appropriation of nature may fail because of the physical properties of the resource to be appropriated. The attempt to liberalize water provision in the United Kingdom, for example, met with a lot of difficulties that its proponents had not foreseen. The reason for the difficulties lay in the bio-chemical properties of water: the problem of transporting drinking water over large distances or mixing water from different sources without risking undesirable chemical reactions that diminish the quality of water. Water thus balks at certain forms of its commodification, which is why Bakker (2003), who analysed this issue, called water an 'uncooperative commodity'.

Tension can occur, too, between different dimensions of physical materiality, in the case of water between nature and the built environment in the form of water infrastructures. Once created, the latter require a minimum flow rate in order to function and thus contribute to generating certain consumption patterns. In times of drought, these patterns and the underlying requirements of the infrastructure system can come into conflict with the requirements of the hydrological balance. Maintaining the functioning of the former can threaten the reproduction of the latter, and vice versa. This tension between nature and the built environment gives rise to conflicts along various lines: between providers and consumers, between different types of consumers and between consumer interests and those of environmental protection (Wissen 2009).

Taking physical materiality into account in order to better understand social processes should not be confused with a deterministic concept of nature or the built environment as external entities that require society to comply with their laws or modes of operation. Even in the form of environmental crisis, nature's materiality does not determine social action. Instead, not only the question of how to deal with the crisis remains contested, but also how to define it, i.e. to determine whose, and what kind of, problem it is. Furthermore, as crisis phenomena like climate change show, physical materiality itself already has a strong social content. It may be not just *physical* in the strong sense of the word, but also socially produced as much as it is, in turn, constitutive for social processes. The merit of political ecology is that it has pointed out this dialectic and further developed a non-essentialist understanding of the 'productive or generative role' (Bakker and Bridge 2006: 9) that ecological processes play in relation to society. In doing so, it has also contributed to correcting influential notions like Neil Smith's (1984) classical account of the production of nature which, in attempting to show how capitalism has inscribed itself into nature, has neglected nature's materiality. As Castree (2000: 29) puts it:

created ecosystems, while intentionally and unintentionally produced by capitalism, possess causal powers of their own and take on agency in relation to the capitalist processes of which they are a medium and outcome. To phrase all of this in Smith's language, nature may indeed be 'produced' but produced nature, in turn, cannot be exploited indefinitely: it has a materiality which cannot be ignored.

### *Space*

A third concept which is central to political ecology is space (cf. the contribution by Dietz *et al.*, Chapter 3 in this volume). The space which interests political ecology is socially produced; it is a physical or institutional manifestation of social forms. In turn, it affects social processes and relations of forces, and it gives them a certain direction and benefits some forces while disadvantaging others. This was famously pointed out by Henri Lefebvre, whose crucial insights into the 'production of space' (Lefebvre 1991 [1974]) have provided an essential basis for critical geographical and political ecology debates. Space, as a social product and as an implication of social processes, is similar to physical materiality. In the form of the built environment, the latter indeed constitutes an important spatial dimension. However, space encompasses more than just the physical dimension (just as physical materiality, as seen above, cannot be reduced to the built environment but also implies nature).

Jessop *et al.* (2008) have proposed a systematization of socio-spatial relations along various dimensions. They distinguish between territory (constituted by borders, which separate an inside from an outside), place (created by face-to-face relations between people and by the meaning which arises from these relations and is attached, for example, to a village, a town or a region), network (which, like territory, is a horizontal dimension of space but, in contrast to territory, does not cover but spans space), and scale (the vertical dimension of space, constituted through processes such as globalization and regionalization, which emerge from territorially-bounded forces and their interaction and, in turn, transform them, for example by inserting them into overarching units like the European Union or a liberalized world market).

Applying these spatial concepts in political ecology contributes to a better understanding of both socio-environmental and broader political processes. The concept of territory has been fruitfully deployed by Neumann in his analysis of national parks and their role in the constitution of the modern state. Taking the Selous Game Reserve in Tanzania as an example, Neumann (2004) shows how the enclosure of local commons and their transformation into national parks implies the constitutive elements of the modern state: drawing borders, controlling resources, creating traditions of progress and environmental protection, delegitimizing native societal nature relations through 'degradation narratives' and eliminating the history and spatiality of local populations. According to Neumann 'these proprietary claims and the process of mapping, bounding, containing and controlling nature and citizenry are what make a state a state. States come into being through these claims and the assertion of control over territory, resources, and people' (Neumann 2004: 202; see also Scott 1998).

Concerning the scalar dimension of space, Swyngedouw has shown that processes of rescaling are closely linked to the transformation of the conditions of access to, and control over, nature. He analyses the rescaling of water resource management as a means of shifting the social relations of forces in Spain at the end of the nineteenth and beginning of the twentieth century. At that time, modernizing forces in Spanish society tried to establish the river basin as a new spatial scale of water resource management. In so doing, they intended to withdraw the water management competences from the municipal, provincial and national scales and thereby to weaken the semi-feudal elites who organized themselves on these scales. Swyngedouw thus demonstrates that:

> nature and environmental transformation are [...] integral parts of the social and material production of scale. More importantly, scalar reconfigurations also produce new sociophysical ecological scales that shape in important ways who will have access to what kind of nature, and the particular trajectories of environmental change.
>
> (2004: 134)[5]

The concepts of place and network have been applied mainly in poststructuralist contributions to political ecology like those of Escobar (who also connects them to the categories of scale and territory). Starting from the assumption that the 'reassertion of place [...] appears as an important arena for rethinking and reworking Eurocentric forms of analysis' (Escobar 2001: 141), Escobar makes visible place-based practices with regard to nature, which are often more reflexive than mere capitalist ones but which can easily be overseen even by critical researchers when they focus on the structuring forces of global capitalism. He elucidates the critique of power and hegemony that is inherent in the practices of the black communities in the Pacific rainforest region of Colombia and gains strength not least through their insertion into transnational networks of social movements (Escobar 2001, 2008; see also Gibson-Graham 2002).

In spite of different spatial concepts and social theories, all these accounts of space have in common that they do not consider a spatial perspective as an end in itself. Instead, analysing spatial configurations should help us to understand socio-environmental conflicts, to excavate relations of power and domination and to identify the possibilities of overcoming them. This is in line with the conceptualization of space in the work of Lefebvre, and it is important in order to avoid the danger of isolating space from social practice. As Margit Mayer puts it, the:

> relevance of a particular spatial form – either for explaining certain social processes or for acting on them – can be measured only from the perspective of the engaged actors. Thus, in order to define criteria for the relevance of (a specific form of) spatiality, we need to start, both in our theoretical endeavors

as well as in political practice, from concrete social processes and practices rather than reifying spatial dimensions.

(2008: 416)

## Towards a political ecology of agrofuels

The concepts of space, materiality and power, as they have been applied and further developed in political ecology, offer several possibilities for enriching the study of energy- and agrofuel-related issues. Indeed, critical research on these issues has to a large part been undertaken from a political ecology perspective and, in doing so, drawn on the concepts as outlined above (or on specific expressions of them). In this section, I discuss the role the concepts play in energy and agrofuels research from a political ecology perspective and how they improve our understanding of ongoing transformations.

As far as *power* is concerned it is needless to say that the expansion of agrofuel production is not just a technical issue, but a strategy driven by economic and political interests. However, there are various mechanisms of power at work and different theoretical perspectives from which they can be analysed. And depending on the concrete mechanism, or the relationship between different mechanisms, as well as depending on the theoretical perspective, the assessment of concrete constellations of social forces and of the potentials of emancipatory movements will differ.

As seen above, an important form of power is rooted in the capitalist relations of production. As far as the energy sector in general and agrofuels in particular are concerned, this power articulates itself not just in the 'normal' capitalist process of extended reproduction, in that it shapes the relation between capitalists and workers or between individual firms and countries. Instead, an important tendency that has been observed recently is the commodification of social relations and parts of nature that were not previously subordinated to capitalist relations of production. This is due to the current fundamental reorganization of the energy system: if the provision of energy becomes increasingly dependent on renewables instead of fossil biomass, control over arable land will gain a new significance (cf. Haberl *et al.* 2011).

The current forms of commodifying land, which have been denoted as *land grabbing* (Borras Jr. *et al.* 2011), *green grabbing* (Fairhead *et al.* 2012) or, if driven by financial capital, as *financialization of nature* (Kaltenbrunner *et al.* 2011; Tricarico 2011), have to be seen in this context. A Marxist theoretical framework for understanding this manifestation of power has been offered by David Harvey (2003). His notion of 'accumulation by dispossession' allows the conceptualization of the recent global 'land rush' (Arezki *et al.* 2012) in the context of the environmental, energy and economic crisis of capitalism. The massive extension of the production of agrofuels then becomes understandable as being driven by the over-accumulation of capital in the global North and by the environmental and energetic limits being approached by the resource- and emission-intensive *imperial mode of living* (Brand and Wissen 2012, 2013) that is dominant in Northern societies and increasingly spreading to Southern ones.

In this sense, Gillon, following Castree (2008), has analysed agrofuel production 'as an "environmental fix", symptomatic of contradictory imperatives of capitalism to both exploit and conserve natural resources for accumulation' (Gillon 2010: 726).

In order to explain whether, and to what extent, agrofuel politics is successful from the point of view of its proponents and in spite of its destructive socio-environmental consequences, it is helpful to modify the structural Marxist concept of power with a Gramscian and/or Foucauldian one. With Gramsci, we can understand how and why local populations are eventually integrated into a hegemonic agrofuels project, for example by offering them new income and employment opportunities and spreading new images of economic success, and how conflicting interests are successfully marginalized. Foucault allows us to explain the subjective dimensions of the project, i.e. for example whether, and how, new entrepreneurial subjectivities are created among local farmers. The integration of smallholders into the agrofuels projects and the attempt to spread an 'entrepreneurial culture' among them have been analysed e.g. for palm oil production in Colombia. Here, so-called 'strategic alliances' between peasants and palm oil companies have been created with strong support by the government and with the objective 'to legitimize and ratify a massive land grab by paramilitary-backed oil palm companies' in Afro-Colombian territories (Ballvé 2012: 617; see also Gómez 2010).

As far as physical *materiality* is concerned, the agrofuels project first has to be understood in the context of capitalism's ecological contradictions. Insofar as capitalism, since its beginning, has been fuelled by fossil energies, it has proven blind to the reproductive necessities of nature. The result is that resources, the development of which required millions of years, have been depleted in a few hundred years and that the specific form in which these resources have been consumed, i.e. utilizing their energetic potential through burning them, has caused climate change (Altvater 2005; Huber 2008). This is one of the most severe manifestations of the material contradictions of capitalism, which are now to be fixed with the partial substitution of fossil energy sources by agrofuels.

A second point which has to be considered when nature's materiality is taken seriously is the way the physical properties of the plants to be grown influence the social and spatial relations of agrofuel production. Some plants, like sugar cane, are grown on big plantations, whereas others may allow small-scale or smallholder production. A question to be raised here is what this means for power relations in agrofuel production and for the chances of weaker actors to articulate their interests. Furthermore, the crops used for the production of agrofuels pose certain requirements regarding their further processing. The fruits of oil palms, for example, have to be processed within 24 hours of being harvested. This gives rise to the industrialization of rural areas, accompanied by a shift in social structures and conflicts, and by the transformation of the built environment through the creation of transport infrastructures (Backhouse 2013a; Pye 2013).

Third, the large-scale industrial production of agrofuels can provoke unintended consequences like diseases, soil degradation and water pollution. In

Colombia for example, oil palm plantations have been devastated by a disease called 'bud rot' (*phytophthora cinnamomi* – PC). Although its cause has not yet been fully understood, there is evidence of a close relationship between the fast spread of the microorganism causing the disease and the large-scale, monocultural production of agrofuels:

> Historically, when the focus of palm cultivation was cooking oil and cosmetics, Colombia always had manageable amounts of PC [...]. But all that changed this decade when palm oil became desirable as a cleanburning substitute for diesel, and acreage doubled.
>
> (Kraul 2010)[6]

There is thus evidence that the specific materialities of the plants utilized may balk at large-scale agrofuel production. Instead of functioning as an environmental fix for the material contradictions of capitalism, the agrofuels project provokes its own contradictions, which threaten its viability and may give rise to new social conflicts and shifts in power relations.

In *spatial* terms, the agrofuels project is one of the most recent expressions of a spatial decoupling of resource extraction and resource consumption and thereby of an unequal distribution of the damages and benefits caused by certain patterns of producing and appropriating nature. In order to maintain, or modernize, fossilist consumption patterns in the global North, ecosystems in the global South are being destroyed, people are being evicted from their traditional places and the capacity of Southern countries to adapt to climate change is being undermined (Dauvergne and Neville 2009; Pye 2009). New frontiers are being created, both in an ecological sense (indirect land-use change as a consequence of an expanding agrofuel production) and in a social sense (new strategies of exclusion and integration), and new territorial units are being produced and inserted into global agricultural and financial markets (Holt-Giménez and Shattuck 2009).

A spatial perspective helps, first, to make these processes visible and to describe them in detail. Second, it helps to explain them by placing them in the context of broader tendencies of capitalist development (like the de- and re-territorialization of capital in the course of the current economic crisis). Third, it sheds light on the problems and requirements of a progressive politicization of the agrofuels issue. In this context, Pye (2010) has identified a mismatch between global campaigns against agrofuels and the interests of local smallholders and plantation workers. Whereas the former focus on the socio-environmental consequences of agrofuels production, such as its threat to biodiversity, climate justice and the global climate itself,[7] the latter are more concerned with land rights and employment conditions. A spatially informed political ecology perspective thus reveals that the *metabolic* rift, for which the spatial decoupling of the production and consumption of agrofuels stands, is accompanied by a *scalar* rift in the manner the resulting problems are politicized.

## Conclusions

The objective of this chapter was to identify crucial concepts that have been applied in political ecology and to discuss their use-value for an analysis of the agrofuels project. The latter can be understood as a technical fix for the environmental contradictions of capitalism. Analysing it from a political ecology perspective allows us to explain the different dimensions of *power and domination* on which the extension of agrofuel production is based. It helps us to understand whether, and under what conditions, the production of agrofuels may become a dominant or even hegemonic strategy and thus gain the character of a societal 'project', i.e. shape discourses, integrate subaltern forces and inscribe itself into the state apparatus.

However, the success of agrofuel production as a project depends not only on the social relations of forces, but also on the *materiality* of the plants from which agrofuels are extracted. This is a crucial insight of more recent political ecology discussions: nature is not only socially produced, but it also produces society, i.e. it constitutes an essential factor of social processes. As such it may both strengthen and undermine social relations of power and domination. It is important to take this into account when studying the agrofuels issue from a political ecology perspective: to carefully identify not only the hegemonic potentials but also the material contradictions (e.g. in the form of diseases), which the agrofuels project intensifies, which severely question its functionality as a technical fix for the crisis of fossil capitalism and which may become the starting-point for the politicization of this seemingly technical issue.

A *spatial* perspective – besides contributing to identifying how social relations of power and domination are inscribed into the built environment, how they create new enclosures and territories and how they insert places and regions into new scalar configurations and networks – helps to further elucidate the conditions for a successful emancipatory politicization of the agrofuels project. As the findings and concepts discussed above reveal, these conditions essentially consist of building up multi-scalar alliances which aim to strengthen the territorial rights of agricultural producers, as a means of combating both the environmental and the social destructiveness of the agrofuels project.

## Notes

1 See particularly the special issue of *The Journal of Peasant Studies*, 37, 4 (2010).

2 A fruitful extension of the planetary boundaries concept by the notion of 'social boundaries' has been developed by Raworth, who emphasizes that the 'biggest source of planetary boundary stress today is the excessive consumption levels of roughly the wealthiest 10 per cent of people in the world, and the production patterns of the companies producing the goods and services that they buy' (Raworth 2012: 19).

3 See, for this, the section on the commodity fetish in the first volume of *Capital* (Marx 1988 [1867]: 85–98) and the concept of reification in Lukács' essay 'Reification and the Consciousness of the Proletariat' (Lukács 1988 [1922]: 170–355).

4 For an excellent reconstruction of the elder Critical Theory from the perspective of societal nature relations, see Görg (2003).

5 For similar treatments of the relationship between nature and scale see McCarthy (2005), Köhler (2008), Neumann (2009) and Wissen (2011).
6 Similar problems have been observed in Brazil (see Backhouse 2013a).
7 Life cycle analyses have revealed that the savings of $CO_2$ emissions through agrofuel consumption could be overcompensated for by the $CO_2$ intensity of their agro-industrial production (see McMichael 2008).

## References

Agarwal, Bina (1998) 'The Gender and Environment Debate', in Roger Keil, David V.J. Bell, Peter Penz and Leesa Fawcett (eds) *Political Ecology. Global and Local*. London: Routledge, 193–219.

Agrawal, Arun (2005) *Environmentality. Technologies of Government and the Making of Subjects*. Durham: Duke University Press.

Altvater, Elmar (2005) *Das Ende des Kapitalismus, wie wir ihn kennen. Eine radikale Kapitalismuskritik*. Münster: Westfälisches Dampfboot.

Arezki, Rabah, Klaus Deininger and Harris Selod (2012) 'The Global Land Rush. Foreign Investors are Buying up Farmland in Developing Countries', *Finance & Development* 49(1), 46–49.

Backhouse, Maria (2013a) 'Die Landfrage im Kontext der Palmölexpansion in Pará/Brasilien', in Hans-Jürgen Burchardt, Kristina Dietz and Rainer Öhlschläger (eds) *Umwelt und Entwicklung im 21. Jahrhundert. Impulse und Analysen aus Lateinamerika*. Baden-Baden: Nomos, 135–149.

Backhouse, Maria (2013b) *Palmölproduktion in Pará – Eine neue, grüne Landnahme? Fair Fuels? Working Paper 6*. Berlin.

Bakker, Karen (2003) 'From Public to Private to … Mutual? Restructuring Water Supply Governance in England and Wales', *Geoforum* 34(3), 359–374.

Bakker, Karen and Gavin Bridge (2006) 'Material worlds? Resource Geographies and the "Matter of Nature"', *Progress in Human Geography* 30(1), 5–27.

Ballvé, Teo (2012) 'Everyday State Formation: Territory, Decentralization, and the Narco Landgrab in Colombia', *Environment and Planning D: Society and Space* 30(4), 603–622.

Biro, Andrew (ed.) (2011) *Critical Ecologies: The Frankfurt School and Contemporary Environmental Crises*. Toronto: University of Toronto Press.

Blaikie, Piers and Harold Brookfield (1987) 'Defining and Debating the Problem', in Piers Blaikie and Harold Brookfield (eds) *Land Degradation and Society*. London: Routledge, 1–26.

Borras Jr., Saturnino M., Jennifer C. Franco, Cristobal Kay and Max Spoor (2011) *Land Grabbing in Latin America and the Caribbean Viewed from Broader International Perspectives*. Santiago de Chile: Food and Agriculture Organization of the United Nations.

Brand, Ulrich and Markus Wissen (2012) 'Global Environmental Politics and the Imperial Mode of Living. Articulations of State-Capital Relations in the Multiple Crisis', *Globalizations* 9(4), 547–560.

Brand, Ulrich and Markus Wissen (2013) 'Crisis and Continuity of Capitalist Society-nature Relationships. The Imperial Mode of Living and the Limits to Environmental Governance', *Review of International Political Economy* 20(4), 687–711.

Bryant, Raymond L. and Sinéad Bailey (1997) *Third World Political Ecology*. London: Routledge.

Buckel, Sonja (2008) 'Zwischen Schutz und Maskerade – Kritik(en) des Rechts', in Alex Demirovic (ed.) *Kritik und Materialität*. Münster: Westfälisches Dampfboot, 110–131.

Budds, Jessica (2008) 'Whose Scarcity? The *Hydrosocial* Cycle and the Changing Waterscape of La Ligua River Basin, Chile', in Michael K. Goodman, Maxwell T. Boykoff and Kyle T. Evered (eds) *Contentious Geographies. Environmental Knowledge, Meaning, Scale*. Aldershot: Ashgate, 59–78.

Castree, Noel (2000) 'Marxism and the Production of Nature', *Capital and Class* 24(3), 5–36.

Castree, Noel (2008) 'Neoliberalising Nature: The Logics of Deregulation and Reregulation', *Environment and Planning A* 40(1), 131–152.

Dauvergne, Peter and Kate J. Neville (2009) 'The Changing North–South and South–South Political Economy of Biofuels', *Third World Quarterly* 30(6), 1087–1102.

Dempsey, Jessica and Morgan M. Robertson (2012) 'Ecosystem Services: Tensions, Impurities, and Points of Engagement within Neoliberalism', *Progress in Human Geography* 37(6), 758–779.

Dietz, Kristina (2011) *Der Klimwandel als Demokratiefrage. Sozial-ökologische und politische Dimensionen von Vulnerabilität in Nicaragua und Tansania*. Münster: Westfälisches Dampfboot.

Ehrlich, Paul (1968) *The Population Bomb*. New York: Ballantine Books.

Escobar, Arturo (2001) 'Culture Sits in Places: Reflections on Globalism and Subaltern Strategies of Localization', *Political Geography* 20, 139–174.

Escobar, Arturo (2008) *Territories of Difference*. Durham: Duke University Press.

Fairhead, James, Melissa Leach and Ian Scoones (2012) 'Green Grabbing: A New Appropriation of Nature?', *The Journal of Peasant Studies* 39(2), 237–261.

Fogel, Cathleen (2004) 'The Local, the Global, and the Kyoto Protocol', in Sheila Jasanoff and Marybeth Long Martello (eds) *Earthly Politics. Local and Global in Environmental Governance*. Cambridge, MA: MIT Press, 103–125.

Folke, Carl (2006) 'Resilience: The Emergence of a Perspective for Social–Ecological Systems Analyses', *Global Environmental Change* 16(3), 253–267.

Gibson-Graham, J.K. (2002) 'Beyond Global vs. Local: Economic Politics Outside the Binary Frame', in Andrew Herod and Melissa W. Wright (eds) *Geographies of Power. Placing Scale*. Oxford: Blackwell, 25–60.

Gillon, Sean (2010) 'Fields of Dreams: Negotiating an Ethanol Agenda in the Midwest United States', *The Journal of Peasant Studies* 37(4), 723–748.

Goldman, Michael (1998) 'Allmacht und Allmende. Die "Commons"-Debatte und der Aufstieg der globalen Ressourcenmanager', in Michael Flitner, Christoph Görg and Volker Heins (eds) *Konfliktfeld Natur. Biologische Ressourcen und globale Politik*. Opladen: Leske and Budrich, 87–118.

Goldman, Michael (2004) 'Eco-governmentality and Other Transnational Practices of a "Green" World Bank', in Richard Peet and Michael Watts (eds) *Liberation Ecologies. Environment, Development, Social Movements*. London: Routledge, 166–192.

Gómez-Baggethun, Erik and Manuel Ruiz-Pérez (2011) 'Economic Valuation and the Commodification of Ecosystem Services', *Progress in Physical Geography* 35(5), 613–628.

Gómez, Andrés (2010) 'La alianza productiva y social de María la Baja en Colombia. Un contraste de lo encontrado con los discursos del desarrollo local', *Eutopia* (1), 85–99.

Görg, Christoph (2003) *Regulation der Naturverhältnisse. Zu einer kritischen Theorie der ökologischen Krise*. Münster: Westfälisches Dampfboot.

Haberl, Helmut, Marina Fischer-Kowalski and Fridolin Krausmann (2011) 'A Socio-metabolic

Transition towards Sustainability? Challenges for Another Great Transformation', *Sustainable Development* 19(1), 1–14.

Hardin, Garrett (1968) 'The Tragedy of the Commons. The Population Problem Has no Technical Solution; It Requires a Fundamental Extension in Morality', *Science* 162, 1243–1248.

Harvey, David (2003) *The New Imperialism*. Oxford: Oxford University Press.

Holt-Giménez, Eric and Annie Shattuck (2009) 'The Agrofuels Transition: Restructuring Places and Spaces in the Global Food System', *Bulletin of Science Technology & Society* 29, 180–188.

Horkheimer, Max and Theodor W. Adorno (1990 [1944]) *Dialektik der Aufklärung. Philosophische Fragmente*. Frankfurt am Main: Fischer.

Huber, Matthew T. (2008) 'Energizing Historical Materialism: Fossil Fuels, Space and the Capitalist Mode of Production', *Geoforum* 40(105–115).

Jessop, Bob, Neil Brenner and Martin Jones (2008) 'Theorizing Sociospatial Relations', *Environment and Planning D: Society and Space* 26, 389–401.

Jessop, Bob and Ngai-Ling Sum (2006) 'Towards a Cultural Political Economy: Poststructuralism and the Italian School', in Marieke de Goede (ed.) *International Political Economy and Poststructural Politics*. New York: Palgrave, 157–176.

Kaltenbrunner, Annina, Susan Newman and Juan Pablo Painceira (2011) *The Financialisation of Natural Resources: Understanding the New Dynamics and Developing Civil Society Answers to It*. Financialisation of natural resources – understanding the new dynamics and developing civil society responses, Paris.

Köhler, Bettina (2008) 'Die Materialität von Rescaling-Prozessen. Zum Verhältnis von Politics of Scale und Political Ecology', in Markus Wissen, Bernd Röttger and Susanne Heeg (eds) *Politics of Scale. Räume der Globalisierung und Perspektiven emanzipatorischer Politik*. Münster: Westfälisches Dampfboot, 208–223.

Kraul, Chris (2010) 'Disease Lays Waste to Colombia Oil Palms', *Los Angeles Times*, 7 April 2010.

Lefebvre, Henri (1991 [1974]) *The Production of Space*. Malden, MA: Blackwell.

Lukács, Georg (1988 [1922]) *Geschichte und Klassenbewusstsein. Studien über marxistische Dialektik*. Darmstadt: Luchterhand.

McAfee, Kathleen (2012) 'The Contradictory Logic of Global Ecosystem Services Markets', *Development and Change* 43(1), 105–131.

McCarthy, James (2005) 'Scale, Sovereignty, and Strategy in Environmental Governance', *Antipode* 37(4), 731–753.

McMichael, Philip (2008) 'Agro-fuels, Food Security, and the Metabolic Rift', *Kurswechsel* (3),14–22.

Mann, Geoff (2009) 'Should Political Ecology be Marxist? A Case for Gramsci's Historical Materialism', *Geoforum* 40, 335–344.

Martín, Facundo (2013) Latin American Political Ecology and the World Ecological Crisis. Recent Developments, Contributions and Dialogues with the Global Field. Paper presented at the 8th Pan-European Conference on International Relations. Warsaw, 18–21 September 2013.

Marx, Karl (1988 [1867]) *Das Kapital. Kritik der politischen Ökonomie. Erster Band*. Berlin: Dietz Verlag.

Mayer, Margit (2008) 'To What End Do We Theorize Sociospatial Relations?', *Environment and Planning D: Society and Space* 26, 414–419.

Meadows, Donella H., Dennis Meadows, Jorgen Randers and William Behrens (1972) *The Limits to Growth*. New York: Universe Books.

Millennium Ecosystem Assessment (2005) *Ecosystems and Human Well-being: Synthesis*. Washington, DC: Island Press.

Neumann, Roderick P. (2004) 'Nature – State – Territory. Toward a Critical Theorization of Conservation Enclosures', in: Richard Peet and Michael Watts (eds) *Liberation Ecologies. Environment, Development, Social Movements*. London: Routledge, 195–217.

Neumann, Roderick P. (2009) 'Political Ecology: Theorizing Scale', *Progress in Human Geography* 33(3), 398–406.

Newell, Peter and Matthew Paterson (2010) *Climate Capitalism. Global Warming and the Transformation of the Global Economy*. Cambridge: Cambridge University Press.

Paulson, Susan, Lisa L. Gezon and Michael Watts (2003) 'Locating the Political in Political Ecology: An Introduction', *Human Organization* 62(3), 205–217.

Peluso, Nancy Lee and Christian Lund (2011) 'New Frontiers of Land Control: Introduction', *The Journal of Peasant Studies* 38(4), 667–681.

Pye, Oliver (2009) 'Biospritbankrott: Europäische Klimapolitik, Palmöl und kapitalistische Naturverhältnisse in Südostasien', *PROKLA. Zeitschrift für kritische Sozialwissenschaft* 39(3), 441–457.

Pye, Oliver (2010) 'The Biofuel Connection – Transnational Activism and the Palm Oil Boom', *The Journal of Peasant Studies* 37(4), 851–874.

Pye, Oliver (2013) 'Migration, Netzwerke und Alltagswiderstand: Die umkämpften Räume der Palmölindustrie', *Peripherie* (132), 466–493.

Raworth, Kate (2012) *A Safe and Just Space for humanity. Can We Live within the Doughnut?* Oxfam Discussion Papers. Oxford: Oxfam.

Robbins, Paul (2004) *Political Ecology. A Critical Introduction*. Malden, MA: Blackwell.

Rocheleau, Dianne, Barbara Thomas-Slayter and Esther Wangari (1996) 'Gender and Environment', in Dianne Rocheleau, Barbara Thomas-Slayter and Esther Wangari (eds) *Feminist Political Ecology. Global Issues and Local Experiences*. London: Routledge, 3–23.

Rockström, Johan, Will Steffen and Kevin Noone (2009a) 'Planetary Boundaries: Exploring the Safe Operating Space for Humanity', *Ecology and Society* 14(2).

Rockström, Johan, Will Steffen and Kevin Noone (2009b) 'A Safe Operating Space for Humanity', *Nature* 461, 472–475.

Sayer, Andrew (1992) *Method in Social Science. A Realist Approach*. London/New York: Routledge.

Scott, James C. (1998) *Seeing Like A State. How Certain Schemes to Improve the Human Condition Have Failed*. New Haven: Yale University Press.

Smith, Neil (1984) *Uneven Development. Nature, Capital and the Production of Space*. Oxford: Blackwell.

Swyngedouw, Erik (2004) 'Scaled Geographies: Nature, Place, and the Politics of Scale', in Eric Sheppard and Robert B. McMaster (eds) *Scale and Geographic Inquiry. Nature, Society and Method*. Oxford: Blackwell, 129–153.

Tricarico, Antonio (2011) 'The "financial enclosure" of the commons. Financialisation of natural resources – understanding the new dynamics and developing civil society responses', Conference in Paris, 28–29 October. Available at: www.un-ngls.org/gsp/docs/Financialisation_natural_resources_draft_2.pdf (accessed: 8 August 2014).

Vayda, Andrew P. and Bradley B. Walters (1999) 'Against Political Ecology', *Human Ecology* 27(1), 167–179.

Wissen, Markus (2009) 'Wassermangel im Überfluss – zum Spannungsverhältnis von Infrastruktur- und Wasserhaushaltsproblemen', in Christoph Bernhardt, Heiderose

Kilper and Timothy Moss (eds) *Im Interesse des Gemeinwohls. Regionale Gemeinschaftsgüter in Geschichte, Politik und Planung*. Frankfurt am Main: Campus, 115–151.

Wissen, Markus (2011) *Gesellschaftliche Naturverhältnisse in der Internationalisierung des Staates. Konflikte um die Räumlichkeit staatlicher Politik und die Kontrolle natürlicher Ressourcen*. Münster: Westfälisches Dampfboot.

# 3 Territory, scale and networks

## The spatial dynamics of agrofuels

*Kristina Dietz, Bettina Engels and Oliver Pye*[1]

### Introduction

The socio-spatial implications of agrofuels are substantial. The increase in agrofuels consumption in Europe for example (EurObserv'ER 2012) is closely linked to the boom in flex crops and agrofuels production in Asia, Latin America and sub-Saharan Africa, thus indicating the interconnected socio-spatial dynamics of agrofuels across world regions and national boundaries (Danker *et al.* 2013). Because of this growing spatial decoupling of consumption and production at the global scale, much of the critique of agrofuels is spatialized as a North–South exploitative relation, symbolized by images that show Northern car owners 'tanking' Southern land conflicts or rainforest destruction. In their analysis of the spatial restructuring of the 'agrofuels transition', Eric Holt-Giménez and Annie Shattuck (2009: 181) argue that 'the poor, food-insecure countries of the Global South are being called upon to supply ethanol and biodiesel for the overconsumption of liquid fuel in the affluent North'. In this chapter, we demonstrate that this argument, although correct in its politicization of certain agrofuel-related conflicts, is too one-dimensional to provide an accurate understanding of the interrelated and complex socio-spatial dynamics of the 'agrofuels project' (McMichael 2008: 14).

Most crops used for agrofuels are produced in agroindustrial complexes relying on large-scale, spatially concentrated monoculture plantations, modern technologies and, not least, capital investment. The spatial expansion of these modes of agricultural production into areas hitherto sparsely opened up for capital, results in new spatial demarcations, new spatially mediated modes of exercising power and authority, of inclusion and exclusion, and in the massive transformation of socio-ecological landscapes. These transformations are behind the land conflicts that accompany agrofuels expansion in the South. But the new global production networks that drive the expansion are not only created by Northern-based corporations. New transnational 'South–South–North' relations and networks with powerful private and state actors from the 'South' emerge within these transnational spaces, thereby changing the structures of global political economy (Dauvergne and Neville 2009, 2010; Hollander 2010).

In addition, investment decisions are not just made by (trans-)national agroindustrial companies, national, supra- or sub-national parliaments and governmental entities, but increasingly at the scale of globalized financial markets. Speculative trading in palm oil, ethanol or soy futures not only impacts global food prices, it also results in spatial changes and restructuring. Decisions made by hedge funds, pension funds, banks and other financial institutions influence whether and where capital is circulated for new plantations, thereby evoking spatial transformations. Financial markets are thus themselves a scale of regulation interacting with other socio-spatial dynamics of agrofuels.

Many of the socio-spatial dynamics of agrofuels result from their characteristics as flex crops. Sugar, rape seed, palm oil and soybeans can be used interchangeably for food and fuel, so agrofuels from these sources exaggerate existing tendencies of agri-food capitalism and, as Philip McMichael emphasizes (Chapter 6 in this volume), are contributing to a crisis of the 'corporate food regime'. For this reason, Ben White and Anirban Dasgupta (2010: 599) argue that 'agro-fuels capitalism [is not] essentially different from other forms of capitalist monocrop production'. But this only holds true for the impact of agrofuels on agrarian transformation. At the regional and local scale of production, it indeed makes little difference to indigenous peoples, small-scale farmers, or workers in the agribusiness plantations whether the crops they produce, or that displace them, end up as fuel or food. But at other scales, agrofuels are characterized by new power relations, political projects, and discourses. A look at the national and supra-national scale shows that agrofuel politics are underscored by new discourses of national development, energy security and climate change. Transnationally, agrofuels link together state actors, agribusiness and chemical and energy industries, creating new knots of power within a global network that is plugged into financial markets.

This chapter explores the spatial dynamics of the political ecology of agrofuels beyond a one-dimensional and state-centred analysis of North–South relations. Its aim is twofold: outlining the analytical potential that spatial categories have for politico-ecological research on agrofuels; and tracing the distinctive socio-spatial dynamics that are created through agrofuel politics. We start from the assumption that spatial categories enable a comprehensive analysis that can link the macro-structures of the global political economy of agrofuels to concrete, place-based struggles. We further assume that an analysis of the socio-economic and political dynamics of agrofuels through a spatial perspective will provide insights to different aspects of social inequalities that accompany the 'agrofuels project'.

We start with general reflections related to our theoretical foundations and the relevance of 'space' within the field of political ecology. We then proceed to sketch three spatial dynamics that we see as crucial for an encompassing understanding of the complex social and political dynamics of the 'agrofuels project': territorialization, global production networks and the global scale of financialization. The chapter concludes, first by summarizing why 'space' and spatial categories matter for a political ecology of agrofuels and second, by discussing the socio-political implications of our analytical approach. The material for this article comes from both literature reviews and our own research conducted in different world regions and sub-national contexts on territorialization effects, struggles over land, global production networks that span different world regions and the finanzialization of agrofuels.[2]

## Theoretical foundations: spatial perspectives

Space is relevant in the social production, use, and regulation of agrofuels and the spatial dynamics are at the same time socially produced and contested. Core analytical categories are: *place* (socially constructed locations, contingent and 'filled up' with historical experiences and social meanings) (Harvey 1996; Massey 1991, 1994; Escobar 2008); *scale* (the vertical dimension of space, socially produced and politically contested) (Herod 2011); *territory* (border demarcations as a means of exercising power, categorizing and in – or excluding through – space)[3] (Peluso and Lund 2011; Vandergeest and Peluso 1995; Sack 1986; Taylor 2003); and *network* (transversal structures, forms of interspatial interconnections between places, things, actors and institutions) (Latour 2005; Castells 2010; Sheppard 2002). Instead of privileging one dimension over others, Bob Jessop *et al.* suggest conceptualizing different dimensions as 'mutually constitutive and relationally intertwined dimensions of socio-spatial relations' (Jessop *et al.* 2008: 389). We share this relational and multidimensional approach of theorizing socio-spatial relations. However, a relational spatial perspective on agrofuels is not an end in itself. Rather, it provides analytical tools that can help us to understand and explain social processes, the (re-)production of social power relations and social inequalities or the creation of new geographies of power and hegemony. It is in this sense that Margit Mayer claims that the:

> relevance of a particular spatial form … can be measured only from the perspective of the engaged actors. Thus, in order to define criteria for the relevance of (a specific form of) spatiality, we need to start, both in our theoretical endeavours as well as in political practice, from concrete social processes and practices rather than reifying spatial dimensions.
>
> (2008: 416; see also Wissen in this volume)

Starting from these considerations, a politico-ecological approach builds upon a non-essentialist concept of space, arguing against both the deterministic idea of space as an external 'container' encompassing social processes, and against purely constructivist approaches that deny any autonomist materiality of space. The latter assume space to be, first and foremost, produced by ideas, discourses and ascriptions. However, accepting that physical materiality is indeed relevant for socio-spatial transformation processes does not necessarily result in essentialist conceptions of space and nature. Nor does it mean assuming that material constellations have the same or a similar meaning for all subjects, and it definitely does not mean that material constellations determine social action. Following Marx (Marx 2007 [1867]: 198), we start from the premise that the physical materiality of nature and space as such is always socially produced. Space is, at the same time, socially produced and productive, meaning that it can indeed structure, in one way or another, social action. But only in social practices, particularly in the appropriation of nature, does materiality become socially meaningful.

Let us take the example of the oil palm: its fatty acid starts dissolving between 12 and 24 hours after the fruit is harvested. In order to ensure processing shortly after harvest, a specific 'built environment', as David Harvey (1982: 233) calls it, is needed: mills, streets, bridges, etc. Investment in such immobile elements and technical infrastructure binds capital and changes the general conditions of production and consumption. The social forms of appropriating nature and spatial orderings are thus reconfigured (ibid.: 233–234). The capitalist logic of accumulation demands, for instance, that palm oil mills are run 24 hours a day to produce oil. As a consequence, plantations must be large enough to be profitable. This example demonstrates how the oil palm's materiality, mediated through capitalist nature appropriation and valuation, matters in socio-spatial terms, in the way it enters and may structure global production networks (see Bridge 2008), in the way labour is organized and in the way landscapes are transformed via exclusive patterns of land use and control. Referencing 'the material' is thus a way of 'acknowledging the embeddedness of social action' (Bakker and Bridge 2006: 18). Nevertheless, acknowledging that materiality makes a difference in the way social relations unfold entails acknowledging that 'things', be they plants, soil or minerals, 'are not pregiven substrates that variably enable [or] constrain social action, but are themselves historical products of material, representational and symbolic practices' specific in time and space (ibid.) Other social practices and forms of nature appropriation thus produce different spatial effects and have different social impacts, though their material base might be the same.

## Territory, territoriality and territorialization

Territorialization can be understood as processes of spatial-administrative reorganisation (Vandergeest and Peluso 1995). Driven by various actors, government agents, private companies, landowners, peasants or indigenous peoples, they aim at establishing control over natural resources and human beings, within or beyond a state's territorial borders and thus changing socio-spatial relations of power. Territoriality, Peter Vandergeest and Nancy Peluso argue, is a central element in understanding state–society relations (ibid.). Robert David Sack defines territoriality as 'the attempt by an individual or group to affect, influence, or control people, phenomena, and relationships, by delimiting and asserting control over a geographic area' (Sack 1986: 19). Territoriality thus refers to the inclusion and exclusion of people within certain geographic borders. Political rulers territorialize power in order to achieve different goals. The enforcement of taxes and access to valuable natural resources are pivotal. Most states further use territorialization to spy on and control their citizens. State authority and domination are secured through territorial control – whereby local actors might accept or ignore state practices of territorial control, or fight against them (Berry 2009: 24). Territory thus cannot be reduced to a fix and static resource (Featherstone 2004: 703), but it is, at the same time, a material reference for authority as well as for the construction of collective identity (Perreault 2013).

Early globalization theorists (e.g. Taylor 1996) argued that in the era of globalization, territorial boundaries may lose their functionality for political authority. These might have been premature considerations, as argued by David Newman (2010). Territory does not lose its relevance, neither for the performance of authority and domination, nor for the construction of collective identities. On the contrary, territorial references are central for identity-related inclusion and exclusion; for the construction of 'Self' and 'Other' (ibid.). Political identities refer, often (not always, nonetheless), to territorially defined spaces, though these are not necessarily linked to nations and states but, possibly more frequently, to other 'imagined communities' such as ethnic, indigenous and autochthonous groups. It is perfectly clear that these identity constructions also play an essential role for place-based struggles related to agrofuels. For instance, most struggles over land between local groups are, at the same time, conflicts about whose claims to land are seen as legitimate, which, in turn, is closely linked to the social negotiation of citizenship and belonging, inclusion and exclusion (Lund 2011). Struggles over land and territorial control are probably the most widespread and visible place-based struggles in the field of agrofuels. A case in point is struggles related to the expansion of oil palm plantations in Colombia (see Marin-Burgos, Chapter 9 in this volume).

Oil palm plantations increased immensely after the government of Alvaro Uribe (2002–2010) adopted norms and policies to foster the national production of agrofuels. While in 2001 oil palms were cultivated on around 160,000 ha of land, this area had tripled by the year 2011, arriving at 430,000 ha (Fedepalma 2012: 23). Today, more than 40 per cent of Colombia's total palm oil production is absorbed by the domestic biodiesel industry, exceeding both the exportation rate and the rate purchased by the traditional domestic industry including edible oils and fats businesses, soap and food companies (ibid.: 26). These developments must be considered against the backdrop of the historically rooted land contentions in Colombia and ongoing conflicts that are primarily 'expressed and conducted as territorial struggles' (Ballvé 2013: 239).

Uribe's presidency was marked by the government imperatives of counterinsurgency and containment of illicit crop production (coca fields). In this context, the government and international donor institutions (the World Bank, European Union and USAID) identified palm crops as a profitable and legal alternative to substitute the production of coca and thus fight 'terrorism'. A relatively rapid expansion of oil palm plantations can be observed in those regions of the country's hinterland that for a long time have been imagined as 'stateless' and have won notoriety as sites of narco-driven, paramilitary, military or guerrilla violence (e.g. the Middle Magdalena Valley, the Pacific Lowlands of Urabá, the Caribbean coastal region) (Ballvé 2012). In many of these regions, as a result of oil palm expansion, contestation of territorial control emerged via either the violent appropriation of fertile land or the introduction of new private property and labour regimes. The latter, thereby, aims at 'including' peasants in an agroindustrial production pattern of palm oil via 'strategic alliances'. Peasant, 'black' or indigenous subjects are thus transformed into rural entrepreneurs (Cárdenas 2012: 329).

A salient example is the case of the municipality of María la Baja in the Montes de María region in north-western Colombia (Coronado Delgado and Dietz 2013). Between 2001 and 2010, the area cultivated with palm crops in María la Baja more than tripled. Currently, over 50 per cent of the arable farm land is covered with palms. Prior to the expansion of oil palms, María la Baja suffered from paramilitary violence and mass displacements. Between the end of the 1990s and the mid 2000s, more than one-third of the local population (17,500 inhabitants) was displaced, and more than 21,000 ha of land were abandoned (ibid.: 105). When oil palm cultivation was introduced in 2001, a new era of capitalized rural development began, with adverse socio-spatial effects. Farmers possessing legal titles to land of a reasonable size can become strategic, crop-producing allies of the local palm oil companies that support the production technologically, run the mills and commercialize the oil. Contracts mostly last for 20 years, implying the exclusive purchase of the farmers' yields by the corresponding company and the exclusive sale of the company's technical services to the farmers (Gómez 2010: 95). Those who cannot prove legal land titles and those who are landless or have been formally displaced, are excluded. Territory is thus produced via strategic alliances between companies, farmers and state agents – the latter promoting these alliances through formal institutions and incentives – and via the limitation of other, often collective uses of land (Coronado and Dietz 2013: 109). In two Afro-Colombian communities located within the municipality of María la Baja, communal paths became impassable for local residents because of palm cultivation. In other cases cattle that had crossed palm cultivation areas disappeared or were found dead, and trees and bushes from neighbouring areas that hindered the undisturbed development of palm trees were cut (ibid.).

The example of María la Baja shows how the expansion of agrofuel crops at the regional and local scale is accompanied by strategies of territorialization. Palm oil companies and landowners growing oil palms use territorialization strategies (claiming of buffer zones, introduction of private property rights, fencing of plantations, employment of armed guards to secure property boundaries, etc.) to prevent 'others' from trespassing, or from producing other types of crops (in this case coca) and hence to secure control over land use, economic gains and political processes.

## Global production networks

At the global scale, the agrofuels boom is characterized by a dynamic reconfiguration of the spatio-temporal dimensions of production and consumption, and of economic spaces of resource and capital flow. According to Harvey, geographical expansion and concentration in capitalist societies are both to be regarded as products of the contradictory dynamics of capital accumulation in space (Harvey 2001a: 246). Meanwhile, the imperative to accumulate produces the concentration of production and of capital and, at the same time, the need for the realization of value leads to a further expansion of the market. But the expansion of market structures is inherently characterized by the contradictory tendencies toward equalization and differentiation of patterns of production and consumption in space and time (Smith 2010 [1984]: 133). Thus capitalist development in general equalizes spatial differences and at the same time produces spatial inequality. These logics of uneven development in capitalism likewise apply to the agrofuels sector. In order to solve crises of accumulation, especially those of over-accumulation, or to benefit from local advantages (e.g. cheap labour, cheap means of production, path dependencies) capital and labour are spatially redistributed at different scales. This leads to a spatialization of bust and boom, whereby old centres of accumulation decay and new ones are produced (ibid.: 150). In this respect, the concept of 'spatial fix' is pivotal. It refers to 'capitalism's insatiable drive to resolve its inner crisis tendency by geographical expansion and geographical restructuring' (Harvey 2001b: 24). The agrofuels boom represents such a spatial fix, first by redirecting agricultural production in the United States and Europe towards agrofuels production, and then by increasing territorial expansion in the South.

A cursory look at global production figures reflects these processes of geographical restructuring at the regional, national and global scales. At the same time, it contradicts the notion that there is a division of labour in which the Global South produces agrofuels for the 'affluent North'. Rather, production and not just consumption of agrofuels to date is concentrated in the North, not in the South. With more than 50 billion litres in 2011, the USA is by far the world's largest producer of ethanol, while the European Union accounts for 68 per cent (9.2 billion litres in 2011) of agrodiesel production (Danker *et al.* 2013: 11–12; Murphy *et al.* 2012: 19). In Brazil, the only major producer of agrofuels in the South, ethanol production reached an amount of 20 billion litres in 2011. In contrast to the USA or the EU, ethanol has here been part of a 'full-fledged national biofuel region' since the 1970s, with 'flex-fuel car development' and nationally mandated blending policies (Mol 2007: 303) taking precedent over exports.

However, as higher agrofuel targets are set in the North (for the EU see Brunnengräber, Chapter 5 in this volume), additional flex crop production together with – in some cases – fuel refining is physically expanding in Southeast Asia and Latin America, most of which is being brought to Europe, China or the US via transnational supply chains. Today, nearly one-quarter of the European overall consumption of biodiesel originates from Argentina and Indonesia (Danker *et al.* 2013). At the same time, agrofuel consumption in the Global South has itself increased since obligatory blending targets have been adopted in many Asian, Latin American and African countries (BiofuelsDigest 2012; Pradhan and Ruysenaar 2014).

Using the concept of the nation state to compare 'national' production figures does little to uncover the transnational spatial dynamics of agrofuel production. Mol (2007: 303) suggests seeing biofuels as an emerging 'global integrated biofuel network (GIBN)' made up of relatively stable production regions connected by networks of flows. Thereby, he focuses on trade liberalization and governance through certification and labelling. However, his analysis largely ignores the 'organizational transformation of the production process' itself, that is characterized by 'transnational production networks, of which multinational corporations are an essential component' (Castells 2010: 122). Rather than 'national biofuel regions' connected by trade flows, agrofuels are increasingly part of global production networks (GPN),[4] 'whose interconnected nodes and links extend spatially across national boundaries' (Coe *et al.* 2008b: 274). The spatial dynamics of agrofuels do not end with the production of agricultural crops – and their location – but continue along the transnational production networks to the mills, the agrofuel refineries, the distribution networks and the transportation networks connecting these places. As Coe *et al.* (2008b: 278) point out, production and distribution in these GPNs is grounded in the environment and 'fundamental environmental interaction … occurs at every point in the network'. GPNs thus reshape the spatial materiality of agrofuels.

Global agrofuel production networks are increasingly controlled by transnational corporations (TNCs). According to Holt-Giménez and Shattuck (2009: 183), five corporations control nearly 50 per cent of ethanol production in the US and most of the plants under construction are 'owned by large corporations' (ibid.). Agribusiness giants – i.e. the 'ABCD companies' ADM, Bunge, Cargill and Louis Dreyfus – (Murphy *et al.* 2012) are key players because they 'dominate the grain market in both food and fuel crops' (Holt-Giménez and Shattuck 2009: 185). These agribusiness giants use their position in the US to transnationalize operations to other parts of the world. For example, ADM controls one-fifth of ethanol production in the US, with a capacity of 1.8 billion gallons, but also has agrodiesel operations in Germany, Brazil, India and Indonesia to the tune of 450 million gallons (Murphy *et al.* 2012: 44). Similarly, Cargill has a global reach and extensive investments in Brazil, while Louis Dreyfus produces 1.5 million[3] of ethanol, mainly in Brazil, and has recently 'acquired a 200,000-tonne-per-year refinery for biodiesel in Wittenberg in Germany, one of the largest of its kind in the world' (ibid.: 46).

Regarding the dominant role of the ABCD corporations in the USA, Holt-Giménez and Shattuck conclude that 'concentration of ownership of global agrofuels production by U.S. agribusiness is proceeding apace' (Holt-Giménez and Shattuck 2009: 185). However, this is somewhat misleading and underestimates the role of TNCs from the South. For example, Murphy *et al.* depict Cargill as a major palm oil player that 'sources palm oil in Indonesia from its own plantations' (Murphy *et al.* 2012: 19). Although Cargill is the biggest importer of palm oil in the US, and a moderately sized processor with 12 refineries, it only has four plantations in Indonesia (Rainforest Action Network 2010). This makes it a tiny player in the production of palm oil compared to the transnationals from

Singapore, Malaysia and Indonesia. For example, the Malaysian state corporation Sime Darby owns 71 plantations and 25 mills in Indonesia alone (and many more in Malaysia).[5] Other major players from the South include the Chinese agribusiness corporation COFCO, the Singapore-based Olam International and the Hong Kong-based Noble Group (Murphy *et al.* 2012: 41–44). But companies involved in the agrofuel business from either the North or the South are not just transnational in their performances, that is, controlling transnational production chains or investing in plantations and mills across national borders. They are also becoming more and more transnationalized themselves.

A good example of a new South–South–North TNC involved in agrofuels is Wilmar International, based in Singapore and owned by the Malaysian Kuok Group, the Indonesian millionaire Sitorus and ADM (10 per cent). Now the biggest processor of palm oil and producer of palm oil agrodiesel in the world, the corporation owns over 250,000 ha of palm oil plantations in Malaysia, Indonesia, Uganda and West Africa. It controls transnational production networks that link nearly 100 refineries with 14 oleochemical and 14 speciality fats factories in Indonesia, Malaysia, China, Vietnam, Europe and Africa. By 2012, Wilmar had built eight agrodiesel factories with a total capacity of two million tonnes (Wilmar International 2013: 26).

But the Global Integrated Biofuels Network is not only characterized by agribusiness TNCs. A crucial difference between agrofuels and 'flex crop' production in general is the role of oil corporations, as agrofuels merge with the 'conventional fossil fuel scape' (Mol 2007: 310). In 2012, *Business Week* proclaimed that 'Big Oil has become the biggest investor in the race to create green fuels' (Wells 2012). Chazan (2009) cites Katrina Landis, head of BP's Alternative Energy Division, as stating that 'Oil companies have a natural affinity for the biofuels business' and explains that 'biofuels made the cut, in part because they fit nicely into the company's existing infrastructure of refineries, pipelines and distribution networks' (ibid.). Agrofuels have fused agribusiness, chemical and oil corporations together, for example, in partnerships between 'ADM with both Monsanto and Conoco-Phillips; BP with DuPont and Toyota, as well as with Monsanto and Mendel Biotechnology; Royal Dutch Shell with Cargill, Syngenta, and Goldman-Sachs; and DuPont with British Petroleum and Weyerhauser' (Holt-Giménez and Shattuck 2009: 185).

The key role of oil corporations is linked to the fact that – in contrast to the food industry, where the retailers are the dominant players – they control the distribution networks, as agrofuels are blended into conventional petrol and diesel, which are then sold at petrol stations. The connection between food production, agrofuel refineries and petrol station distribution networks creates new and spatially specific global production networks.

One example of these new GPNs can be seen in the construction by the Finnish oil corporation Neste Oil of two of the largest agrofuel refineries in the world, in Singapore and Rotterdam. The Singapore refinery, completed in 2010 and costing over 500 million dollars to build, can produce 800,000 tonnes of agrodiesel a year, 'enough for 10 million cars to run continuously' (Neste Oil

2013) with a 10 per cent blend of 'green fuel.' The Rotterdam plant is of a similar size, making the oil corporation 'the world's largest producer of biodiesel' (Greenpeace undated). Neste Oil sources the bulk of its feedstock for the plants from palm oil and, according to Neste Oil (2013), is the first company to be awarded an RSPO-RED Supply Chain certificate in line with the EU's Renewable Energy Directive (RED). According to Greenpeace (undated) 'most of the palm oil for Neste Oil has been bought from IOI Group, a Malaysian corporation responsible for the ongoing destruction of rainforests, peatlands and habitats of orang-utan, for illegal plantings, land-grabbing and land conflict'.

## The financialization of agrofuels

If the production of agrofuels takes place within transnational networks, capital movement and investment decisions that drive them happen at a global scale of the 'space of flows' of 'globally integrated financial markets' (Castells 2010). At the global scale, agrofuels as flex crops accelerate existing tendencies of the 'financialization in the global food system' (Clapp 2014: 2) and at the same time are becoming an object of financial speculation in their own right, a tendency that looks set to increase in connection with the emergence of a global carbon market. Financialization is understood as a process by means of which global financial markets, institutions, instruments and actors gain ever more influence over nature in general and food and crop production in particular, e.g. through derivates, futures, mortgages etc. (Tricarico and Löschmann 2012). Hence, financial markets have themselves become a new scale of regulation in the agrofuels sector, interacting with other socio-spatial dynamics already mentioned above. According to Eric Swyngedouw, scales or scalar configurations are the 'outcome of sociospatial processes that regulate and organize social power relations'. That means scales, e.g. the global scale of climate negotiations, the European scale of agrofuels politics, the local scale of the mills and plantations, etc. are social constructions with concrete and powerful meanings, as they become 'arenas around which sociospatial power choreographies are enacted and performed' (Swyngedouw 1997, 2004: 132). Scalar reconfigurations, that is the emergence of new scales of political regulation and contestation, can produce substantial changes or lead to the perpetuation of existing socio-spatial power relations at and between different scales. Related to political ecological questions, scalar reconfigurations 'shape in important ways who will have access to what kind of nature, and the particular trajectories of environmental change' (ibid.). It is in this sense that the emergence of the global financial market as a new scale of regulation of agrofuels can be conceptualized as a politico-ecological process.

The increasing influence of 'banks, finance houses, insurance companies, sovereign wealth funds, private equity consortia, hedge funds, superannuation funds and other financial agencies' (Burch and Lawrence 2009: 268) in the food industry has led to a financialization of the food system. The involvement of financial institutions takes many forms, ranging from investment in farmland by hundreds of 'investment, super-annuation and hedge funds' (ibid.: 271), direct investment into agri-food processing by banks, land-grabbing by sovereign wealth funds, 'agricultural investment funds set up by hedge fund and private equity managers' and futures trading in commodities (ibid.: 271–273).

As Jennifer Clapp (2014: 1–18) explains, the strengthening of the global networked scale of financial markets has led to a shift of power towards financial actors:

> Deregulation has enabled the entry of new actors, or middlepersons, into agrifood commodity chains. Banks, institutional investors and new financial investment arms of the ABCDs are now active participants in agrifood supply chains, bringing with them considerable decision-making power based on their financial weight. Along with new actors, new types of relationships between actors, in the form of abstract commodity derivatives, have also changed the shape and culture of those commodity chains. Financial investors are not necessarily buying agricultural commodities in their physical form, but rather are seeking profits in and around the commodity chain, through new kinds of financial derivatives that are abstractions from the physical commodity.
>
> (Clapp 2014: 9)

Agrofuels contribute in several ways to the financialization of the agri-food regime. First, they create a 'demand shock' (Friends of the Earth 2012), increasing interest in land for the production of agrofuels. Sovereign wealth funds from countries such as 'Abu Dhabi, Norway, China, Singapore, Saudi Arabia, Libya and South Korea' invest in land in order to increase their food and also energy security via agrofuels (Burch and Lawrence 2009: 272). Second, agrofuels are integrated into increasingly speculative trading, for example in the form of commodity index funds (CIFs). CIFs speculate with options on commodity futures linked to the prices of bundled commodities on indexes such as Standard and Poor's Goldman Sachs Commodity Index (GSCI) and the Dow Jones American International Group (AIG) Index. According to Isakson (2014: 10) 'speculation in CIFs exploded following financial deregulation, ballooning from US$13 billion in 2003 to US$317 billion in 2008'.

Besides adding to agricultural commodity speculation, agrofuels are part of other investment funds that are emerging in anticipation of profitable ventures in the 'green economy'. Hundreds of equity funds like HgCapital, Impax, InfraRed, BlackRock, Calvert and Foresight, and pension funds such as Calvers, have invested billions of dollars in 'renewable energy funds'. Although returns have dropped, prompting some commentators to predict investors being reluctant to invest more in the 'much-hyped fad' (Johnson 2013), agrofuels have remained more profitable than other types of renewable energy, such as solar power. At the same time, banks such as the World Bank, SEB and Nikko Asset Management and companies like the French energy corporation EDF, have issued 'green bonds', which include agrofuels and which are attracting increased interest.

Agrofuels are also the subject of financial speculation in their own right. In the US, the 2005 Renewable Fuel Standard (RFS) established mandatory levels of agrofuels that have to be blended into petrol and diesel and that rise gradually up to 2020. It also established a market for agrofuel credits, so that refineries or importers of petrol and diesel that did not blend as much as required could offset by buying from other producers. These credits are traded under 38- digit numbers called Renewable Identification Numbers, or RINs. After blending, these RINs can be 'detached, and can then be bought and sold like other commodities' (Yacobucci 2013: 6) on the Chicago Mercantile Exchange (CME).

Enter the financial institutions: after RIN prices had stagnated during the first years, they rose dramatically from 0.7 to 1.16US$ between January and July 2013 (ibid.: 8). An article in the *New York Times* uncovered that banks and financial institutions such as Morgan Stanley, JPMorgan Chase, Citigroup and Barclays were registered to trade with ethanol credits and that traders had 'amassed millions of the credits just as refiners were looking to buy more of them to meet an expanding federal requirement' (Morgenson and Gebeloff 2013). The article also uncovered that in 2006, Morgan Stanley had bought TransMontaigne, a petroleum corporation with 21 refineries that generates RINs that are then traded via the Morgan Stanley Capital Group. In another scandal, three companies (Clean Green Fuels, Absolute Fuels and Green Diesel) were accused of fraudulently generating 140 million biodiesel RINs in 2010 and 2011 (Yacobucci 2013: 11). Creating and selling what is ultimately a 38-digit number without any real agrofuel production is a systemic risk in this kind of credit permit market, echoing similar scandals in the European Emissions Trading System (ETS) (Gilbertson 2011). The ETS, too, is a multibillion dollar market characterized by 'trading through spot, futures and options contracts' (ibid.). The signs are that – when aviation companies are included in the scheme – agrofuels will become more intimately linked to financial speculation in the EU too.

## Analysing agrofuels through spatial perspectives – what do we see and learn?

Our motivation was to explore the analytical potential that different spatial categories have for politico-ecological research and action on agrofuels. Our verdict is that a spatial perspective provides tools that enable us to understand the complex power relations and linkages that transcend national boundaries and characterize the 'agrofuels project'. At the same time, it succeeds in displaying and explaining the multiple ways in which social inequalities are produced and reproduced through agrofuels themselves and the politics and discourses that create and regulate them. Thus spatial categories, such as the ones considered in this chapter, provide tools to go beyond initial politico-ecology approaches that differentiate in a more one-dimensional manner between 'place-based and non-placed-based actors' (e.g. O'Brien 1985; see also, Bryant 2001). As we have tried to show, we cannot explain social processes and transformations with one spatial dimension alone. Through the relational perspective introduced by the

TPSN (Territory, Place, Scale, Networks) approach, we can analyse dynamic and context-dependent socio-spatial relations that reflect distinctive orderings and new geographies of power and hegemony (Jessop *et al.* 2008: 395). Different actor constellations and socio-spatial dimensions are, thereby, equally important. One such dimension is territory. Analysing the expansion of agrofuels through a spatial perspective highlighting territory provides us with tools to both situate the agrofuels project in the contexts of concrete social and political dynamics and struggles and understand how social inequalities of access as well as unequal social power relations are (re-)produced via bounded spaces. Territorialization via the expansion of crop production for agrofuels thereby serves specific purposes and claims, be they private economic interests, the consolidation and exercise of state power, the assertion of specific territorial rights or the repression of certain crops and patterns of production against others. In so doing, societal nature relations are being changed as the emerging agrofuel landscapes in one specific place are being linked to the global scale of financial regulation.

Another dimension related to processes of territorialization is GPNs. A GPN perspective that transcends methodological nationalism can uncover spatial processes that cannot solely be understood by a comparison between countries (which country produces, consumes, imports, exports, etc.) A GPN perspective places companies and inter-company relations at the centre of the analysis and underlines the manifold variations of organizational forms – state-owned companies and capital, transnational agribusiness firms, (national) oil companies, newly emerging South–South–North companies, private–public partnerships, etc. (Bridge 2008). It further highlights different levels of vertical integration and transnational linkages between concrete places of crop production, the companies' headquarters in urban metropolises of the US, Europe, Singapore, Malaysia, China or Brazil, and transportation hubs like the harbour at Rotterdam. In so doing, such a perspective draws attention to relations of power, dominance and dependence within and across the network, and it reveals how these change over time and space.

The role of GPNs and increasing finanzialisation highlight the significance of scale in the agrofuels project. While territorialization imposes on specific places and creates place-based struggles, these are in turn related in manifold ways to national, transnational and global scales. National and supra-national (i.e. EU) legislation and subsidies supporting agrofuels are still a key scale of regulation because they provide the incentives for capital to invest. The national scale is also crucial for the discursive, regulative and repressive power needed to extend and uphold agrofuel expansion in the national territory. But the linkages created by the transnational scale of GPNs mean that agrofuel policies at the national scale drive agrofuel production in other parts of the world. In turn, at the global scale, financial markets create a 'space of flows' in which financial institutions invest and speculate largely without government regulation. Power concentration thereby shifts from the national and transnational scale to the global, and 'green investment' becomes a powerful global narrative that shapes agrofuel policies at the national scale.

## Conclusion

We have shown in this chapter that the spatial dynamics connected to the agrofuels project are characterized by a combination of accelerating transformations within the corporate food regime and by shifting power constellations that bring together actors outside of agribusiness. On the global scale of the deregulated financial markets, agrofuels add to the financialization of the food regime by increasing investment and speculation by banks, hedge funds, sovereign wealth funds, commodity index funds and the like. As flex crops, they also connect agricultural commodities to speculation and capital flows that focus on energy and 'renewables'. And, in the form of RINs and as offset credits on the carbon market, agrofuels become themselves an object of global investment flows and speculation. These agrofuel-related financial flows fuse new coalitions of actors together, connecting financial institutions with oil corporations, airlines, chemical corporations and the automotive industry with agribusiness giants in North and South.

As capital flows through the financial markets in search of its 'spatial fix', it immobilizes in the form of petrol stations, refineries, fleets of tankers, mills and plantations. The geographical decoupling of production and consumption results in new spatio-temporal concentrations as well as in differentiations across and within national boundaries. Growing palm oil landscapes in Indonesia, Colombia and Brazil, soy landscapes in Argentina, as well as the emergence of new labour regimes, often based on (transnational) labour migration (see Pye, Chapter 11 in this volume), are specific expressions of these spatial dynamics. Space is thus simultaneously absorbed and produced by the agrofuels industry; thereby new spatial disparities related to the provision of technical and social infrastructure are produced.

Although an analytical perspective that starts from concepts like spatial fix or uneven development illuminate fundamental transnational spatio-temporal dynamics of the agrofuels project at a global scale, it also bears the risks of one-dimensionalism and economy-centrism. In order to understand the complexity of the socio-spatial dynamics of agrofuels, we need to understand capitalism's inherent logics in the production of space, but this is by no means sufficient. As we have seen, the expansion of agrofuels is mediated by historically- and context-specific political and legal decisions, by narratives and scaled frames and by a multitude of actors operating at different scales. Thus, a '*complex inquiry*' of the socio-spatial dynamics of the political ecology of agrofuels requires what Jessop *et al.* call a 'reflexive attention to combining different dimensions of socio-spatial analysis with other features of the research object in question' (Jessop *et al.* 2008: 392).

The politicization of agrofuels starts in place-based struggles that are about economic justice and land rights, cultural integrity, autonomy and environmental justice. These place-based struggles usually have to deal with discourses of national development that have been refined and 'greenwashed' by global discourses of win–win scenarios and green economies. As a counter-discourse, small-scale farmers involved in the network Via Campesina argue that 'another agrarian transition is possible', one that 'responds not to the logic of capital but to the redistributive logic of *food sovereignty*' (Holt-Giménez and Shattuck 2009: 186). As place-based struggles connected with activists in the North, new transnational campaign alliances around the negative socio-ecological consequences of the agrofuels boom emerge, thus creating new transnational political spaces of contention (Pye 2010). Campaign coalitions have mobilized, in part successfully, against pro-agrofuel regulation at the national and supranational (EU) scale.

But the spatial analysis of agrofuels shows us that place-based struggles addressing national policies are not enough to deal with the new power concentrations of the global production networks and global financial markets. A response to agrofuels expansion cannot just be about food sovereignty, because agrofuels are related to 'interlocking crises of the climate, energy, food, and finances' (Rosset 2009: 189). Developing strategies that can intervene at the scale of global production networks and at the scale of the global financial markets will be a crucial part of connecting the struggle against agrofuels to these broader challenges.

## Notes

1 We would like to thank Bernd Belina and Achim Brunnengräber for their valuable comments on an earlier draft of this chapter.
2 The main body of research was conducted in the context of three different research projects: Kristina Dietz conducted research in Colombia as well as on transnational production networks and resource flows in the context of the project 'Fair Fuels? Between Dead End and Energy Transition: A Socio-ecological Multilevel Analysis of Transnational Biofuel Policy', funded by the German Federal Ministry of Education and Research (2009–2013). The research in Colombia consisted of the analysis of the territorializing effects of expanding palm oil plantation for biodiesel in two different sites: one in the central 'Magdalena Medio' region, province of Santander, municipality of San Alberto, and another in the Caribbean region, provinces of Sucre and Bolívar, municipality of María la Baja. Bettina Engels conducted research in Ethiopia and Burkina Faso in the context of the project 'Local Conflict Dynamics: Environmental Change, Food Crisis, and Violent Conflicts in sub-Saharan Africa', funded by the German Foundation for Peace Research. Oliver Pye is involved in ongoing research on the palm oil industry, including research with migrant palm oil workers in Malaysia in the context of the German Research Foundation funded project on 'The Making of Social Movements under Conditions of Precarization and Transnationality in Southeast Asia' (2009–2013), and work on transnational campaigning around palm oil and agrofuels.
3 We owe this wording/idea Bernd Belina. Thank you!
4 Global production networks analysis is similar to global commodity chain (GCC) and global value chain (GVC) analysis, in that it looks at transnational production networks but pays more attention to 'multi-actor and multi-scalar characteristics of transnational production systems through intersecting notions of power, value and embeddedness' (Coe *et al.* 2008a: 267). As Coe *et al.* point out, 'the term transnational production network is more accurate. However, current usage suggests that it is better to retain the term "global"' (Coe *et al.* 2008b: 274).
5 See www.simedarbyplantation.com/Sime_Darby_Plantation_in_Indonesia.aspx (accessed: 14 March 2014).

## References


Bakker, Karen and Gavin Bridge (2006) 'Material worlds? Resource geographies and the "matter of nature"', *Progress in Human Geography* 30(1), 5–27.

Ballvé, Teo (2012) 'Everyday state formation: territory, decentralization, and the narco landgrab in Colombia', *Environment and Planning D: Society and Space* 30(4), 603–622.

Ballvé, Teo (2013) 'Territories of life and death on a Colombian Frontier', *Antipode* 45(1), 238–241.

Berry, Sara (2009) 'Property, authority and citizenship: land claims, politics and the dynamics of social division in West Africa', *Development and Change* 40(1), 23–45.

BiofuelsDigest (2012) 'Biofuels mandates around the world: 2012' (www.biofuelsdigest.com/bdigest/2012/11/22/biofuels-mandates-around-the-world-2012/, accessed: 23 April 2014).

Brenner, Neil (2004) *New State Spaces. Urban Governance and the Rescaling of Statehood*. Oxford: Oxford University Press.

Bridge, Gavin (2008) 'Global production networks and the extractive sector: governing resource based development', *Journal of Economic Geography* 8(3), 389–419.

Bryant, Raymond L. (2001) 'Political ecology: a critical agenda for change?', in Castree, Noel and Bruce Braun (eds) *Social Nature. Theory, Practice, and Politics*. Malden, Oxford: Blackwell, 151–169.

Burch, David and Geoffrey Lawrence (2009) 'Towards a third food regime: behind the transformation', *Agriculture and human values* 26(4), 267–279.

Cárdenas, Roosbelinda (2012) 'Green multiculturalism: articulations of ethnic and environmental politics in a Colombian "black community"', *The Journal of Peasant Studies* 39(2), 309–333.

Castells, Manuel (2010) *The Rise of the Network Society*. Chichester: Wiley-Blackwell.

Chazan, Guy (2009) 'Big oil looks to biofuels' (http://online.wsj.com/news/articles/SB10001424052970204731804574386960944758516, accessed: 12 March 2014).

Clapp, Jennifer (2014) 'Financialization, distance and global food politics', *The Journal of Peasant Studies*, DOI: 10.1080/03066150.2013.875536.

Coe, Neil M., Peter Dicken and Martin Hess (2008a) 'Introduction: global production networks—debates and challenges', *Journal of Economic Geography* 8(3), 267–269.

Coe, Neil M., Peter Dicken and Martin Hess (2008b) 'Global production networks: realizing the potential', *Journal of Economic Geography* 8(3), 271–295.

Coronado Delgado, Sergio and Kristina Dietz (2013) 'Controlando territorios, reestructurando relaciones socio-ecológicas: La globalización de agrocombustibles y sus efectos locales, el caso de Montes de María en Colombia', *IberoAmericana* 49, 93–116.

Danker, Hans-Christian, Kristina Dietz, Nicola Jaeger and Wiebke Thomas (2013) *Die Globalisierung der Agrarkraftstoffe. Produktion, Handel und Akteure*. Berlin, Fair Fuels? Working Paper 7.

Dauvergne, Peter and Kate J. Neville (2009) 'The changing North–South and South–South political economy of biofuels', *Third World Quarterly* 30(6), 1087–1102.

Dauvergne, Peter and Kate J. Neville (2010) 'Forests, food, and fuel in the tropics: the uneven social and ecological consequences of the emerging political economy of biofuels', *The Journal of Peasant Studies* 37(4), 631–660.

Escobar, Arturo (2008) *Territories of Difference: Place, Movements, Life, Redes*. Durham, London: Duke University Press.

EurObserv´ER (2012) *Biofuels Barometer*. Paris, Berlin: EurObserv'ER.

Featherstone, David (2004) 'Spatial relations and the materialities of political conflict: the construction of entangled political identities in the London and Newcastle Port Strikes of 1768', *Geoforum* 35(6), 701–711.

Fedepalma (2012) *Anuario Estadístico. La agroindustria de la palma de aceite en Colombia y en el mundo, 2007–2011*. Bogotá: Fedepalma.

Flood, Chris (2013) 'Green bonds take root in maturing market' (www.ft.com/intl/cms/s/0/1fb827d6–5789–11e3–86d1–00144feabdc0.html#axzz2vNYeGpdt, accessed: 10 March 2014).

Friends of the Earth (2012) 'Farming Money. How European banks and private finance profit from food speculation and land grabs' (www.foeeurope.org/sites/default/files/publications/Farming_money_FoEE_Jan2012.pdf, accessed: 11 March 2014).

Gilbertson, Tamra (2011) 'Fraud and scams in Europe's Emissions Trading System' (http://climateandcapitalism.com/2011/05/05/fraud-and-scams-in-europes-emissions-trading-system/, accessed: 11 March 2014).

Gómez, Andrés (2010) 'La alianza productiva y social de María la Baja en Colombia. Un contraste de lo encontrado con los discursos del desarrollo local', *Eutopia* 1, 85–99.

Grajales, Jacobo (2011) 'The rifle and the title: paramilitary violence, land grab and land control in Colombia', *The Journal of Peasant Studies* 38(4), 771–792.

Grajales, Jacobo (2013) 'State involvement, land grabbing and counter-insurgency in colombia', *Development and Change* 44(2), 211–232.

Greenpeace (undated) 'Neste Oil – biodiesel driving rainforest destruction' (www.greenpeace.org/finland/en/What-we-do/Neste-Oil–driving-rainforest-destruction/, accessed: 29 April 2014).

Hall, Ruth (2011) 'Land grabbing in Southern Africa: the many faces of the investor rush', *Review of African Political Economy* 38(128), 193–214.

Harvey, David (1982) *The Limits to Capital*. Chicago: The University of Chicago Press.

Harvey, David (1996) *Justice, Nature and the Geography of Difference*. Cambridge, Oxford: Blackwell.

Harvey, David (2001a) *Spaces of Capital. Towards a Critical Geography*. New York: Routledge.

Harvey, David (2001b) 'Globalization and the "Spatial fix"', *Geographische Revue* 3(2), 23–30.

Herod, Andrew (2011) *Scale*. London: Routledge.

Hollander, Gail (2010) 'Power is sweet: sugarcane in the global ethanol assemblage', *The Journal of Peasant Studies* 37(4), 699–721.

Holt-Giménez, Eric and Annie Shattuck (2009) 'The agrofuels transition: restructuring places and spaces in the global food system', *Bulletin of Science Technology & Society* 29(3), 180–188.

Isakson, S. Ryan (2014) 'Food and finance: the financial transformation of agro-food supply chains', *The Journal of Peasant Studies*, DOI: 10.1080/03066150.2013.874340.

Jessop, Bob, Neil Brenner and Martin Jones (2008) 'Theorizing sociospatial relations', *Environment and Planning D: Society and Space* 26(3), 389–401.

Johnson, Steve (2013) 'Private equity retreats from renewables "fad"' (www.ft.com/intl/cms/s/0/ef1b2248–94bb-11e3–9146–00144feab7de.html#axzz2vNYeGpdt, accessed: 21 March 2014).

Latour, Bruno (2005) *Reassembling the Social: An Introduction to Actor-Network-Theory*. Oxford: Oxford University Press.

Lavers, Tom (2012) 'Patterns of agrarian transformation in Ethiopia: state-mediated commercialisation and the "land grab"', *The Journal of Peasant Studies* 39(3–4), 795–822.

Lund, Christian (2011) *Land Rights and Citizenship in Africa*. Discussion Paper 65. Uppsala: Nordiska Afrikainstitutet.

McMichael, Philip (2008) 'Agrofuels, food security, and the metabolic rift', *Kurswechsel* 3, 14–22.

McMichael, Philip (2009) 'The agrofuels project at large', *Critical Sociology* 35(6), 825–839.

Marston, Sallie A. (2000) 'The social construction of scale', *Progress in Human Geography* 24(2), 219–242.

Marx, Karl (2007 [1867]) *Das Kapital. Kritik der politischen Ökonomie*. Erster Band. Berlin: Karl Dietz.

Massey, Doreen (1991) 'A Global Sense of Place', *Marxism Today* (June), 24–29.

Massey, Doreen (1994) *Space, Place, and Gender*. Minneapolis: University of Minneapolis Press.

Mayer, Margit (2008) 'To what end do we theorize sociospatial relations?', *Environment and Planning D: Society and Space* 26, 414–419.

Mol, Arthur P. J. (2007) 'Boundless biofuels? Between environmental sustainability and vulnerability', *Sociologia ruralis* 47(4), 297–315.

Molano Bravo, Alfredo (2009) *En Medio del Magdalena Medio*. Bogotá: CINEP.

Morgenson, Gretchen and Robert Gebeloff (2013) 'Wall St. exploits ethanol credits, and prices spike (www.nytimes.com/2013/09/15/business/wall-st-exploits-ethanol-credits-and-prices-spike.html?pagewanted=all&_r=1&, accessed: 06 March 2014).

Murphy, Sofia, David Burch and Jennifer Clapp (2012) 'Cereal secrets: the world's largest commodity traders and global trends in agriculture' (www.oxfam.org/sites/www.oxfam.org/files/rr-cereal-secrets-grain-traders-agriculture-30082012-en.pdf, accessed: 06 March 2014).

Nalepa, Rachel A. and Dana Marie Bauer (2012) 'Marginal lands: the role of remote sensing in constructing landscapes for agrofuel development', *The Journal of Peasant Studies* 39(2), 403–422.

Neste Oil (2013) 'Neste Oil's Singapore refinery – the world's largest and most advanced' (www.nesteoil.com/binary.asp?path=1;41;540;2384;18010;21058&field=FileAttachment, accessed: 29 April 2014).

Newman, David (2010) 'Territory, compartments and borders: avoiding the trap of the territorial trap', *Geopolitics* 15(4), 773–778.

O'Brien, Jay (1985) 'Sowing the seeds of famine: the political economy of food deficits in Sudan', *Review of African Political Economy* 33, 23–32.

Peluso, Nancy Lee and Christian Lund (2011) 'New frontiers of land control: introduction', *The Journal of Peasant Studies* 38(4), 667–681.

Perreault, Tom (2013) 'Nature and nation: hydrocarbons, governance, and the territorial logics of "resource nationalism" in Bolivia', in Bebbington, Anthony and Jeffrey Bury (eds) *Subterranean Struggles: New Dynamics of Mining, Oil, and Gas in Latin America*. Austin: University of Texas Press, 67–90.

Pradhan, Shishusri and Shaun Ruysenaar (2014) 'Burning desires: untangling and interpreting "pro-poor" biofuel policy processes in India and South Africa', *Environment and Planning A* 46(2), 299–317.

Pye, Oliver (2010) 'The biofuel connection – transnational activism and the palm oil boom', *The Journal of Peasant Studies* 37(4), 851–874.

Pye, Oliver (2013) 'Migration, Netzwerke und Alltagswiderstand: die umkämpften Räume der Palmölindustrie', *PERIPHERIE* 33(132), 466–493.

Rainforest Action Network (2010) 'Cargill's problems with palm oil' (http://understory.ran.org/wp-content/uploads/2010/05/Cargills_Problems_With_Palm_Oil_low.pdf, accessed: 11 March 2014).

Rosset, Peter (2009) 'Agrofuels, food sovereignty, and the contemporary food crisis', *Bulletin of Science Technology & Society*, 29(3), 189–193.

Sack, Robert (1986) *Human Territoriality. Its Theory and History*. Cambridge: Cambridge University Press.

Sheppard, Eric (2002) 'The spaces and times of globalization: place, scale, networks, and positionality', *Economic Geography* 78(3), 307–330.

Smith, Neil (2010 [1984]) *Uneven Development. Nature, Capital and the Production of Space*. London, New York: Verso.

Swyngedouw, Erik (1997) 'Neither global nor local. "Glocalization" and the politics of scale', in Cox, Kevin (ed.) *Spaces of Globalization: Reasserting the Power of the Local*. New York, London: Guilford, 137–166.

Swyngedouw, Erik (2004) 'Scaled geographies: nature, place, and the politics of scale', in Sheppard, Eric and Robert B. McMaster (eds) *Scale and Geographic Inquiry. Nature, Society, and Method*. Oxford: Blackwell, 129–153.

Taylor, Peter J. (1996): 'On the nation-state, the global, and social science', *Environment and Planning A* 28(11), 1917–1995.

Taylor, Peter J. (2003) 'The state as container: territoriality in the modern world-system', in Brenner, Neil, Bob Jessop, Martin Jones and Gordon Macleod (eds) *State/Space. A Reader*. Malden, Oxford: Blackwell, 101–113.

Tricarico, Antonio and Heike Löschmann (2012) 'Finanzialisierung – ein Hebel zur Einhegung der Commons', in Helfrich, Silke and Heinrich-Böll-Stiftung (eds) *Commons. Für eine neue Politik jenseits von Markt und Staat*. Bielefeld: transcript, 184–195.

Vandergeest, Peter and Nancy Lee Peluso (1995) 'Territorialization and state power in Thailand', *Theory and Society* 24, 385–426.

Wells, Ken (2012) 'Big oil's big in biofuels', *BloombergBusinessweek*, 10 May (www.businessweek.com/articles/2012–05–10/big-oils-big-in-biofuels, accessed: 14 March 2014).

White, Ben and Anirban Dasgupta (2010) 'Agrofuels capitalism: a view from political economy', *The Journal of Peasant Studies* 37(4), 593–607.

Wilmar International (2013) 'Wilmar International Annual Report 2012' (http://media.corporate-ir.net/media_files/IROL/16/164878/Wilmar_International_Limited_AR_2012.pdf, accessed: 29 April 2014).

Yacobucci, Brent D. (2013) 'Analysis of renewable identification numbers (RINs) in the Renewable Fuel Standard (RFS)' (https://www.fas.org/sgp/crs/misc/R42824.pdf, accessed: 12 March 2014).

# 4 The gendered political ecology of agrofuels expansion

*Ben White, Clara M. Park and Julia*

> Look at all the sheanut trees you have cut down already, and consider that the nuts that I collect in a year give me cloth for the year and also a little capital. I can invest my small income in a sheep and sometimes in a good year I can buy a cow. Now you have destroyed the trees and you are promising me something you do not want to commit yourself to. Where then do you want me to go? What do you want me to do?[1]

## Introduction and starting-points

The use of plant materials and agricultural (by-)products for fuel is nothing new in human history; it is as old as the domestication of fire. Wood and charcoal, crop residues and animal dung have been used as fuels for centuries and still are used in many countries (Magdoff 2008: 34–35). Women (and male and female children) are often those most involved in the collection of these materials from nature. Using land for fuel production in this way does not in itself threaten the food security of individuals or communities. In this chapter, however we focus not on these age-old 'subsistence' biofuels, but on the emergence of corporate ventures aiming to produce fuel commercially from plant materials – on a large scale – to replace fossil fuels.

As noted in the editors' introduction to this volume, most crops used in commercial agrofuel production – with the exception of jatropha – are 'flex crops', and those who grow them or live in communities affected by their expansion may not even know whether the crops they see around them are destined for final use as fuel, food, cosmetic or other products. This raises the question of whether 'agrofuels' is a relevant analytical category for an analysis of the social, economic, political and cultural dimensions and implications of their production. Is there any important difference to local communities between a crop produced as agrofuel feedstock, and other crops grown corporately in large-scale, corporate, monocrop schemes, such as tobacco, mangos or cotton?

With this caution in mind, our chapter explores the gender dimensions of the large-scale, corporate production of agrofuels destined for commercial use and consumption in distant places. Combining gendered political ecology and agrarian studies perspectives, we draw on six recent case studies from African and

Asian countries to examine the direct and indirect implications of flex crops or agrofuels expansion for shifts in gendered power and labour relations, access to resources and livelihoods, incorporation in new production regimes and gendered responses to these changes. The six case studies are from Indonesia (oil palm), Lao PDR (jatropha), Sierra Leone (sugar cane and oil palm), Tanzania (jatropha), Zambia (sugar cane) and Uganda (oil palm).

Our study is guided by some of the basic ideas of feminist political ecology.[2] Political ecology applies basic political economy questions and concepts to the production of environmental change, inspired by a notion of environmental justice. Feminist political ecology inserts gender relations and gender justice as critical variables in political ecology, focusing particularly on the intersection of gender with class and other dimensions of difference in shaping processes of resource access and control and ecological change, and the struggles of men and women to sustain ecologically viable livelihoods (Rocheleau *et al.* 1996). It draws particular attention to questions of gendered knowledge, gendered resource access and control, and the engagement between local struggles and more global issues, while avoiding the seductions of 'earth mother myths' (Leach 2007).

The three crops on which our chapter focuses, like virtually all common crops entering wider (national and global) economic circuits, can be efficiently farmed (cultivated), with high per-hectare yields, on either small-scale or large-scale farm units. Corporate agriculture then does not necessarily mean 'corporate farming', if we make a distinction between *farming* – 'what farmers do', production on the land and 'their social and ecological conditions and practices, labour processes and so on' – and *agriculture*, a much broader notion embracing 'farming together with all those economic interests, and their specialised institutions and activities, "upstream" and "downstream" of farming that affect the activities and reproduction of farmers' (Bernstein 2010: 65, 124; 2013: 22). But small-scale farmers producing for distant markets require larger-scale units to take care of downstream (and in many cases upstream) activities. Cultivators are therefore, like it or not, engaged with the corporate[3] sector in one form or another, in complex commodity chains.

Considering gender in research on agrofuels expansion means considering problems in gender relations not only on (small-scale) farms but also in different positions in differentiated agrarian labour regimes, and at different points in agro-commodity chains. Women and men (and male and female children) may be involved in agrofuel production systems in many different capacities. They may be direct producers on their own account, or unpaid family workers in family farms (including contracted farms); they may be wage workers on the farms of others (larger farms or industrial plantations); they may be actors (own-account, unpaid family workers, wage workers) in the upstream and/or downstream entities in agro-commodity chains; and they may be consumers of food and other agricultural products which they have not themselves produced, and providers of care and food in households where one or more members are involved in agrofuels production.

This means that rather than assuming shared experiences, interests and struggles among women (and men), we need to be sensitive to differences both between and within rural communities in the gendered experience of incorporation. Currently available research (for example Behrman *et al.* 2012) has rarely explored these differences in detail. We include in this critique the case studies discussed in the rest of this chapter; still, these case studies do allow us to detect some broad contrasts in gendered processes and experiences on the ground.

After briefly introducing the six case studies in the next section, we draw on them to explore differences between and within communities in women's changing experience of access to/control of land and other resources; division of labour and participation in production; access to food and the household food situation; and voice and participation in decision-making processes at the household and community level.[4]

## Cases and methods

The Indonesia case study concerns the expansion of oil palm plantations in Sanggau District, a sparsely populated region of West Kalimantan covering some 13,000$^2$ km, in which the indigenous Iban Dayak population have traditionally cultivated subsistence crops and commercial rubber in mixed gardens, combined with the collection of forest products for subsistence and for sale. The first oil palm plantation was established in the 1980s, and by 2008 there were 20 oil palm companies operating in Sanggau's 15 sub-districts. Four of these corporations are based on foreign investment, 15 on domestic private investment, and one is the state-owned company, PTPN XIII. The plantations combine wage labour and contract farming. The case study is based on a period of one month's field research by Julia in 2008 (reported in Julia and White 2012), and further qualitative research conducted by Julia in October 2011.[5]

The Lao PDR, Sierra Leone, Tanzania and Zambia reports are all part of a Food and Agriculture Organization of the United Nations (FAO) programme, which aims to investigate the gender and social equity implications of land-related investments.[6] All cases, carried out between 2011–2013, are based on about one month of fieldwork, consisting of interviews with national government ministries and organizations, local government offices, development partner and civil society stakeholders and focus group discussions with local farmers and agricultural workers involved in the investments. The research teams were generally made up of a senior researcher supported by two national research assistants.

The Lao PDR study was conducted in the provinces of Vientiane, Vientiane Capital and Borikhamxai. The jatropha case presented here is just one of the eight companies analysed. The company, the agriculture and biofuel division of a large Korean conglomerate, cleared vast tracts of land in Vientiane Province to start a jatropha plantation after having signed a Memorandum of Understanding with the Ministry of Planning and Investment in 2008 and while waiting to get its land concession. It also hired both salaried employees and wage labourers from the area.[7] Local households, regardless of the number of family members actually engaged in the work, received an annual lump sum payment to maintain a specific area of jatropha plantation over a whole season. However, when the research team visited the area in November 2011, the company had already discontinued its operations and the plantation had fallen into neglect, with all those who had been hired having lost their jobs.

The Sierra Leone case was conducted in four different investment sites. For this chapter, two cases in the Northern Province were analysed: a Swiss corporation producing sugar cane for ethanol and the for-profit company of a Dutch philanthropic foundation producing palm oil. The sugar cane company, established in 2008, is part of a Swiss corporation involved in several ventures in the energy sector in West Africa. The company leased 55,000 ha of land in two districts in the Northern Province and, after having reverted some land it did not need back to the communities, as of 2013 controlled 44,000 ha of land comprising 94 villages in three chieftaincies, of which it planned to utilize about 12,000 ha for the large-scale production of sugar cane for conversion to ethanol for export. An ethanol production plant was under construction and production was expected to start in November 2013. The palm oil company, established in 2006, operates a plantation and a processing facility for the cultivation and processing of oil palm on 164 acres of land located in Tonkolili District. The land, all customary land under the control of the Paramount Chief of Gbonkolenken, was leased from five landowning families for 25 years.[8] In addition, the company has 1,500 registered outgrowers operating within a 10-mile radius from the company's premises. Only 5 per cent of these outgrowers were women (Barley and Kondoh 2013: 12).

The Tanzania case involved a company established in Arusha District in 2005, owned by a holding whose sole shareholder was a Dutch foundation. The company produced jatropha crude oil, as well as seed cakes, for sale on the domestic market, sourcing jatropha seeds from farmers who already had trees on their land as hedge plants through a system of collectors supervised by company field officers. The farmers were discouraged from planting new trees. At the time of the fieldwork, the company had 13 permanent employees, around 70 collectors (around 80 per cent men) and some 5,000 participating farmers, of whom 65 per cent were women. The company had also just started implementing a group system of collection in collaboration with local partner organizations. In this system, the farmers are organized in groups and provided with capacity building in negotiation skills, record-keeping and group organization. The company went into voluntary liquidation in 2012. Soon after, its former General Manager was able to secure limited funding to continue the collection of seeds and the sale of oil and seed cakes.[9]

The sugar cane company analysed in the Zambian case operates in Mazabuka District in the Southern Province, where it combines own-production and contract farming on company leased land. Originally a project of the Commonwealth Development Corporation, the company has now been privatized and is owned by the investment arm of a smallholder farmer association and two other private shareholders. Of the 4,315 ha of land, leased from the government on a 99-year lease, the company subleases about a quarter to its outgrowers free of charge. At the time of the fieldwork, the company had about 160 outgrowers – of whom 43 were women – and 362 employees, divided between fixed term (63) and seasonal workers (299). Participation in the scheme ensures farmers an average of 6.5 ha of land under 14 year renewable sub-leases, in addition to between 0.5–1 ha of land per household for dwelling, domestic food production and income generation. The company accounts for about 8 per cent of Zambia's national cane production.

The Uganda case focuses on the gendered experience of oil palm expansion in Kalangala District, southern Central Uganda. The Vegetable Oil Development Programme of Oil Palm Uganda Ltd. incorporated local people as both wage labourers and contract farmers. Claudia Piacenza's qualitative study was based on two months of field research and interviews with 28 farmers (with both male and female farmers as respondents) and 16 other informants (Piacenza 2012).

## Access to resources and livelihoods

> Women are regarded as being particularly vulnerable when common land is diverted to biofuels feedstock production. Because of their limited control over private resources they draw on common property resources for many goods and services for meeting household needs, such as food, fuel, building materials, and medicines.
>
> (Clancy 2008: 422)

Women's land rights often consist of secondary use and access rights granted through male relatives, and are thus subject to change along with any changes in the conditions of men's rights over land. Various studies have documented how state-codified, individual and 'household-based' forms of land allocation in contract farming schemes often annihilate women's customary rights to land (Jacobs 2010).

In West Kalimantan, the expansion of oil palm plantations undermined Dayak women's customary land rights through the formal system of smallholder registration based on 'Family Heads' (*Kepala Keluarga*), whereby the male 'head of the family' is registered as the smallholder. When the conversion to oil-palm was made and households surrendered (on average) 7 ha of land under customary tenure to get 2 ha of contract-farming land in formal ownership tenure, the new contract-farming plots were nearly always registered in the name of the male household head. So in a single bureaucratic stroke, women lost rights to land and produce (Julia and White 2012). In an Iban Dayak community located on the Indonesia–Malaysian border, besides losing their customary residential land and being relocated into the plantation compound, many households were forcibly dispossessed of their agricultural lands (rice fields, fruit and rubber orchards) and access to rattan, one of the main livelihood sources of the community. Two women recalled:

> I had my four rubber orchards flattened to the ground. How should I eat now?
>
> The clearance [of my paddy fields] was done early in the morning. Initially, it was said to be for road construction. No compensation was given at the time of clearance [2009], but now it was paid, just recently [2011]. [I] just took what was paid.
>
> (Fieldwork Julia 2011)

The loss of their lands and other sources of livelihood left women in these households with no other livelihood option than wage work on the plantation, with daily payments ranging from Rp32,200 (approximately US$4) as daily labourers to Rp41,200 (approximately US$5) as permanent contract labourers. The women are mostly recruited as daily labourers, while the men have a wider range of jobs available, from daily labourer (usually harvester) to public relations officer. Due to the destruction of customary protected forests for oil palm expansion, people have lost access to rattan, which was particularly important for landless households who were most dependent on extracting forest products. One female head of household explained that she had to take up other kinds work such as rubber sharecropping, collecting agar wood and wage work as a log carrier; she has to work longer hours with less income compared to when she could collect rattan.

In the case of the Swiss corporation in Sierra Leone, the communities are free to access and use the land until the land is physically occupied and put into sugar cane production. Already, however, both women and men have lost access to land they cultivated with mixed food and cash crops, including rice, ground nuts, peppers, onions, aubergines, cucumbers and water melons. Most families also had tree crops, particularly pineapple and oil palm. Women, in this patriarchal culture, cannot access positions as chiefs and are less likely to participate in decision-making. Women and men farm land that is regarded as the 'husband's land', while women have individual plots for the production of seasonal cash crops, which they access through their husbands or through *bora*, symbolic payments made to other male landowners. Women in focus group discussions reported having lost access to such lands in many cases (Winsborg 2013: 44). When land was allocated to sugar cane production, the lease payments and additional benefits all targeted the (male) head of household. The perception of company managers was that a few powerful individuals were controlling the benefits and that women, in particular, were disadvantaged. However, company staff felt they 'cannot interfere in local culture' (ibid.: 45).

This is confirmed by findings from another case study of a Dutch company involved in oil palm cultivation and processing in the Northern Province of Sierra Leone. The deals, consisting of 164 ha of land, were signed by the heads of five landowning families who received the cash payments and then distributed them to the family members. Male focus group participants in Matopie lamented that while their lands had been taken, the payments received were too small to have any meaningful impact on their lives. Overall, women received less than men, with some getting as little as SLL2,000 (equivalent to US$0.46). Only widows, being recognized as household heads, were given an equal share. The women, in particular, said that despite their desire to engage in business activities, they could not do so with the small amounts of money received (Barley and Kondoh 2013: 14–15).

In the Ugandan Kalangala oil palm project, where oil palm now covers one-third of Bugala Island, only a few female household heads were able to acquire land for oil palm cultivation. Women's limited capacity to demonstrate their legitimate status as landowners (through land title or occupancy certificates) reduced their possibility of acquiring full oil palm grower status, although they are often engaged in farming activities on their husband's plots (Piacenza 2012: 10, 14). Later however the Kalangala Oil Palm Growers' Trust (KOPGT) relaxed the definition of land 'ownership' required for registration and thereby facilitated women's inclusion, so that 35 per cent of registered smallholders are now women (Piacenza 2012: 17). This case shows a full spectrum of gendered experiences, in which women's initial landholding status and intra-household bargaining position can lead to highly contrasting outcomes. In one case a woman, whose husband started growing oil palm, did not even know how much land he now controlled or how much income he gained, and she struggled to provide decent food and clothing for her children; initial inequalities had been exacerbated with the introduction of oil palm. In another case, while the husband started to grow oil palm and acquired extra land in his own name, the couple later bought land together and put it in the wife's name so she that could register with the KOPGT as a grower, with access to her own bank account; she maintains her own mixed garden and uses hired workers in her oil palm gardens, and they have kept a part of their land in forest to ensure wood fuel supplies (Piacenza 2012: 22–24).

Some of the cases did not involve the acquisition of large tracts of land and dispossession of local people. However, all were reported as having significant gendered impacts on the control of land and its produce, particularly with respect to the collection of, and access to, non-traditional forest products (NTFPs). For instance in the jatropha investment case in Lao PDR, before 1998, local women and men had been farming upland rice under shifting cultivation in the plantation area, which was officially state forest. After the government banned shifting cultivation and allocated the area to the village as communal land, the women used it for collecting NTFPs for household consumption and sale. Since the jatropha plantations started, however, access to NTFPs and the income from it have diminished. Women in particular were affected, although both women and men got some cash income from working as casual labourers for the company when the plantations started operating (Daley *et al.* 2013: 31).

## Division of labour and access to employment

Does biofuel production create employment opportunities that enable poor women to move out of poverty? There are two issues particularly relevant here. The first is poor women's ability to participate in paid labour markets

> and second is the levels of remuneration that they would receive in those markets.
>
> (Clancy 2008: 423)

In 2012, the sugar company in Sierra Leone employed 1,460 staff, of which over 1,000 were temporary. In spite of the company's claims of gender parity objectives in recruitment, only 10 per cent of the staff were women. Hiring was managed through the local authorities, who 'when they come up with the lists, bring mainly men'. In spite of disappointment over short-term jobs, employment was perceived as the main benefit arising from involvement with the company and the most important compensation for loss of land. Income was primarily used for food, particularly rice; other uses included school fees, savings for education and clothes. Some women highlighted their (young adult) children's employment as a source of household income and relief from the drudgery of subsistence farming, including the possibility of taking food on credit. A group of women believed that food security in the village had improved, but widespread recourse to loans for food purchase suggests that food security was under pressure (Winsborg 2013: 46).

Likewise, at the oil palm company, there were only three women employees out of a total of more than 100 (both permanent and casual). The three women who were permanent employees were assigned to soap making and cleaning duties at the processing site; these tasks were allocated lower salaries than the tasks assigned to men (Barley and Kondoh 2013: 18). The elders said that they had thought that the company's priority would be to provide jobs for their children. However, this had not been the case and they noted, 'our children have all gone to mine gold' (ibid.: 16); in several communities in the investment area, a considerable number of youths had migrated to work in mining, which in turn led to labour shortages on small farms, particularly those operated by widows (ibid.: 19).

Employment opportunities that are unsustainable in the long term do not provide real livelihood alternatives. The jatropha investment in Lao PDR is one example, having already discontinued its operations when the research team visited the area in 2011. Both women and men who had worked as casual labourers for the company wished that it had resumed operations. However, they also had negative or neutral opinions about the company, mainly because of the worsening of their food situation, particularly due to women's diminished access to NTFPs, as already noted. The village head explained that when the jatropha plantations came, jobs were created at the start to clear and plant the land, and many households lost their own crops or missed other opportunities because they spent their time working for the company, so when the company stopped operating they lost livelihoods and harvests. 'We welcome investment here', he explained, 'but these should be good companies with stable financial situations and operations. These companies should also support the community and have a good vision for sustainable development' (Daley *et al.* 2013: 37).

The Zambian sugar company had a total of 364 employees, among them 63 fixed-term staff (six of them female in mainly administrative and secretarial positions) and about 300 seasonal wage workers (17 per cent female) engaged in the labour-intensive phases of cane growing, which comprise planting, fertilizer and herbicide application, and irrigation. While daily rates are standardized, women are generally engaged in lighter tasks, such as replanting and weeding, which are more limited in quantity and duration. Moreover, traditional gender divisions of labour in agriculture tend to confine women to the production of staple crops or domestic food crops. Cash cropping of cane sugar, or seeking work in cane production is considered to be inappropriate for women (FAO 2013: 23).

Whilst crucial for food collection and production, access to land also affects women's chances of participating on their own account in contract farming. This was evident in the Sierra Leone oil palm investment, where only 5 per cent of the company's 1,500 outgrowers were women. These women had become outgrowers by virtue of the fact that their husbands (the registered outgrowers) had died and they had become heads of households. The company selected prospective outgrowers based on the amount of land they had available to cultivate the new varieties of oil palm (Barley and Kondoh 2013: 19). In all the communities visited, women were not entitled to independent land rights and had limited control over household resources and decision-making. The men were emphatic about women not being entitled to own land. In Matopie, the men stated, 'we the men own the land and make decisions for the women' (ibid.: 16).

According to focus group participants, participation in the outgrower scheme has reduced their workload by cutting some of the activities in the processing of palm fruits. Both men and women agreed that it was the women who benefitted the most from this reduction. The women appreciated having more time available for undertaking other chores as a result of the reduction in their workload. However, women in Petifu Mayopoh complained about the unavailability of by-products of palm kernel oil and 'Enoi' (the chaff of the kernel fruit, used as cooking fuel). In terms of income, it was the men who benefitted most from the sale of palm oil, as the income was received and controlled by them. But men lamented that now they were selling the oil to the mill, their wives knew how much money they were making as they were aware of the amount and price of the fruits being sold to the mill; thus, now they are more accountable to their wives while before, they could report any amount (ibid: 20).

The Zambian sugar cane company chose its outgrowers in a competitive selection process, with successful applicants being allocated 4–6 ha of standing cane field and a further 0.5–1 ha of residential and home-garden land, free of charge. Initially, it was mostly men who got the land. Subsequently, through a succession clause embedded in the contract, more women were able to access the scheme. At the time of the fieldwork, 43 out of 160 farmers were women. Of these, 51 per cent were single, young, never-married women who had all entered the scheme through inheritance. Married women constitute 30 per cent of all female outgrowers, while widows or surviving female spouses of original male scheme members represent 19 per cent (FAO 2013: 25). Focus group discussions with male and female outgrowers in Mazabuka District, where the company operates, revealed that in male-headed households where women were the main (registered) members of the scheme, women had a greater say over how the income was used than when the man was the registered member (FAO 2013: 27). 'They believed that through their involvement they have challenged negative attitudes about their capabilities and their rights. They also felt they were able to articulate their problems and identify solutions when necessary, particularly when in group settings.' (ibid.: 29)

In the Tanzania (jatropha) case, although the income the farmers made from selling jatropha seeds was minimal, people in focus group discussions said that they were doing it as a complementary income-generating activity in addition to, and alongside, their own farming. The company's local collectors, supervised by the company's field officers, were given cash payments in advance in order to buy seeds for cash directly from the farmers. The seeds were then stored at collection sites, where company trucks could pick them up and take them to the factory for processing (Daley and Park 2012: 13). Participants in focus group discussions depended heavily on their own food crop production and their food security suffered when maize and bean harvests were bad. Jatropha collection seemed to benefit women in particular, because of men's willingness to leave jatropha as a women and children's crop given its minimal profitability. In fact, it was usually the women, supported by the children, who picked and dried the seeds and took them to the collection sites where they were paid. The money went into paying for small household items, which women were in charge of purchasing in any case, such as bars of soap, cooking oil, kerosene and so on.

> I am very happy to have Diligent [name of the company] here as now there's another crop I can get money from and I am happy to have that possibility to support my children.
>
> (Young widow)

> I feel good because I can get money now for buying exercise books for school without asking my father.
>
> (Young girl still at secondary school, living with her father after her mother died)

Some male farmers said that if prices were higher, they would put more effort in jatropha collection, as they do for coffee, but because of the low returns they were happy to leave the task to women:

> I am happy … because there is now less fighting in our family with my wife, as she gets money from the jatropha seeds which cater to her small needs so other money coming into the family can be saved for other things.
>
> (Old man – all quotations, Daley and Park 2012: 16)

The company estimated that 65 per cent of the approximately 50,000 farmers selling seeds through this collection system were women, while only 20 per cent of the 70 collectors were women. The higher returns on being a collector made it more attractive for men to be involved.

These examples indicate that different women have different experiences with different forms of incorporation in corporate farming. Some women appeared to be more independent as a result of their engagement with the investments. Others welcomed the extra cash coming from casual labour on plantations or minimal income from sale of jatropha seeds.

Wage labour represents another form of incorporation into corporate agriculture. The case studies suggest, as can be expected, that those who were landless or had less land were more likely to engage as wage labourers on plantations and as casual labourers for other farmers. For instance, in Lao PDR both women and men in an isolated village with few employment opportunities welcomed casual labour on the jatropha plantation, although this was short-lived (Daley *et al.* 2013: 37).

Overall, women who are engaged in wage work tend to be in non-permanent, worse-paid jobs that are often segregated by sex, task and crop (Daley *et al.* 2013; FAO 2013). Women were absent or under-represented in managerial or supervisory roles. Consequently, women were, on average, earning less than men per month and, as casual workers, had worse employment conditions compared to their male counterparts in permanent roles. Women were also responsible for most domestic work, so as wage workers they experienced an increase in their overall workload.

In all cases across countries and crops, there was a distinct division of labour, whereby women had specific tasks in farming and agriculture, were responsible for the collection of NTPFs, fuelwood and water and bore more or less entirely (together with children) the weight of care and domestic tasks. As wives of contract farmers, women had increased workloads in farming and agriculture but also in domestic activities (Daley *et al.* 2013). However, in some cases, as for instance in oil palm operations in Sierra Leone, women in outgrowing families benefitted from a reduced work burden, thanks to the streamlining of processing activities.

There were also cases of successful female outgrowers who were, in some instances, more productive than their male counterparts. The following testimony is from a prosperous Zambian female sugar cane farmer, who is also self-sufficient in maize, owns a car, has built a house with rooms for her two servants, and is building a house in the nearby town:

> I am able to grow maize on the 1 ha of dwelling space to meet my food security requirements and generate additional income for my household. With proper management and care, we are able to feed ourselves for the whole year.
>
> I have built a three bedroom house with a two-room servant quarters. I belong to a group of farmers which has 50 members, but only four of us are women. Yet female farmers generally perform better with respect to productivity and investments in household welfare improvements. They invest

> more in household goods and family education. As you can see, I own a car and have bought a 30 × 30 metres residential plot in Mazabuka. I want to rent the new house so that I diversify my risk portfolio. In addition, I have employed 25 irrigation workers which most male farmers fail to do.
>
> (FAO 2013: 29)

## Food security and food sovereignty

Eide (2008: 4) notes that biofuel production is likely to weaken access to adequate food, or to the resources by which vulnerable people feed themselves, in at least three ways. First, by contributing significantly to the increase in food prices; second by causing land concentration for plantation-type production and third, because biofuel production causes a number of environmental problems, reduces biodiversity and leads to competition for water.

In Zambia, the sugar cane company made 0.5–1 ha of marginal lands available to its outgrowers as domestic plots. Interviews with the company's extension officer and individual farmers indicated that many farmers had utilized the land to produce food for their own consumption. Contrary to men, women tended to use these plots for household food crops rather than cash crops due to time constraints. Thus, despite being constrained by their overall labour burden in their ability to produce and sell cash crops, female-headed households appeared to be faring better in terms of availability of food (FAO 2013: 28).

In Sierra Leone, women in the focus group in Manewa said:

> The bad is more than the benefits. They don't keep the promises: employment, money and food. The money is small compared to what they said. We can't see the benefits. They take the land but are deviating from the contract. This is very grave in our hearts. We suffer more. We do most of the household chores. We buy for the household. We don't have husbands who give us money for food. They may just give rice, then we have to go and get the sauce. Before we used resources on their farms. Men control the money. It is a problem but it is a tradition.
>
> (Winsborg 2013: 44)

In the same study, women in Romaro also reported having lost access to wild (red) palm oil for household use and having to purchase it. They said that the time spent collecting firewood had increased substantially, and that they now had to spend up to seven hours every other day to meet household needs (Winsborg 2013: 44).

In the Tanzania case, people engaged in corporate agriculture reported greater benefits in terms of cash income than improvements in the household food situation (Daley and Park 2012). Seventy-one per cent of respondents said they were better off in terms of cash income, while only 49 per cent said they were better off in terms of their food situation. In this case, the majority of negative responses were linked to incomes from involvement with investments not being enough for families to keep up with increasing food prices – a point confirmed by the Indonesian example below. However, about half of the focus group participants reported improvements in their family food situation (Daley and Park 2012: 30). Having more cash does not automatically translate into better access to food, as the Indonesian case shows. In the Iban Dayak community, women who lost their land have also become wage labourers on the oil palm plantation. In general, households are increasingly becoming more dependent on cash for their food supply and on the plantation as the main source of cash economy. But this is seen in a very different light by women in different positions. One woman noted that:

> [We] produced our own in the past, then, one should obtain and produce by oneself in order to have them [the food]. Now, we must purchase in order to obtain food as lands have all been condemned. None can be planted anymore. Like it or not, it's only money that talks now.

Conversely, a woman whose husband was employed as public relations officer at the plantation stated her appreciation for the greater ease of life made possible by the new availability of purchased foods:

> In the past, if we wanted to buy fish or other types of meat, we had to travel as far as Seluas [the sub-district town]. Now, those things come by themselves ... really, people deliver them [to the village].[10]

Women in small-scale farmer households also complained about the rising price of food and other goods. Thus, although they had more cash, their purchasing power was weaker and they reported experiencing difficulty in meeting other basic needs, such as health and education, for their children.

## Gendered political responses from below: resistance, negotiation and protest

Most of the case studies that we have used in this chapter do not provide sufficient information on women's political responses to the expansion of corporate flex crop/agrofuel production in their communities. To avoid any impression that women are passive victims (or beneficiaries) of these changes, we provide some isolated details on women's agency and political responses.

First, it is clear that in many of the cases lack of access to land and patriarchal power relations have restricted women's participation in community decision-making and their interaction with the investors. Julia and White report difficulties faced by Iban Dayak women, who were excluded from the village consultations over the establishment of the oil palm plantation; customarily women are not allowed to participate in community politics and 'the voice of the men was considered to be the unanimous voice of the villagers' (Julia and White 2012: 1012).

In Sierra Leone, the preparation of the investment involved a significant process of consultation; villagers recalled several meetings held about the project. Some women respondents reported difficulties in having their voices heard during these meetings and in challenging the authorities (including the Paramount Chief, section chiefs, etc.) who were proposing the deal, although this varied from village to village. The Liaison Committee in Manewa said that 'everyone' participated in village consultation meetings, men, women and youths. However, women were marginalized in village consultations and decision-making; many women were busy cooking during the consultation and only a few joined the discussion once they were done with their duties.

> First the company came. They called a meeting. A few were at the first meeting but women only listened. Addax came and requested land. The women there also supported. There were about fifteen women at that meeting. They told [us] that the youth will be employed and all will benefit. It was the first time a foreign company came, so we could not say no, because of the money. We were happy that employment would give money to buy food, so we clapped. They came for about seven meetings in the village. They bought food for the meetings, for everyone in the village. Women did the cooking, so some of us missed the first parts of the meetings. Land owners took the decision; we only listened. We did not talk among each other, because men do not allow us. We were not allowed to even vent our opinion. We do not own the land, we just come through marriage, we are told. But we talked with each other in other fora, at the well and so on.
>
> (Women's focus group, Manewa, Winsborg 2013: 41)

In the Indonesian oil palm case, despite the cultural taboo on women's participation in public fora, women have been involved in resistance, both alongside men and individually. When villagers protested against the state-owned plantation company PTPN XIII by blockading and harvesting a part of the company's nucleus zone, only five women were involved directly in the protest. The military were brought in to confront the protesting villagers, who were all taken into police detention. After negotiation between the male villagers and the police, the women were released and sent back to Anbera hamlet, while the male villagers were jailed for a day. Some women also voiced their protest in village meetings with the company. One woman, for example, stated her objection to the violation of promises made by the company at the sub-district office:

> The company promised us that after our land had been planted with oil palm trees, it would be given to the farmers but ten years after, it was not cleared, not maintained. So we re-occupied it. Our rubber trees had been cleared down without compensation, so I took hold of the plantation plot as the compensation of our land, because it belonged to us. The rubber trees were

> compensated with the oil palm trees, so don't say that we steal. I said this at the Yanba sub-district office.
>
> (Mrs Ayokng)

At the district level, women members of the Oil Palm Farmers Union (SPKS) engaged in open protest against the malpractices of the oil palm plantation companies in the district. Mrs Rini, a Iban Dayak woman who lives in a neighbouring village, was known for her strong resistance against the plantation company on behalf of both her own family and the community:

> As a teacher working in the village and originally from the village Kampuh, and as indigenous Hibun, my husband and I feel that it is our obligation to react collectively. We and other teachers in the area as well as the other communities here are members of the cooperative. Our land was taken with empty promises, but none of our *adat* leaders have reacted. They became the guardians of the company, not any more our leaders. By joining SPKS, step by step we will gain back our cooperative and we can negotiate with the company to gain a better position.
>
> (Mrs Rini, quoted in Sirait [2009])

The practice of scavenging for fallen oil palm kernels *(berondol)* on the plantation, which is a part- or full-time occupation for many local women, may also be considered as an expression of 'everyday' resistance. Although the company leaves the *berondol* to rot on the ground, the *berondol* seekers are considered as thieves by the plantation company, and women have to evade the plantation's security officials or police who patrol the area. The women are often intimidated, being told they will 'will be sent to Sanggau'[the District capital] or 'will be jailed in Sanggau' if they are caught. Such intimidation, and the sexual harassment that often accompanies it, has created intense fear among some women (Julia and White 2012).

## Concluding reflections

What do these findings mean for a feminist political ecology of agrofuels? First, in regions of corporate expansion of flex/agrofuel crops, paths of differentiation and class formation are diverse, depending on the initial conditions, the terms of incorporation and entitlements of different individuals and household members. So while some women consider the opportunity to engage with corporate agribusiness to be beneficial, for others it has meant more work and few benefits. Second, some women do not have the capital and the resources needed, or are prevented from participation by discriminatory land allocation and employment recruitment practices. The case studies suggest that women's access to the means of production, particularly land, and control over what to produce, are still largely constrained by patriarchal relations of power operating at the community and household level.

At the same time, however, there are also important differences in women's experiences both within and between communities, shaped by the perceptions and opportunities that different (classes of) women have vis-à-vis diverse forms of incorporation. The case studies we have presented are undeniably inadequate for a thorough analysis. However, they do support the argument that 'women are not all the same' to begin with, in terms of resource control and position within the household and community, but also in relation to what they may want and expect for themselves and their families by engaging (or not) in corporate agriculture. Issues around gender inequalities, patriarchal relations and class-based differences need to take into account the diverse positions and roles of different groups, and women in different positions within those groups.

Finally, in the event that gender justice and empowerment for rural women could be achieved, to assume that all rural women would choose (small-scale/family) farming as opposed to engagement with corporate agriculture is quite a leap of faith. The incorporation of gender interests and gender analysis in the study of agrofuels expansion and its consequences requires recognition and exploration of differences in women's experiences, interests and responses, both between and within communities. The current, continuing dearth of careful empirical work on these dimensions of corporate agrofuel or 'flex crop' expansion should be recognized, as a matter of some urgency, in research agendas for the coming years.

## Notes

1 A woman in Alipe village, Ghana Northern Region, addressing Mr. Finn Nyberg, Director of Land Acquisition for Biofuel Africa about the company's destruction of forests to establish 'the largest jatropha plantation in the world' (Nyari 2008).

2 Recognizing that the case studies in this chapter permit exploration of only some of the core elements in a feminist political ecology framework (without, for example, allowing analysis of gendered knowledge), we have chosen to call our analysis a 'gendered political ecology'.

3 'Corporate' here can mean private, public or cooperatively owned corporate entities.

4 Three of these six case studies (those from Indonesia, Lao PDR and Zambia) have also been discussed in the authors' recent article on gender and food sovereignty (Park *et al.* n.d.). Our summaries of these three studies in this chapter inevitably reflect a certain degree of (permissible) 'self-plagiarism'.

5 The 2011 study, which has not been published, was commissioned by the Economic and Social Empowerment Commission of Pontianak Archdiocese.

6 Under the FAO programme case studies were conducted between 2011 and 2013 in Tanzania, Sierra Leone, Ghana, Zambia, Lao PDR, Mozambique and the Philippines.

7 Several informants reported that the company had not followed the correct procedures and that this had caused delays in getting the concession. One informant at the district level informed the research team that the company's request was pending further investigation before any documents could be issued. Another informant added that the company had bypassed the district authorities and gone to the villages using a well-known local man as a broker.

8 With the exception of 10 acres for an experimental farm, which were leased for 15 years.

9 Email communication with Jan Gaevart, General Manager of Diligent Ltd, 11 November 2013.

10 These quotations are from Julia's unpublished 2011 field research (see note 5 above).

## References


Barley, Kenyeh and Francess Kondoh (draft) (2013) *The Gender and Equity Implications of Land-Related Investments on Land Access and Labour and Income-Generating Opportunities: Two Additional Case Studies of Selected Agricultural Investments in Sierra Leone*. Rome: FAO.

Behrman, Julia, Ruth Meinzen-Dick and Agnes Quisumbing (2012) 'The gender implications of large-scale land deals', *The Journal of Peasant Studies* 39(1), 49–79.

Bernstein, Henry (2010) *Class Dynamics of Agrarian Change (Agrarian Change and Peasant Studies Series)*. Black Point, Nova Scotia: Fernwood Publishing.

Bernstein, Henry (2013) 'Food sovereignty: a skeptical view'. International Conference on Food Sovereignty: A Critical Dialogue, Yale University, September 2013.

Clancy, Joy (2008) 'Are biofuels pro-poor? Assessing the evidence'. *European Journal of Development Research* 20(3), 416–431.

Daley, Elizabeth and Clara Park (2012) *The Gender and Equity Implications of Land-Related Investments on Land Access and Labour and Income-Generating Opportunities. A Case Study of Selected Agricultural Investments in Northern Tanzania*. Rome: FAO.

Daley, Elizabeth, Martha Osorio and Clara Park (2013) *The Gender and Equity Implications of Land-Related Investments on Land Access and Labour and Income-Generating Opportunities. A Case Study of Selected Agricultural Investments in Lao PDR*. Rome: FAO.

Eide, Asbjørn (2008) *The Right to Food and the Impact of Liquid Biofuels (Agrofuels)*. Right to Food Studies. Rome: FAO.

FAO (2013) *The Gender and Equity Implications of Land-Related Investments on Land access, Labour and Income-Generating Opportunities A Case Study of Selected Agricultural Investments in Zambia*. Rome: FAO.

Jacobs, Susie (2010) *Gender and Agrarian Reforms*. London: Routledge.

Julia, and Ben White (2012) 'Gendered experiences of dispossession: oil palm expansion in a Dayak Hibun community in West Kalimantan', *The Journal of Peasant Studies* 39 (3–4), 995–1016.

Leach, Melissa (2007) 'Earth mother myths and other ecofeminist fables: how a strategic notion rose and fell', *Development and Change* 38(1), 67–85.

Magdoff, Fred (2008) 'The political economy and ecology of agrofuels', *Monthly Review*, July–August, 34–50.

Nyari, Bakari (2008) 'Biofuel land grabbing in Northern Ghana' (www.biofuelwatch.org.uk/files/biofuels_ghana.pdf, accessed: 1 August 2014).

Park, Clara, Ben White and Julia (n.d.) 'We are not all the same: taking gender seriously in food sovereignty discourse' (under review, *Third World Quarterly*).

Piacenza, Claudia (2012) 'Negotiating gendered property relations over land: palm oil expansion in Kalangala District, Uganda', International Conference on Global Land Grabbing II, Cornell University, October 2012.

Rocheleau Diane, Barbara Thomas-Slayter and Esther Wangari (1996) *Feminist Political Ecology: Global Issues and Local Experiences*. New York: Routledge.

Sirait, Martua. T. (2009) Indigenous Peoples and Oil Palm Plantation Expansion in West Kalimantan, Indonesia. The Hague: University of Amsterdam and Cordaid Memisa.

Winsborg, Poul (Draft report) (2013) *'Nah All Man Business': Gender and equity implications of land-related investments for access to land, employment and other income opportunities: Two case studies in Sierra Leone*. Rome: FAO.

# 5 Bridging the gap with agrofuels

## Energy hunger, energy scarcity and climate change in the European Union

*Achim Brunnengräber*[1]

### Introduction

Economic growth, competition and economic stability in the anthropocene[2] are highly dependent on cheap, sufficient and secure energy sources. The anthropocene is a destructive, man-made epoch, which started with the Industrial Revolution and is characterized by the congruence of fossil fuels and capitalism. Fossil fuels suit the capitalist mode of production perfectly and have contributed to changing the planet profoundly. However, against the background of peak oil,[3] climate change and changing geo-economic constellations on the global scale, the need for alternatives to fuel the fossilist energy system – I call it 'fossilism' – has become more and more pressing. Agrofuels have become part and parcel of the consolidation and prolongation of a critical, eco-unfriendly fossilism, which needs the increase in energy input to produce the global output of goods and services corresponding to our consumption and mobility patterns. Based on this assumption, my aim is to show that because of the existing power constellations in the energy, transport, automobile and agriculture sectors, agrofuels – despite all criticism – will continue to serve as a technological 'fix' for a globally fragmented fossilism. Fossil fuels have an adverse impact on the global climate. It was hoped at the beginning of the agrofuel boom that they would help to mitigate that impact through the reduction of greenhouse gas (GHG) emissions. In so doing, they were supposed to lead us into a sustainable energy future by 'feeding' and facilitating green economic growth. However, blending gasoline and diesel with ethanol and biodiesel which are made from corn, wheat, palm oil or soy has become a highly controversial issue in both the Global North and the Global South.

This chapter focuses on the European Union (EU) and its member states, which are struggling to redefine future blending targets for agrofuels. In 2009, the EU established a 10 per cent target to be met by 2020. But since a number of studies have shown that the growth of agrofuel production has caused negative climate and other ecological side-effects, as well as negative social effects, scientists, politicians, interest groups and civil society organizations have begun to question this target. Some of them are demanding a 5 per cent cap for agrofuels in conventional fuels, e.g. the German Ministry for the Environment, while the agrofuel industry is lobbying for the consolidation, and even a further increase, of the 10 per cent target. Others, such as the former UN Special Rapporteur on food rights, Jean Ziegler, have called the agrofuels policy of the EU an 'idiocy'[4] because of the impact on global food prices, land use changes, and ongoing deforestation, especially in the Global South, and have pleaded for a ban on agrofuels altogether. The argument is that under current conditions, the production of agrofuel crops is already displacing food crops, pushing up food prices, and the use of some energy crops is actually leading to higher carbon emissions than that of fossil fuels if direct and indirect changes in land use are fully accounted for (UNEP 2009, 2014). Blending agrofuels seems to be less promising now than at the beginning of the millennium. In 2012, the production of agrofuels dropped slightly worldwide, even though some countries increased their production and markets for renewable energy sources continued to expand.

Despite all these criticisms there is still strong support for increasing production due to other reasons, which play an important role in the EU's energy policies.[5] The overall aim is to bridge the gap between the different policy objectives: reducing (GHG) emissions, securing energy supply, supporting growth and fostering competitiveness. Can agrofuels play a role in this? Is a fusion between energy security and climate protection possible and will it result in opportunities for a green economy in the EU? It will be argued in this chapter that the coming together of climate protection and energy security, as aimed at by the European Commission (EC), cannot be achieved with the strategies and measures which have been taken so far. But why is support for agrofuels growing despite the lack of clear benefits and harsh criticism from various actors? The narratives about the advantages of agrofuels and the discursive strategy to promote agrofuels for energy security reasons (see Vogelpohl, Chapter 14 in this volume) are no longer important. The EU is no longer the key player in this field because the empowered agrofuels industry and individual member states are now dictating the way forward. Even if there are no incentives coming from the EC, the member states and strong lobby groups continue to support agrofuels on a high level. Now that the political start-up phase of the agrofuel economy is over, profit-driven forces are writing a neoliberal agrofuels agenda.

This chapter is based on a critical political and economic perspective, taking into account recent conceptual debates in the field of political ecology. My theoretical framework is the 'ambiguous post-neoliberalizing of nature' (Brand 2009), meaning the new forms of appropriation and changes in investments in nature, spatial and political reconfigurations, and the restructuring of industrial agricultural production. This is not a fateful process with a determined end, but is highly questioned and quite controversial, accompanied by conflicts, contradictions and struggles, and it has not only a material but also a symbolic and discursive dimension.[6] A discourse is a pattern of concepts that give meaning to physical and social reality, but if materialist and profit-driven interests, structures and actors are strong enough, they can temporarily counteract any critical discourse (for a similar discussion about the neoliberalizing of climate change, see

Brunnengräber 2013). However, the political ecology of agrofuels as an area of conflict is increasingly coming under pressure as the contradictions become more and more obvious.

To bring the different symbolic and material dimensions together, I shall start by reflecting on the crucial changes in the architecture of world energy and the emergence of a fragmented energy world. The extraction of gas and oil in the United States, the increased use of hydraulic fracturing (fracking) and the deepening of South–South relations are indications of this. I shall then present a brief overview of EU policy on agrofuels and look at the EU's concepts to show how they programmatically try to interlink agrofuel policy with energy security and climate change. After that, I shall explore the EC's back-down in its support of agrofuels, which are now strongly promoted politically by individual member states and the agrofuels industry. Following that, I shall argue that growth and competition has priority over the reduction of GHG emissions (see also Park *et al.* 2008) and I shall present findings on the social and environmental consequences of EU agrofuel policies in the South before moving, finally, to the role of agrofuels in the capitalist energy system. Summarizing, I will show how a fragmented world energy market and a shift to a new fuel took place without a shift in patterns of consumption or economic growth. Fossilism can be prolonged into the future.

## The fragmentation of the energy world

The financial crisis of 2008 changed the world dramatically. The budgetary problems of EU member states had dire consequences for climate policy and the energy sector – and vice versa. Energy production and consumption became central issues in the revitalization of the economy. The UN Conference on Climate Change in 2010 in Cancún, agreed to commit to a maximum increase in the planet's average temperature of two degrees above the pre-industrial level. On a rhetorical level, the UN still supports this target, but no further progress was made at the conferences which followed in Durban (2011), Doha (2012) and Warsaw (2013), and no new legal obligations will be implemented before 2020. It is obvious that the UN Framework Convention on Climate Change (UNFCCC), signed in 1994 and the Kyoto Protocol, which became effective in 2005, still do not have the power to regulate and bring governments towards the desired aim (Altvater and Brunnengräber 2011a, 2011b). Since 2000, an average annual increase of 2.9 per cent has meant that global GHG emissions reached a total of 34.5 billion tonnes in 2012 and, 'despite many countries taking new actions, the world is drifting further and further from the track it needs to follow' (IEA 2013b: ii).

Cheap and sufficient energy is the basis of fossilism. It is helping to bring member states back to growth and prosperity. Furthermore, 'the rise of unconventional oil, gas and renewables is transforming our understanding of the distribution of the world's energy resources' (IEA 2013a: 3). However, the situation differs among countries and regions. Major importers are becoming exporters.

Due to the dynamic changes in many regions of the world, global energy demand will grow by more than one-third by 2035, triggered mainly by the enormous need for energy in China, India and the Middle East, which account for 60 per cent of the growth. The IEA projects that EU reliance on imported oil will increase from around 80 per cent today to more than 90 per cent by 2035. 'This increases the EU's vulnerability to supply and energy price shocks' (EC 2014: 11). Demand is barely rising in the OECD (Organization for Economic Cooperation and Development), but there is a pronounced shift towards gas and renewables (IEA 2012a). Powerful industrial countries like the EU member states, China and India are doing a lot to secure a stable supply of oil and gas. As a result, new supply regions must be found (import diversification) and the energy mix must be changed (resource diversification) to reduce dependency on the oil-producing countries.

Paradoxically, the country which did not ratify the Kyoto Protocol, the United States, has been making some steps towards decoupling GHG emissions from energy consumption. In 2012, emissions decreased by 4.0 per cent compared to 2011. The use of unconventional shale gas produces fewer GHG emissions. The IEA has called the new period the 'Golden Age of Gas'. It is estimated that the reserves will be enough for another 250 years. Furthermore, the extraction of unconventional oil will turn the United States into a main oil producer and by 2035 it will no longer need to import fossil fuels (IEA 2013a; 2012b). Simultaneously, agrofuels are becoming less important in the United States and, for competitive reasons, more important in the EU.[7] Nevertheless, the impacts of fracking on the environment, on human settlements and health are unclear and very risky. 'Knowledge deficits' concerning the technology have brought the German Advisory Council on the Environment to make a clear statement: fracking 'can currently not be allowed' (SRU 2013: 43). The German government demanded a moratorium on fracking in 2013, but in the meantime, in the United States, the cost-effective combination of horizontal drilling and hydraulic fracturing is booming.

National availability means that the new resources are of vital importance geo-strategically. The United States produces its own cheap energy and therefore has advantages on the world market. However, this is not the case in the EU, where the debate about the environmental and economic merits of fracking is still going on. The member states cannot agree on a common position on whether or not to allow fracking. If it is not a part of the diversification strategy of the EU, there is a need for alternatives. Agrofuels are such an alternative and an answer to the gap between the EU's energy hunger and energy scarcity:

> Agricultural and energy markets have become increasingly interlinked not only through the supply side as energy is an important input but also on the demand side as the demand for biofuels has shifted outward the demand for several crops including maize, wheat, sugar and various oilseeds as feedstocks.
>
> (OECD and FAO 2013: 22)

The EU's strong fixation on low energy prices and its adherence to fossil fuels as a precondition for economic stability and growth (EC 2014) are confronted with this new circumstance on the world market. Can agrofuels be a solution? At first glance they cannot: they are systematically connected with negative effects on food production and security, natural resources, including land, soil and water and rural development. However, the global interest in agrofuels as a way of reducing dependency on fossil fuels is increasing. Emerging Southern economies like Brazil and India have the potential to support and finance the industry and encourage investment in less developed and less powerful countries. 'This is seen in the increasing influence wielded by developing countries over the direction of agricultural and energy production, a trend reinforced by their rapidly growing populations and energy demands' (Dauvergne and Neville 2009: 1099). Agrofuels are deepening and strengthening South–South relations, but are not diminishing the influence of Western economies on Southern countries. I shall demonstrate the impact of EU agrofuels policy as a clear example.

The crude oil price and the opportunity to shift to alternatives is a strong driving force. Liquid agrofuels provided about 3.4 per cent of global road transport fuels, with a small but increasing use in the aviation and marine sectors (a 0.8 per cent estimated agrofuel share of global final energy consumption in 2011) (REN21 2013: 16):

> The average annual growth rate in the period from the end of 2007 through 2012 was nearly 11% for ethanol and 17% for biodiesel. Although biodiesel production continued to expand in 2012, it was at a much slower rate of growth, whereas ethanol production peaked in 2010 and has since declined.
> (REN21 2013: 18)

Growth is continuing and is strongly supported by several governments. In the EU, 'UK, Spain, Germany, Italy, Poland and France account for 70% of the additional biofuel demand between 2008 and 2020' (Bowyer 2011: 10). The worldwide competition for arable land is increasing in China, the United States and the oil countries of the Middle East:

> Biofuel blending mandates were identified at the national level in 27 countries and in 27 states/provinces. Despite increasing pressure on major markets such as Europe and the United States, due to the growing debate about the overall sustainability of first generation biofuels, regulatory policies promoting the use of biofuels have existed in at least 49 countries as of early 2013.
> (REN21 2013: 15)

To sum up, agrofuels play a decisive role in times of multiple crisis and in the age of fossilism; a fragmented energy system such as that of the EU helps to diversify supply, improve energy security and encourage competition.

## EU strategies for coherent energy and climate policies

The promotion of agrofuels within the EU began when oil prices dramatically increased in 1973 and was controversial from the beginning. The intention was to reduce dependency on energy imports from the OPEC and other oil-producing countries. 'Also, if oil price rises put new constraints on agriculture, they would also open the possibility of new outlets for products of agricultural origin which could be used as raw material for energy production' (CEC 1980: 16). It took a long time, till 2003, before the first EU Directive covering agrofuels was approved (EPC 2003). The EU Directive 2003/30/EC set a minimum percentage of agrofuel to replace diesel or petrol for transport purposes in every member state. It set a minimum share of 2 per cent agrofuels in road transport fuels beginning in 2005, increasing to 5.75 per cent by 2010, and projected a 10 per cent target by 2020 (EPC 2003). To help to meet this target, the 'EU Biofuel Strategy' (COM (2006) 34 final), consisting of different policy actions for agrofuels, was adopted (EC 2006).

In order to interlink climate with energy policies and to counteract the political and economic risks involved in the dependency of the EU on energy imports, the EC wished to start a 'new industrial revolution' (EC 2007: 5). Low-carbon growth was therefore to be accelerated, energy consumption diversified and competitiveness maximized. This was to be achieved by increasing the share of renewable energies in total energy consumption by 20 per cent and reducing GHG emissions by 20 per cent, both by 2020 (EC 2008a). The EU wished to be a pioneer and model for sustainable development in the twenty-first century (EC 2007, 2008). Meeting the 20 per cent target would require growth in electricity, agrofuels, and heating and cooling (EC 2007: 14). In order to reach this goal 'a 10% share of energy from renewable sources in transport by 2020' was necessary (EC 2013b: 2; EU Directive 2009/28/EC).[8]

It was the aim of the EC to bridge the gap between energy security and climate protection. In a 'clear framework', the EU wished to 'integrate different policy objectives such as reducing greenhouse gas (GHG) emissions, securing energy supply and supporting growth, competitiveness and jobs through a high technology, cost effective and resource efficient approach' (EC 2013a: 2). Agrofuels are a part of this narrative dimension of the political strategy of the EU: they reduce the consumption and dependency on fossil fuels in the transport sector, protecting the climate and establishing a new export market for developing countries; *ergo* a win–win–win solution.

Beside this narrative, the EC also pinpointed the economic problems and interests of the EU: 'The EU is becoming increasingly exposed to the effects of price volatility and price increases on international energy markets and the consequences of the progressive concentration of hydrocarbons reserves in few hands' (EC 2007: 4). This is one of the reasons for supporting agrofuels, despite the fact that they are now more expensive than fossil fuels and other forms of renewable energy. 'Over the next 15 years they are the only way to significantly reduce oil dependence in the transport sector' (EC 2007: 15). Already in the EU

Strategy '20 20 by 2020', it is clearly formulated that the central aim is the safeguarding of the prosperity of the EU (EC 2008a: 2; 2008b). This is necessary because the growing competition to access fossil fuels and the worldwide increase in the prices of oil, coal and gas are causing considerable political conflicts. The fragmented energy world and the growing dependency on imported energy are recognized by the EU (EC 2013a: 7).

In 2009, the EU adopted the Renewable Energy Directive, which stipulated that the share of energy from renewable sources in the transport sector must amount to at least 10 per cent of final energy consumption in the sector by 2020. The directive also established an environmental sustainability criterion with which agrofuels consumed in the EU have to comply. Biofuels and bioliquids used in electricity, heating and cooling must contribute to a reduction of at least 35 per cent of GHG emissions in order to be taken into account. From 1 January 2017, their share of GHG emissions savings will be increased to 50 per cent. Biofuels and bioliquids should not be produced using raw materials from land with high biodiversity value or with high carbon stock (and cross-compliance rules for domestically produced agro/biofuels are essential). However, the negative impacts have already led to a debate on the correctness of EU agrofuel policy. In October 2012, the EC published a proposal to limit the share of agrofuels in final energy consumption in transport to 5 per cent, and in December 2013 the European Council failed to agree on a compromise. The draft of a European Council Directive, which was released at the end of 2013, stated that the share of energy from agrofuels should be no more than 7 per cent of the final consumption of energy in transport in 2020.[9] However, conflicting interests and positions in the energy strategy within the EU are hindering a common decision.

The debates have been shaped by the EU's neoliberal agenda and the strong interests of individual member states. Privatization, marketization and regulation are the ideological background of the EU's agrofuels policy (Brunnengräber 2011; Levidow 2013: 212f.) Any plans to back down on the quotas are therefore 'heavily opposed by biofuel supporters … especially in Member States with currently high biofuel shares such as France and Germany' (Grethe *et al.* 2013: 13). In its policy framework for climate and energy in the period from 2020 to 2030, the EC announced an increased 'flexibility for Member States to define a low-carbon transition', that 'food-based biofuels should not receive public support after 2020' and that first-generation agrofuels should play a limited role in decarbonizing the transport sector (EC 2014: 3f., 6).The European Biofuels Technology Platform (EBTP) formulated the consequences of the ongoing negotiations:

> The continued uncertainty in European biofuels policy is deterring investment in the industry, making it harder for demonstration and flagship plants to secure the funding needed for commissioning. This potentially puts on hold the creation of thousands of new jobs in the European Bioeconomy.[10]

Domestic political interests and the interests of European corporations are taking over the shaping of the political economy of the EU's agrofuel policy.[11]

## The primacy of economic growth and competition

Agrofuels have many advantages for a capitalist system based on fossil fuels. In comparison to solar and wind power they are very compact, they can easily be transported and stored, they are highly suited for global trade and are sought after by 'multiple suppliers, which can be kept in competition which each other' (Levidow 2013: 220). The automotive industry benefits greatly from agrofuels, as these do not contradict their huge invested capital, subsidised road transportation and individual mobility structure, and it is argued that this benefits society as a whole. Agrofuels are part of the accumulation dynamism of capitalist societies – at least until the peak of its output of fossil fuels has been reached (peak oil) (Altvater 2011: 129). Agrofuels can postpone this peak into the future, another important reason why they are promoted by EU member states.

Inevitably, these processes are open to the influence of powerful actors – electricity, gas and oil companies, industry and retailing, to name but a few. 'Their scope and ability to appeal to so many interest groups' is also a big advantage of agrofuels (Smith 2010: 92), as they bring together many actors along the production chain from seeds to generic crops, from harvesting to processing, from transportation to storage and from natural science to lobbies and consultancies. In this way, a range of actors have been supporting agrofuels for different reasons. Financial investors can enforce their profit strategies in the politically created markets in order to offer innovative and profitable financial instruments. A part of the over-accumulated global liquidity – which can no longer be invested in real estate and properties – can be invested in the emerging, profitable agribusiness sector. The plants produced are flex crops (see Hunsberger, Chapter 8 in this volume): they can be used for cars, cows and cooking; they are flexibly available for various buyers and product uses. Traders can sell them in different sectors and in countries in which they can fetch a high price.

Renewable energy sources require favourable EU and state policies in order to become cost-competitive. A framework of economic incentives (e.g. blending quotas, tax reductions) was needed to support agrofuels, due to their higher production costs compared with fossil fuels. In the long run, 'the objective of a more sustainable energy system must go hand in hand with the need for a fully liberalised and integrated energy market capable of mobilising and allocating investment efficiently' (EC 2013a: 7). However, carbon credits, offsetting and accounting are also part of the neoliberal environmentalism, like tariffs, targets and preferences. In addition, the World Bank, the International Monetary Fund and the World Trade Organization all support the global agrofuel business. A frenzy of (fragmented) activities persists globally.

Supporters in the EU are the European Biodiesel Board (EBB) and ePure, which are both members of the industry-led European Biofuel Technology Platform (EBTP).[12] Their 'aim is to promote the use of biodiesel in the European Union, at the same time, grouping the major EU biodiesel producers' (www.ebb-eu.org) and 'to ensure the formation of policies that promote the beneficial uses of ethanol in Europe. We pursue this by representing our member's interests to the European institutions, industry stakeholders, the media and general public' (www.epure.org/what-we-do). The 'Mission of the European Biofuels Technology Platform is to contribute to: the development of cost-competitive world-class biofuels value chains, to the creation of a healthy biofuels industry, and to accelerate the sustainable deployment of biofuels in the EU' (www.biofuelstp.eu). The agrofuels lobby is trying to become influential in the negotiations at EU level, and the European Parliament is not totally opposed to the lobby's interests.

The EU's research agenda has also strongly supported the development of agrofuels.[13] In 2005, the Biofuels Research Advisory Council (Biofrac), 'effectively a pro-biofuel lobby' (Franco *et al.* 2010: 663), was established by the EC in order to prepare the groundwork for innovative technologies. The Seventh Framework Programme for Research and Technological Development (FP7) funded the Technology Platform on Biofuels to develop and implement a strategic research agenda (SRA): 'The SRA is heavily influenced by private capitals pursuing profit and not necessarily public interest, and includes a vision that up to 25% of road transport fossil fuels will be substituted by biofuels by 2030' (Ponti and Gutierrez 2009: 498).

The EU welcomes the interest of the European agrofuels industry and research institutions in industrial monocultures in the South, as it is expected to strengthen the economic and technological competitiveness of internal market forces in the world (Levidow 2013: 216). However, this provokes conflicts where the rights of small farmers to access land, water and seed, and their right to the preservation of local knowledge as the basis for food and notional sovereignty, is questioned. The effective interplay of scientists, politicians and corporations is promoted as an impetus for better agrofuels governance. Socio-ecological necessities are subordinated to the objectives of strategic, economic and energy security policy rationale.

Since the negative effects of the first-generation agrofuels are now obviously costly, and because only a fraction of the plants are converted to fuel, the EU (FP7) started favouring second-generation agrofuels, which are manufactured from biomass residues, or non-food, to more resource-efficient energy crops in order to make the impacts of agrofuels on food security and land grabbing less problematic. The technologies for large-scale production are still under development, however, and so the new generation is not yet integrated into the agrofuels market. If there is no strong support and regulation, market forces will decide which energy source is more profitable and should be used. The prices of first-generation agrofuels are projected to continue to rise over the coming ten years against the background of expectedly high crude oil prices and continuing agrofuels policies around the world. The EU and its member states are not totally opposed to this development: on the contrary, agrofuels are a way to reduce dependency on fossil fuels. Without cheap fuels, it seems unrealistic to expect that there can be economic growth and that the EU can play a leading role in a world of free trade.

## Social and environmental consequences

The EU's positive narrative on agrofuels had its equivalent in the South. The aim was to extend the benefits of agrofuels to the developing countries by creating jobs, opening up potential export markets, diversifying the production structure and promoting rural development. This euphoric phase is definitely over and must be replaced by a more realistic perspective on the complex agrofuel sector (UNEP 2014). Strong criticism and serious concerns of scientists, expert committees, NGOs and politicians were the driving force behind the negative agrofuels narrative (Giampietro *et al.* 2005; WBGU 2009; Bowyer 2011; Grethe *et al.* 2013):[14]

> Potentially harmful impacts of renewable energy growth have also been highlighted, with watchdog organizations citing the danger of land uptake for renewables (particularly biofuels) by governments and foreign investors as a threat to the social and economic prosperity of local communities.
>
> (REN 2013: 24)[15]

From a watchdog point of view, the EU energy, climate and agribusiness policy will have a considerable impact on societies and ecosystems.

*First*, a rise in food prices. The worldwide price increases of basic foodstuffs have already shown dramatically the real and future conflicts between malnutrition and food sovereignty, especially in the South, and energy security in the North. Agrofuels:

> overall now account for a significant part of global use of a number of crops.... With such weights of biofuels in the supply-demand balance of the products concerned, it is not surprising that world market prices of these products (and their substitutes) are substantially higher than they would be if no biofuels were produced.
>
> (FAO *et al.* 2011)

A study commissioned by the Heinrich Böll Foundation and Oxfam indicates that EU's agrofuels policy results in 16 per cent higher prices for plant oils, 10 per cent higher prices for oil seeds and about 2.6 per cent higher global crop prices on average (Grethe *et al.* 2013: 7). The effects of agrofuels policies in other countries such as the United States are not considered. IFPRI calculated that the prices of basic foodstuffs, which are of great importance above all in African, Latin American and Asian countries, will continue to show an upward trend until 2020 due to increasing demand: maize by 41 per cent, oilseed by 76 per cent and cassava by 135 per cent (IFPRI 2006: 8). The food crisis, with riots in Haiti, Yemen, Zimbabwe, Indonesia, Mexico and Bangladesh, would then only be the forerunner of a development revealing the destructive power of competition between foodstuffs for humans, animals and motors (OECD and FAO 2013).

*Second*, higher GHG emissions. The discussion around the reduction of emissions by using agrofuels has been controversial (Crutzen *et al.* 2008) and the reduction is obviously impossible if direct and indirect land use change is fully taken into account, but the EU's 'existing risk governance apparatus has failed to acknowledge the inherent difficulties of both identifying and interpreting knowledge pertaining to this issue' (Palmer 2012: 495). Nevertheless, environmental campaigners and many EU parliamentarians have raised concerns about the additional carbon contribution of indirect land use change caused by first-generation agrofuels:[16]

> This land use change increases stress on water resources and habitats and accelerates the release of soil carbon into the atmosphere, potentially undermining efforts to reduce greenhouse gas emissions that many governments hope to achieve through the use of renewable fuels and alternative energy sources, like biofuels.
>
> (IFPRI 2011; see also Leopoldina 2012: 24f.)

If the complete life cycle of the fuels from planting through production to transport is taken into account, the result is often no better – and perhaps even worse – than in the case of conventional fossil fuels (IFPRI 2006; IFPRI 2010). An assessment by the Institute for European Environmental Policy (IEEP) concluded that use of additional conventional agrofuels up to 2020 on the scale anticipated in the 27 NREAPs (National Renewable Energy Action Plans) would lead to between 81 per cent and 167 per cent more GHG emissions than meeting the same need through fossil fuel use (Bowyer 2011: 2; see also Grethe *et al.* 2013: 22ff.) Nevertheless, the carbon footprint in the EU is improving because of the shifting of GHG production and the related environmental impacts of energy farming to developing countries under the 'guise of rural development policies and aid' (Ponti and Gutierrez 2009: 498).

*Third*, land grabbing and other social repercussions. The EU does not have the land base needed to meet its growing demand for agrofuels, therefore it must turn to the South for its supplies. 'If EU policy would stick to its current biofuels targets, about 85% of politically driven EU demand for biofuels in 2020 would directly or indirectly stem from imports' (Grethe *et al.* 2013: 8). The conversion of rainforests, peatlands, savannas or grassland to produce agrofuels creates a 'biofuel carbon debt' (Fargione *et al.* 2008: 1235; see also Oxfam 2011). In spite of that, a window of opportunities opens for an extraordinary variety of actors, such as commodity traders, speculators, hedge funds and brokers, consultants and agrofuel energy companies. They have been at the forefront of a global rush for land that has forced hundreds of thousands of smallholder farmers off their lands and taken away their livelihoods and water supplies. The subsidy-driven demand for food-based agrofuels drives farmers (typically in developing countries) to sell their land or to clear forests and other carbon sequestering landscapes (see Backhouse, Chapter 10 in this volume). The dispossession is deepening the 'metabolic rift': the separation of agriculture from its biological base (McMichael 2008: 15). Fairhead *et al.* call it 'green grabbing', to show how 'green' credentials are used to justify appropriations of land for food or fuel. Appropriation, dispossession and valuation of nature are strategies to favour energy farming and to attract investments (Fairhead *et al.* 2012: 238). A further conflict connected with the expansion of agrofuel production is that it creates more competition with nature conservation.

*Fourth*, ecological degradation and environmental devastation. The demand for additional land to accommodate EU agrofuel ambitions means expanding cropland, which bears the risk of forest clearcuts, land clearance, plundered peatlands, ploughed prairies and displacement of other crops. It can also lead to the expansion of monocultures and the intensive use of pesticides and, as a consequence, to the loss of variety of species. Intensive farming practices are associated with:

> a) the establishment of large-scale plantations and the impairment of smallholder agriculture, b) potentially higher levels of e.g. fertilizer run-off into the water cycle and thus affecting overall water quality, and c) higher emissions of the GHG nitrous-oxide due to higher fertiliser use.
>
> (Grethe *et al.* 2013: 29)

'In current production conditions, it takes on average roughly 2,500 liters of water to produce one liter of biofuel' (FAO n.d.).

'In other words, the European impact on agricultural land is being displaced to developing countries' (UNEP 2009: 63). Since 2009, this should have been reacted to by binding criteria for the preservation of the variety of species and certain forms of land use, but doubts as to the enforceability of these criteria are justified. The EU's environmental sustainability criteria:

> could be challenged in the WTO, and they will be defended successfully only if the EU can show that they are non-discriminatory and scientifically based and that they have been imposed only after meaningful negotiations, with the EU's main suppliers, to develop international standards.
>
> (Swinbank 2009: x)

Examples such as the Roundtable Discussion on Sustainable Biofuels or the Roundtable Discussion on Sustainable Palm Oil show that multinational firms massively violate the basic principles of the criteria – even though they now integrate sustainability criteria and environmental and social standards into their corporate strategies. Furthermore, a plenitude of international certification schemes makes a comparative perspective difficult, if not even impossible, and plans for reporting carbon emissions savings to the EC are heavily opposed by agrofuel proponents. Against this background, certification for sustainability is not enough to overcome the menacing climate problem and the food crisis it brings in its wake; they legitimize the existing energy path. As Levidow states, 'The EU's political accountability is reduced to carbon accounting: in turn it is channelled into expert debates over modelling methods and uncertainties' (Levidow 2013: 211).

To summarize, agrofuels are in many ways problematical because of the intended and unintended impact on societies and the environment. The transition to agrofuels in the growing global transport sector 'causes fundamental contradictions between EU policy assumptions and practices in the real world' (Franco *et al.* 2010: 661; cf. also WBGU 2009): first, the admixture prolongs the use of fossil fuels, which can hardly be described as a climate-friendly strategy. Second, competition arises between 'food, feed and fuel', between food for hungry mouths, feed for animals and fuel for empty tanks, because the same sources of energy (maize and soy, sugar cane and palm oil, rape seed and turnips) serve as food for humans and are transformed in agriculture into meat products and into fuels for automobiles. Third, the GHG emissions incurred by the admixture of agrofuels have – in many cases – proved to be similar to, or even higher than, those connected with the use of fossil fuels (UNEP 2014).

Fracking and the changes on the supply side of fossil fuels are other driving forces behind this process. The analysis shows that the aims of the EU climate and energy policy cannot be realized: (1) the promotion of agrofuels can hardly satisfy all the domestic energy needs without harming the South in social and ecological terms, and (2) the reduction of GHG emissions in the EU goes hand-in-hand with exporting the source of emissions to developing countries, where land use changes can be observed. Agrofuels cannot bridge the gap.

## Conclusion – the emergence of a politico-economic agrofuel complex

If climate protection and the reduction of GHG emissions are seriously formulated as targets, the oil age must be overcome; and this must happen before the scarcity of fossil energy sources ensues. The fundamental importance accorded to fossil energies, energy security and low energy prices hardly points in this direction. Neither in international climate policy nor in the EU is there any sign of the political will to conduct the much-needed radical transformation of the energy systems towards renewable and decentralized energy sources and away from fossilism and centralized energy supply structures. However, as the Renewable Energy Policy Network has pointed out:

> growth in biofuels markets, investment, and new plant construction has slowed in several countries in response to a number of factors: lower margins, spiking of commodity prices, policy uncertainty, increased competition for feedstock, impacts of drought conditions on crop productivity, concerns about competition with food production for land and water resources, and concerns about the sustainability of production more broadly. Even so, biofuel blending mandates continue to drive demand.
>
> (REN21 2013: 31)

By linking agrofuel politics to the global context, it could be shown that agrofuels are not an ecological problem but a symptom of the comprehensive geostrategic, social, political and economic crisis of our fossil fuel-driven society. In response to the assumed energy crisis, EU policy takes the growth of the transport sector for granted and this becomes an imperative for extra fuel sources. The politically stimulated agrarian transformation and the agrofuels economy are based on a combination of peak oil, peak soil and peak everything, which lead to a restructuring of the ambiguous neoliberalizing of nature. Obvious contradictions have emerged between the rapidly growing global economy and the Earth's resources. They raise questions of resistance, de-growth, alternative mobility systems and a new understanding of prosperity on the way to a post-fossilistic society.

On the one hand, the struggle against agrofuels is powerful: the 'academy, NGOs, international organizations and the general media have increasingly questioned the true capacity of biofuels to be sustainable, climate- and people friendly' (Grethe *et al.* 2013: 13). There are a lot of other eco-friendly alternatives to agrofuels, such as small dams, solar power, geothermal energy and wind power. Jean Ziegler has called the use of agricultural output for fuel rather than food 'a 'crime against humanity'.[17] The negative accumulation dynamism of capital and the overuse of natural resources are increasingly being criticised because the destruction of ecosystems, the appropriation of land and the socially unequal distribution of income is part of the agrarian transformation. The negative effects are not just a by-product of accumulation. We are dealing with a classical form of the valuation of nature and neoliberal, market-oriented environmentalism (Fairhead *et al.* 2012). In other words, criticism and political counteraction is taking place under rough conditions.

On the other hand, agrofuels represent a new profitability frontier for agribusiness, for investors and energy companies and for governments. It is this 'sum of interests, rationalities and economic imperatives' that explains the strong support for agrofuels (Smith 2010: 122) despite the harsh criticism. EU policies are facilitating and boosting a new global agro-industrial complex and investments for the benefit of multinational corporations based mainly in industrial and emerging countries. The establishment of an agrofuels economy is taking place in several steps. *First*, agrofuel markets are being politically created, with the result that agrofuels are tradeable on financial markets and stock exchanges by specialized actors. Some member states of the European Union strongly support this process. Economic goods are politically transformed from food to feed or fuel. *Second*, as a consequence, parts of the landscape (e.g. rainforest) or land for growing food have to be transformed into an energy source by a sovereign act of the state. *Third*, the flex crops become part of the capital value of multinational firms and dependent on market prices. Profitable transactions, investments and land grabbing can now – without state support – take place on the world markets and specialized financial markets. This process of the neoliberalization of nature is not new, however: historically, capital accumulation has depended on a dispossession process, subordinating labour, land and natural resources.

It will be possible to extend the structures of the fossil fuel energy system, with its individualist mobility and road transport sector, far into the future, taking with them the economic and political relationships of power which they carry. At the same time, agrofuels offer the participating multinational enterprises investors, various consultants and transnational lobby groups an opportunity to develop new areas of business. 'Only multinational agribusiness corporations will profit from agrofuel technologies and even more so if the technology package includes genetic engineering' (Ponti and Gutierrez 2009: 500). On the other hand, neoliberalization means unequal development through the transfer of risk from the Global North to the Global South, the risk of energy insecurity, the risk of food price rises and market risks (Smith 2010).

What are the consequences? 'In spite of the evidence put forward against politically supporting first generation biofuels … the direction followed by the EU biofuel policy seemed surprisingly unaffected until recently' (Grethe *et al.* 2013: 13). But the EU's misguided policy on agrofuels is like a fast-selling item: it is the member states with their powerful lobby support who justify high blending quotas. From a materialist point of view ignoring social standards, climate change or criteria for sustainability are at the forefront. Resource requirements and the world market will still decide whether mono-cropping and chemically-intensive methods will be used in cultivating energy crops for agrofuels. Already a global agro-industrial complex has arisen, consisting of finite energy sources, raw materials, genetic products and renewable energy. Corporations and financiers are investing massively in agrofuel production in the Global South. This attempt can guarantee energy supply, and thus growth, despite all the apocalypse scenarios with regard to peak oil. McMichael has called it an 'agrofuel project', which enables over-consumption in the Global North and the Global South and which deepens the multiple crises (McMichael 2008: 14). It seems clear that economic profits and a high standard of living must be guaranteed at the sites of production and extraction of raw materials in the South (Burchardt and Dietz 2013). The Global North and the Global South can continue driving to work in a fossil fuel constrained world. Rest assured, we are just at the beginning of a new phase of agrarian and energy transformation. The volume of production, and the price, of agrofuels will move in tandem with the crude oil price.

## Notes

1 The author is grateful to Kristina Dietz, David Buchanan and Michael Krug for their helpful comments and suggestions.

2 This geological chronological term serves to mark the evidence and extent of human activities that have had a significant impact globally on our environment (WSSR 2013).

3 Peak oil is the point in time when the maximum rate of petroleum extraction is reached, after which it is expected to enter terminal decline.

4 Seewww.sueddeutsche.de/wissen/subvention-von-biokraftstoffen-in-der-eu-moerderischer-sprit-1.1833999–2, 27.02.2014 (accessed: 5 March 2014).

5 For a literature review, see Ponti and Gutierrez (2009).

6 For a discursive-institutionalist analysis of renewable transport fuel policy with a focus on the UK's flagship agrofuels policy, see Palmer (2010).

7 'The AEO2013 Reference case reflects a less optimistic outlook for advanced biofuels to capture a rapidly growing share of the liquid fuels market than earlier Annual Energy Outlook' (EIA 2013: 6).

8 See also ec.europa.eu/agriculture/bioenergy/index_en.htm (accessed: 22 January 2014).

9 Seewww.theguardian.com/environment/2013/nov/29/food-based-biofuels-rise-european-union-crops (accessed: 5 December 2013).

10 See www.biofuelstp.eu/overview.html and www.ceasc.com/Images/Content/Agra%20CEAS%20study%20on%20EU%20Biofuels%20Investment%20Development%20-%20Impact%20of%20an%20Uncertain%20Policy%20Environment%20-%20December%202013.pdf (accessed: 19 December 2013).

11 See http://corporateeurope.org/blog/heated-lobbying-over-parliament-agrofuel-vote (accessed: 5 March 2014).

12 See www.biofuelstp.eu (accessed: 22 January 2014).

13 For more information about EU support for bioenergy, see http://ec.europa.eu/research/energy/eu/index_en.cfm?pg=research-bioenergy-support; private companies like BP also fund research, see http://energybiosciencesinstitute.org (accessed: 22 January 2014).

14 For an example of transnational activism and campaigns around the negative consequences of agrofuels in Southeast Asia, see Pye (2010, 2013).

15 See also the Study 'Driving to destruction. The impacts of EU's biofuel plans on carbon emissions and land', launched by a coalition of NGOs. www.actionaid.org.uk/sites/default/files/driving_to_destruction.pdf (accessed: 16 December 2013).

16 See www.theguardian.com/environment/2013/nov/29/food-based-biofuels-rise-european-union-crops (accessed: 22 January 2014).

17 See www.theguardian.com/global-development/poverty-matters/2013/nov/26/burning-food-crops-biofuels-crime-humanity (accessed: 17 December 2013).

## References

Altvater, Elmar (2011) 'The "Tragedy of the Atmosphere" or the Doubling of the Carbon Cycle and the Circulation of Capital', in Altvater, Elmar and Achim Brunnengräber (eds) (2011) *After Cancún: Climate Governance or Climate Conflicts*. Wiesbaden: VS Verlag für Sozialwissenschaften, VS Research Energiepolitik und Klimaschutz, 125–143

Altvater, Elmar and Achim Brunnengräber (eds) (2011a) *After Cancún: Climate Governance or Climate Conflicts*. Wiesbaden: VS Verlag für Sozialwissenschaften, VS Research Energiepolitik und Klimaschutz.

Altvater, Elmar and Achim Brunnengräber (2011b) 'With the Market Against Climate Catastrophe – Can That Succeed?', in Altvater, Elmar and Achim Brunnengräber (eds) (2011) *After Cancún: Climate Governance or Climate Conflicts*. Wiesbaden: VS Verlag für Sozialwissenschaften, VS Research Energiepolitik und Klimaschutz, 11–21.

Bowyer, Catherine (2011) 'Anticipated Indirect land use change associated with expanded use of biofuels and bioliquids in the EU. An analysis of renewable energy action plans', Institute for European Environmental Policy (IEEP) (www.ieep.eu/assets/786/Analysis_of_ILUC_Based_on_the_National_Renewable_Energy_Action_Plans.pdf, accessed: 16 December 2013).

Brand, Ulrich (2009) 'Environmental crisis and the ambiguous postneoliberalising of nature', *Development Dialogue* 51 (Uppsala: Dag-Hammarskjöld-Stiftung), 103–117 (www.dhf.uu.se/pdffiler/DD2009_51_postneoliberalism/Development_Dialogue_51.pdf, accessed: 04 March 2014).

Brunnengräber, Achim (2013) 'Multi-Level Climate Governance: Strategic Selectivities in International Politics', in Knieling, Joerg and Walter Leal Filho (eds) (2013) *Climate Change Governance.* HafenCity University Hamburg, Frankfurt: Springer, 67–83.

Brunnengräber, Achim (2011) 'Greening the Economy in the European Union', in: Altvater, Elmar and Achim Brunnengräber (eds) (2011) *After Cancún: Climate Governance or Climate Conflicts*. Wiesbaden: VS Verlag für Sozialwissenschaften, VS Research Energiepolitik und Klimaschutz, 111–124.

Burchardt, Hans-Jürgen and Kristina Dietz (2013) 'Extraktivismus in Lateinamerika – der Versuch einer Fundierung', in Burchardt, H.-J., Kristina Dietz and Rainer Öhlschläger (eds) *Umwelt und Entwicklung im 21. Jahrhundert. Impulse und Analysen aus Lateinamerika.* Baden-Baden: Nomos, 181–200.

CEC (Commission of the European Communities) (1980) *Reflections on the Common Agricultural Policy*, Bulletin of the European Communities, Supplement 6/80, COM(80)800 final. Luxembourg: Office for Official Publications of the European Communities.

Crutzen, Paul, A.R. Mosier, K.A. Smith and W. Winiwarter (2008) 'N2O release from agro-biofuel production negates global warming reduction by replacing fossil fuels', *Atmos. Chem. Phys*. (8), 389–395.

Dauvergne, Peter and Kate J. Neville (2009) 'The changing North-South and South-South political economy of biofuels', *Third World Quarterly* 30(6), 1087–1102.

EC (European Commission) (2006) 'An EU strategy for biofuels', COM(2006)34 final. Brussels.

EC (European Commission) (2007) 'An energy policy for Europe', Communication from the Commission to the European Council and the European Parliament, COM(2007)1 final. Brussels.

EC (European Commission) (2008a) '20 20 by 2020. Europe's climate change opportunity', COM(2008)30 final. Brussels.

EC (European Commission) (2008b) 'Climate Change and international security', Paper from the High Representative and the European Commission to the European Council, S113/08. Brussels.

EC (European Commission) (2013a) 'Green paper. A 2030 framework for climate and energy policies', COM(2013)169 final. Brussels.

EC (European Commission) (2013b) 'Renewable energy progress report', COM(2013)175 final. Brussels.

EC (European Commission) (2014) 'A policy framework for climate and energy in the period from 2020 to 2030', COM(2014)15 final. Brussels.

EIA (2013) 'The annual energy outlook 2013' (AEO2013), prepared by the US Energy Information Administration. Washington, DC. (www.eia.gov/forecasts/aeo/pdf/0383(2013).pdf, accessed: 12 December 2013).

EPC (European Parliament and Council) (2003) 'Directive 2003/30/EC of the European Parliament and of the Council of 8 May 2003 on the promotion of the use of biofuels and other renewable fuels for transport', *Official Journal of the European Union, L123*, 42–46.

Fairhead, James, Melissa Leach and Ian Scoones (2012) 'Green grabbing: a new appropriation of nature?', *The Journal of Peasant Studies* 39(2), 237–261.

FAO (n.d.) 'Water at FAO' (www.fao.org/nr/water/docs/wateratfao.pdf, accessed: 6 December 2013).

FAO (2011) 'Price volatility in food and agricultural markets: policy responses', Policy Report including contributions by FAO, IFAD, IMF,OECD, UNCTAD, WFP, the World Bank, the WTO, IFPRI and the UN HLTF (www.oecd.org/trade/agricultural-trade/48152638.pdf, accessed: 14 January 2014).

Fargione, Joseph, Jason Hill, David Tilman, Stephen Polasky and Peter Hawthorne (2008) 'Land clearing and the biofuel carbon debt', *Science* 319(5867), 1235–1238 (http://climateknowledge.org/figures/Rood_Climate_Change_AOSS480_Documents.pdf, accessed: 20 December 2013).

Franco, Jennifer, Les Levidow, David Fig, Lucia Goldfarb, Mireille Hönicke and Maria Luisa Mendonça (2010) 'Assumptions in the European Union biofuels policy: frictions with experiences in Germany, Brazil and Mozambique', *The Journal of Peasant Studies* 37(4), 661–698.

Giampietro, Mario, Kozo Mayumi and Jerome Ravetz (2005) *The Biofuel Delusion: The Fallacy of Large Scale Agro-Biofuels Production.* London: Routledge.

Grethe, Harald, Andre Deppermann and Sandra Marquardt (2013) 'Biofuels. Effects on global agricultural prices and climate change', Institute of Agricultural Policy and Markets, Universität Hohenheim (Study on behalf of the Heinrich-Böll-Foundation and Oxfam) (www.boell.de/sites/default/files/biofuels_disk_papier2.pdf, accessed: 12 December 2013).

IEA (2012a) *World Energy Outlook 2012.* Paris: International Energy Agency.

IEA (2012b) *Golden Rules for a Golden Age of Gas*, World Energy Outlook. Special Report on Unconventional Gas. Paris: International Energy Agency. (www.worldenergyoutlook.org/media/weowebsite/2012/goldenrules/WEO2012_GoldenRulesReport.pdf., accessed: 6 December 13).

IEA (2013a) *World Energy Outlook 2013*. Paris: International Energy Agency.

IEA (2013b) *World Energy Outlook Special Report 2013: Redrawing the Energy Climate Map*. Paris: International Energy Agency.

IFPRI (2006) 'Bioenergy and agriculture: promises and challenges', International Food Policy Research Institute, 2020 Focus No. 14, November 2006. Washington, DC (www.ifpri.org/sites/default/files/publications/focus14.pdf, accessed: 9 December 2013).

IFPRI (2010) 'Global trade and environmental impact study of the EU biofuels mandate', Washington, DC: International Food Policy Research Institute (www.ifpri.org/sites/default/files/publications/biofuelsreportec.pdf, accessed: 09 December 2013).

IFPRI (2011) 'Assessing the land use change consequences of European biofuel policies', International Food Policy Institute, Washington, DC (www.ifpri.org/sites/default/files/publications/biofuelsreportec2011.pdf, accessed: 6 December 2013).

Leopoldina (2012) 'Bioenergy – chances and limits', German Academy of Science Leopoldina (Statement) (www.leopoldina.org/uploads/tx_leopublication/201207_Stellungnahme_Bioenergie_LAY_en_final_01.pdf, accessed: 25 February 2014).

Levidow, Les (2013) 'EU criteria for sustainable biofuels: accounting for carbon, depoliticising plunder', *Geoforum* 44, 211–223.

McMichael, Philip (2008) 'Agro-fuels, food security, and the metabolic rift', *Kurswechsel* 3, 14–22 (www.beigewum.at/wordpress/wp-content/uploads/2008_3_014-023.pdf, accessed: 17 December 2013).

Misereor (2011) 'Who feeds the world? The impacts of European agricultural policy on hunger in developing countries' Aachen (www.misereor.org/fileadmin/redaktion/MISEREOR_Who%20feeds%20the%20world.pdf, accessed: 9 Decmber 2013).

OECD and FAO (2013) *Agricultural Outlook 2013–2022.* Paris/Rome: OECD and FAO.

Oxfam (2011) 'Land and power. The growing scandal surrounding the new wave of investments in land', Oxfam Briefing Paper 151, (www.oxfam.org/sites/www.oxfam.org/files/bp151-land-power-rights-acquisitions-220911-en.pdf, accessed: 14 January 2014).

Palmer, James (2010) 'Stopping the unstoppable? A discursive-institutionalist analysis of renewable transport fuel policy', *Environment and Planning C: Government and Policy* 28(6), 992–1010.

Palmer, James (2012) 'Risk governance in an age of wicked problems: lessons from the European approach to indirect land-use change', *Journal of Risk Research* 15(5), 495–513.

Park, Jacob, Ken Conca and Mathias Finger (2008) 'The Death of Rio Environmentalism', in Park, Jacob, Ken Conca and Matthias Finger (eds) *The Crisis of Global Environmental Governance.* London and New York: Routledge, 1–12.

Ponti, Luigi and Andrew Paul Gutierrez (2009) 'Overview on biofuels From a European perspective', *Bulletin of Science, Technology & Society* 29(6), 493–504.

Pye, Oliver (2010) 'The biofuel connection – transnational activism and the palm oil boom', *The Journal of Peasant Studies* 37(4), 851–874.

Pye, Oliver (2013) 'An Analysis of Transnational Environmental Campaigning around Palm Oil', in Pye, Oliver and J. Bhattacharya (eds) *The Palm Oil Controversy in Southeast Asia. A Transnational Perspective.* Pasir Panjang (Singapore): ISEAS Publishing, 179–198.

REN21 (2013) *Renewable 2013. Global Status Report. Renewable Energy Policy Network for the 21st Century*, Paris: REN21 Secretariat. (www.ren21.net/Portals/0/documents/Resources/GSR/2013/GSR2013_lowres.pdf, accessed: 10 December 2013).

Smith, James (2010) *Biofuels and the Globalization of Risk – the Biggest Change in North-South Relationship Since Colonialism.* London: Zed Books.

SRU (2013) 'Fracking for shale gas production. A contribution to its appraisal in the context of energy and environment policy', The German Advisory Council on the Environment (Sachverständigenrat für Umweltfragen), Statement No. 18. Berlin (www.umweltrat.de/EN/TheGermanAdvisoryCouncilOnTheEnvironment/thegermanadvisorycouncilontheenvironment_node.html;jsessionid=32E6C2FF54D065BFDAB2E63A4EFBE439.1_cid325, accessed: 5 December 2013).

Swinbank, Alan (2009) *EU Support for Biofuels and Bioenergy, Environmental Sustainability Criteria, and Trade Policy.* Genf: ICTSD (http://ictsd.org/downloads/2012/03/eu-support-for-biofuels-and-bioenergy-environmental-sustainability-criteria-and-trade-policy.pdf, accessed: 20 December 2013).

UNEP (2009) 'Towards sustainable production and use of resources: assessing biofuels' (www.unep.org/pdf/biofuels/Assessing_Biofuels_Full_Report.pdf, accessed: 27 January 2014).

UNEP (2014) 'Assessing global land use: balancing consumption with sustainable supply', A Report of the Working Group on Land and Soils of the International Resource Panel, Nairobi (www.unep.org/resourcepanel/Portals/24102/PDFs//Full_Report-Assessing_Global_Land_UseEnglish_%28PDF%29.pdf, accessed: 27 January 2014).

WBGU (2009) 'Future bioenergy and sustainable land use', German Advisory Council on Global Change (Wissenschaftlicher Beirat der Bundesregierung Globale

Umweltveränderungen), Berlin (www.wbgu.de/fileadmin/templates/dateien/veroeffent lichungen/hauptgutachten/jg2008/wbgu_jg2008_en.pdf, last access 9 December 2013).

WSSR (2013) *Changing Global Environments, World Social Science Report 2013*. Editors ISSC and UNESCO, OECD Publishing and UNESCO Publishing (www.oecd-ilibrary.org/social-issues-migration-health/world-social-science-report-2013_9789264203419-en;jsessionid=64djevf5fehna.x-oecd-live-01, accessed: 10 December 2013).

# 6 Agrofuels in the food regime

*Philip McMichael*

## Introduction

The recent explosion of biofuels (a questionable response to the energy/climate crisis) is a blunt reminder of the extent to which capitalism externalises its costs. Cost externalisation is one clear consequence of commodity fetishism: wherein the social and ecological impacts of commodity relations are obscured by the price-form. Assigning a price to biophysical processes (as 'natural resources') objectifies them and conceals their socio-ecological relations. As indebted Southern governments compete for biofuel investment finance and Northern governments champion this 'green fuel', the social and ecological consequences of converting crop land and forest into a new profit frontier are hidden behind a façade of market environmentalism. What elsewhere I have called the 'agrofuels project' (McMichael 2008) is at the same time approximating a food-for-fuel regime. Through the lens of food regime analysis, the rush to agrofuels[1] can be seen to be the ultimate demystification of capitalism's subjection of food to the commodity form: deepening the abstraction of food through its conversion to fuel, at the continuing expense of the environment.

Recognition of the contribution of agrofuels to the 2008 food crisis,[2] and the claim that a ton of palm oil produces 33 tons of $CO_2$ – ten times more per ton than petroleum (Rainforest Action Network 2007), emphasises the socio-ecological impact of agrofuels. Not only do agrofuels substitute fuel crops for food crops, but they are also a highly problematic alternative source of energy. As a 'renewable' energy source, they represent an attempt to internalise externalities. But, as suggested, they compound capitalism's externalities – in a combination of artificial offsetting of emissions, releasing more carbon from newly cleared land, and exacerbating food insecurity. Heralded as a form of 'ecological modernisation',[3] they have been revealed as a questionable development, especially insofar as they exacerbate the global food crisis, entwined as it is with the climate and energy crises. As such, the agrofuels phenomenon underlines the *breaching* of neoliberal claims to feed the world through the market via a corporate food regime premised on an unsustainable energy-intensive form of agro-industrialisation. The breaching has three dimensions: failure to deliver on these claims, enabling crossover investment from food to fuel crops, and a violation of trust, as feeding the world claims yield to energy security provisioning for a wealthy minority of humanity.

Accordingly, the agrofuels project reflects a material and epistemic crisis – dramatised by UN Human Rights *Rapporteur* Jean Ziegler's claim in October 2007 that biofuels are a 'crime against humanity' (Borger 2008). It is this claim that anticipates the politicisation of biofuels, in their renaming by social movements and critics as 'agrofuels'. Just as the unravelling of the US-centred food regime by a decade of unruly trade practices led to the renaming of food aid (subsidised food trade) as 'dumping' (Friedmann 2005: 234), so the resort to biofuels as a misguided response to the energy/climate crisis, intensifying the food crisis, is a latent expression of the unravelling of the recent corporate food regime, beginning with its representational crisis (McMichael 2005).

## Food regimes and development

The distinctiveness of the food regime concept is its attention to the significance of food production and consumption relations across historical periods (Friedmann and McMichael 1989; and see McMichael 2009c). These historical periods have been commonly associated with hegemonic moments in the world capitalist economy (British, US, and institutionalised neoliberalism),[4] embodying specific geo-political relations and modes of capitalist development, and accompanying development ideologies. For example, the difference between the first two periods is often characterised as British 'outer-oriented development' and American 'inner-oriented development' – distinguishing a colonial-imperial mode of development from a nation-centred form of development, respectively. Through these historic projects of rule – from the colonial, through the development, to the globalisation project – the ruling powers in each historical moment constructed a ruling *rationale*, whether it was 'civilisation', 'development', or 'globalisation', respectively. The accomplishment of rule in each moment required the construction of 'subjects', 'citizens', or 'consumers' – each social category serving as the ideal vehicle, and product, of development.

Within these moments, food regimes have underwritten projects of rule, whether via *agricultural social forms* animating particular divisions of labour within and across political boundaries (plantations, family farming on settler frontiers, petro-farming, contract farming, and agro-industrial estates), or via the construction and reconstruction of *social diets* as sources of economic and cultural hegemony and political legitimacy. Much has been written about the material and symbolic role of beef, bread, hamburger, tomatoes, chicken, and Chinese noodles as expressing the articulation of cultural and class relations under changing hegemonic conditions (Morgan 1980; Mintz 1986; Rifkin 1992; Friedmann 2000, 2005; Dixon 2002; Pritchard and Burch 2003; Patel 2007; Barndt 2008). But *for the purposes of this essay*, in each project of capitalist development there has been a distinctive problematic regarding the 'agrarian question'. That is, what is the political-economic role of agriculture (and food) in each regime, and what residual and emergent contradictions drive the rise and decline of each food regime and its associated project of development?

As argued elsewhere, the British-centred regime combined residual and emergent contradictions in its juxtaposition of tropical and temperate agricultures, namely, a colonial division of labour and its initial supply of exotic foods for European consumers alongside an emergent nation-centred division of labour between farm and manufacturing sectors pioneered in the settler states (Friedmann and McMichael 1989). Settler farming continued the colonial practice of agro-exporting, focused now on wheat and meat as staple provisions for a formative European proletariat, via a characteristic colonial monoculture deploying an ecologically destructive form of 'soil mining' in developing specialised wheat frontier and livestock pastures (see Crosby 1986; and Friedmann 1978, 2000). Soil mining represented ecological degradation 'at a distance' during the height of this regime,[5] eventually resulting in a catastrophic 'ecological feedback' (Campbell 2009) in the form of the 1930s dust bowl, and its attendant social unrest (cf. Friedmann 2005). The outcome was a new US-centred food regime, based on agricultural subsidies, commodity stabilisation programmes, and petro-farming (Walker 2004). The hallmark of this food regime was its political anchoring in the US farm belt and its agro-industrial form, exported first to Europe through the Marshall Plan, and then to the Third World via the green revolution (an ecological time bomb).

The intensive agricultural methods also had political origins insofar as the agrichemical revolution of the 1950s depended on the conversion of wartime nitrogen production (for bombs) to inorganic fertilizer, which displaced the nitrogen-fixing legumes and manure used previously. Along with mechanisation, the use of inorganic fertilizer increased farm demand for fuel oils, gasoline, and electricity, 'thus increasing agricultural dependence on the energy sector and thereby converting the latter more than ever into a part of agribusiness' (Cleaver 1977: 17). Subsequently, in the name of the UN's Freedom from Hunger campaign (1960), the Food and Agriculture Organization (FAO) provided extension services for the dispersal of surplus inorganic fertiliser across the Third World, deepening agricultural dependence on the energy sector (Cleaver 1977: 28), and deepening petro-farming's ecological degradation through soil mining via chemical fertiliser.

While the agrarian question in the initial food regime focused on the political implications of patterns of proletarianisation of European farmers, overdetermined by the international food trade (Kautsky 1988; McMichael 2009a, 290–3), the agrarian question in the second food regime concerned state pacification of First and Third World farmers via public support, land reforms, and technification. The former agrarian question viewed agriculture through the lens of progressive capital subordination as a backdrop to late nineteenth-century revolutionary politics, while the latter agrarian question focused on publicly managed agro-industrialisation for capitalist farmers, and peasantisation through American-style land reforms (Araghi 1995). The combined processes of industrialisation, proletarianisation, and unionisation during the waning years of the first food regime, followed by world war and depression, imprinted a nexus of social reform displacing the question of agriculture's trajectory from First World revolutionary politics, even as postcolonial politics played out in Third World peasant insurgency – eventually undercut by repression, land reform and a 'green' revolution (Perkins 1997; Patel 2007) that were central tenets of the post-war development project (McMichael 1996).

Arguably, the agrarian question emerging during the current era of the corporate food regime has evolved as an agrarian question of food. Neither simply a question of the political impact of capital's subordination of landed property, nor of political pacification of struggling farmers and peasants in the North and South, today's agrarian question concerns the implications of 'agriculture without farmers',[6] on a world scale (McMichael 2009b; Araghi 2009). While capital and labour relations continue to shape the contours of agrarian transition, the (neoliberal) institutional setting has shifted from a state-centric to a global social landscape. Not only has the scope of the question broadened in an age of increasingly unfettered capital mobility, but also the state system, as a relation of production (Sayer 1987), has been transformed via a combination of privatisation and liberalisation to accommodate transnational capital. This nexus between production *and circulation* relations has been identified by La Vía Campesina as the essence of the early twenty-first century agrarian question in its observation that the 'massive movement of food around the world is forcing the increased movement of people' (2000).

In this statement, dispossession of small producers is linked to the political privileging of capital circuits to enhance market outlets for agro-industrial surpluses and agro-exports. That is, proletarianisation on a world scale for footloose capital is accomplished by agribusiness via a politics of de-peasantisation, expressed in the rise of a global peasant movement (Desmarais 2007; Borras *et al.* 2008). Today's agrarian question is not simply about political tendencies of capitalist development; rather, it concerns the politics of constructing the means of 'accumulation by dispossession' (Harvey 2003) in the agrarian sector *for capital in general*, thereby promoting 'food from nowhere' at the expense of landed food cultures and the natural environment (from soils and water through landraces to livestock species). And here is the point: that biofuels constitute another portal through which capital in general can profit from agriculture.

## The twenty-first century agrarian question

In generating a 'planet of slums' (Davis 2006), neoliberal capitalism has inverted the problematic of the original agrarian question. Because of the uncoupling of urbanisation and industrialisation and its attendant socio-spatial consequences (Araghi 2000), the agrarian question becomes less about the classical question – whether depeasantisation strengthens proletarian organisation, and more about the casualisation of labour worldwide (McMichael 1999), and the dismantling of cultures of social reproduction via small farming for the majority of the world's population. As noted elsewhere, the resulting peasant counter-movement:

> involves developing a praxis premised on a critique of the conditions of global *movement* of capital at this historical moment. It is a class politics with an ethical, historical and ecological sensibility aimed at the machinations of the state system in converting agriculture to a world industry for profit. As such, it concerns questions of rights, social reproduction and sustainability, rather than the questions of teleology, class and accumulation deriving from a productivist understanding of capital and its historical movement.
>
> (McMichael 2009b: 308)

To reformulate the agrarian question as a food question is to acknowledge that the politics of agriculture today is less about chronicling transition than about addressing the crisis of small farming across the world.[7] The food sovereignty movement, combining peasant and farm organisations and associated environmental and urban-alliance movements, is the *political* form of this question. The *technical* form is that represented by the World Bank, in its 2008 *World Development Report*, which reveals a renewed interest in agriculture (after a 25-year hiatus in its reports), and in particular in bankrolling the small farmer as a key to enhancing food production.

What is intriguing here is the re-centring of agriculture from both directions: from peasant mobilisation to promote an ‘agrarian citizenship’, premised on land redistribution and co-operative forms of agro-ecology (Wittman 2009), and from corporate mobilisation, articulated in the Bank’s vision of the ‘new agriculture’: ‘led by private entrepreneurs in extensive value chains linking producers to consumers and including many entrepreneurial smallholders supported by their organisations’ (World Bank 2007: 8). The Bank’s ‘new agriculture for development’ is governed by market intensification, via publicly subsidised agribusiness: ‘The private sector drives the organisation of value chains that bring the market to smallholders and commercial farms’ (ibid.: 8).

In its *World Development Report*, the World Bank identifies ‘two major regional challenges’. The first, in sub-Saharan Africa, views growth in agricultural productivity as ‘vital for stimulating growth in other parts of the economy’, and the second, in Asia, focuses on generating ‘rural jobs by diversifying into labour-intensive, high-value agriculture linked to a dynamic rural, nonfarm sector’ (World Bank 2007: v). The Bank extrapolates future (unsustainable and inequitable) trajectories: ‘To meet projected demand, cereal production will have to increase by nearly 50 percent and meat production 85 percent from 2000 to 2030. Added to this is the burgeoning demand for agricultural feedstocks for biofuels....’ (ibid.: 8).

The world market for biofuels is currently centred in Southeast Asia, where Malaysia and Indonesia are the world’s largest palm oil producers, supplying about 85 per cent of the world market. Meanwhile, Africa, referred to as the ‘Green OPEC’ because of its vast land reserves, is hosting finance from Brazil, Saudi Arabia, and China, the World Bank, USAID, the European Commission, and private corporations to develop biofuels primarily for export. In other words, while the Bank's *World Development Report* advocates biofuels, cautiously, noting that with current technology they have a marginal impact on energy security in particular countries, it nevertheless characterises them unproblematically as 'agricultural feedstocks' subject to 'burgeoning demand'. That is, biofuels represent a logical extension (under peak oil conditions) of an agro-industrial future, in which small farmers are progressively incorporated into food-fuel value chains premised on global 'agriculture without farmers'. These are the relations of subjection against which the food sovereignty movement mobilises, and through which the twenty-first century agrarian question of food emerges.

## Corporate food regime developments

As suggested above, food regimes condition projects of development with residual and emergent contradictions which govern trajectories of subsequent transitions. The corporate food regime is no exception. Bill Pritchard (2009) has argued that the World Trade Organization (WTO) is a hangover from the crisis of the preceding regime, by which he means the WTO emerged as a solution to that regime crisis, but retained some of its mercantilist relations. While his implication is that a 'third food regime' depends on the demise of this institution, arguably the WTO has simultaneously presided over a deepening of agribusiness power as a private regime behind the WTO's multilateral façade (Cutler 2001; Peine 2009; McMichael 2009c). It is this publicly subsidised private regime that has been responsible for constructing the export-oriented 'world farm' (McMichael 2005) around which the new agrarian question revolves. And public subsidies for agribusiness are not going away. In fact this residual contradiction is deepening as a consequence of the combined food and climate crises – which represent the emergent contradictions in the corporate food regime. The most visible aspect of this is the 'global land grab'[8] arising from a combination of *new* mercantilist food security practices, as governments sponsor offshore agriculture to ensure national food security, and offshore investment in land for biofuels production. As suggested, in context of this crisis and the stated misgivings even by authorities[9] still bound to promote biofuels, critics rename them 'agrofuels'.

Assisted by World Bank policy,[10] the land grab is represented as a form of development, insofar as indebted governments in the global South stand to receive foreign investment and hard currency from conversion of their land and forests into agro-export platforms. Biofuels in particular claim a new role in development. In 2001, for instance, President Andrés Pastrana of Colombia sought to lure Malaysian investors for a three million-hectare oil palm project on the grounds that 'progress and social development can reach large areas of Colombia that are ready to join in the cultivation and processing of this primary commodity' (quoted in Escobar 2008: 85).

Echoing the World Bank's challenge to generate employment through rural diversification, Oxfam concludes in its report, 'Bio-fuelling poverty':

> Biofuels need not spell disaster for poor people in the South – they should instead offer new market and livelihood opportunities. But the agro-industrial model that is emerging to supply the EU target poses little in the way of opportunities and much in the way of threats.
>
> (2007: 5)

Oxfam's solution is to propose a set of social principles governing the development of a biofuels industry, one of which is that 'feedstock cultivation does not adversely impact on local communities or indigenous people', without which the EU 'must accept that the ten percent target will not be reached sustainably, and therefore should not be reached at all' (ibid.: 6). The UK Gallagher Report (2008) complements Oxfam's social vision, cautioning against displacing food crops, but noting that alternative energy crops can simultaneously provide new employment and local development opportunities to rural communities.

Arguably, poverty alleviation serves as a proxy for an 'agrofuels project', as a new frontier of green accumulation geared to address the twin problems of peak oil and climate warming. Within the development paradigm, this project gains currency by appealing to an urgent need for alternative, sustainable energy sources. While the criterion of sustainability is open to serious question, and serious abuse, nevertheless it legitimises this project. At the same time, there is a more profound, ontological issue, namely the projection onto the world at large of a development model whose beneficiaries are a minority of the world's population, most of whom consume energy unsustainably, whether they like it or not. In other words, biofuelling poverty, a polite term for the agrofuels project, also means deepening forms of rural dispossession in the name of the market, and on behalf of this minority and its dependence on agribusiness imperialism. It is, perhaps, the apogee of 'global ecology', whereby natural resources are incorporated into a market calculus to sustain unsustainable patterns of profit-making and consumption (cf. Sachs 1993). It is this very incorporation, however, that is revealing the ultimate shortcomings of the development paradigm, and the crisis of the corporate food regime.

The inclusionary reflex – of extending social development to the hinterlands via the biofuel industry – is not without benefits for some (already marginalised rural and forest-dwelling people).[11] And it is important not to assume that many of these people are hapless victims, even when it is clear that they have no choice in the matter. But this is not the point – rather it is to consider the cost of such inclusion in normalising a global process of uneven and combined development whose path-dependence undermines future possibility. Part of this process of erasure includes the elimination of tacit ecological knowledges upon which the survival of the human species might depend in the process of reforming our anthropocentric assumptions and practices. This is not to say that pre-industrial peoples, or those with light ecological footprints, are a necessary and sufficient corrective resource to save the planet. Rather, undermining ecological knowledge reinforces capital's attempts to overcome all barriers to accumulation, in particular the conversion of natural processes (and their discursive representation) into value relations.[12]

In acknowledging the environmental shortcomings of biofuels, the UK Gallagher Report nevertheless proposed continued biofuels production, but only on non-agricultural land, because of 'displacement effects':

> Biofuels have been proposed as a solution to several pressing global concerns: energy security, climate change and rural development. This has led to generous subsidies in order to stimulate supply. In 2003 ... the European Union agreed to the Biofuels Directive...
>
> Five years later, there is growing concern about the role of biofuels in rising food prices, accelerating deforestation and doubts about the climate benefits. This has led to serious questions about their sustainability...
>
> We have concluded that there is a future for a sustainable biofuels industry but that feedstock production must avoid agricultural land that would otherwise be used for food production. This is, because the displacement of existing agricultural production, due to biofuel demand, is accelerating land-use change, and, if left unchecked, will reduce biodiversity and may even cause greenhouse gas emissions rather than savings. The introduction of biofuels should be significantly slowed until adequate controls to address displacement effects are implemented and are demonstrated to be effective.
>
> (2008: 1)

The confusion in this report, and statement, is symptomatic of the developmentalist assumption that energy consumption follows an inexorable trend (either because it is political suicide for governments to break this habit, or because of assumptions about the rise of 'Chindia'). It clearly echoes the World Bank's advocacy of a 'new agriculture for development' – a development projection based on the extrapolation of current trends in resource consumption. Embedded in these projections is the expectation of access to non-agricultural 'idle lands' for biofuel production. The new development discourse re-values such land as a resource for securing new energy supplies to sustain industrial accumulation at the expense of the value these lands have in both sequestering carbon and sustaining livelihoods of so-called 'marginal peoples'. Economic valorisation of hitherto 'unused' habitat represents an attempt to awaken the potential of idle resources through their development.

As the Colombian president claimed, oil palm would bring progress and social development to rural areas. For Colombia, the World Rainforest Movement reports:

> Vast stretches of land are given over to plantations for agrofuel; tropical forests are being cleared to plant thousands of hectares of oil palm, sugar cane and other crops.... In many cases, palm plantations are expanding over the territories of displaced communities.
>
> (Quoted in Lohmann 2009)

A recent report in the *Observer* noted that The Body Shop's main supplier of palm oil (90 per cent) was part of a consortium that used legal and police force to evict over 100 *bona fide* peasant occupiers from land north of Bogotá to grow palm oil (Syal 2009). Not without resistance, hundreds of Afro-Colombians in the Choco coastal rainforests have been illegally dispossessed for oil palm agro-industrial projects administered by the Colombian company Urapalma. Colombian President Uribe Vélez represented palm oil development 'as a strategy for territorial control and paramilitary demobilisation' (FoodFirst 2007: 2). Human rights workers report that:

> Since the beginning of the decade, all the areas of expansion of palm plantations have coincided geographically with paramilitary areas of expansion and presence, to the extent that some of the new plantations being developed have been financed as farming projects for the same demobilised soldiers … who had previously made incursions into these very areas.
>
> (Quoted in Smolker *et al.* 2008: 27)

Southeast Asia concentrates the agrofuels rush. Indonesian cultivation of oil palm has risen from 3.6 million hectares in 1961 to 8.1 million hectares in 2009 (Rist *et al.* 2009). Indonesia is now the world's largest producer of palm oil, with 18 million hectares of cleared forest for timber and future biofuel expansion (Colchester *et al.* 2006: 11–12). Oil palm is a key to rural development strategy, exercised mainly through Nucleus Estate and Smallholder schemes (NES), where farmers allocate a proportion of their land to an oil palm firm's estate plantation, with remaining land retained by farmers, but planted with oil palm by the firm (Rist *et al.* 2009). The Indonesian Department of Agriculture claims that approximately 27 million additional hectares of 'unproductive forestlands' (post-logging and cultivation) are available for conversion into plantations, and Sawit Watch reports almost that 20 million hectares are proposed for biofuel development by local governments (Rist *et al.* 2009: 25).

Land laws are key to these projections. The land tenure system of indigenous (mostly Dayak) groups rests on communal ownership regulated by customary law (*adat*), whereby farmers gain land rights through land clearance and cultivation, 'originally through swidden agriculture although this system has almost ceased due to land shortage caused by the expansion of oil palms' (ibid.: 104). While the Indonesian constitution recognises 'customary law communities' (a status weakened under the New Order, but potentially restored though inconsistently applied under Regional Autonomy laws since 1999), legally the state has the right to control and allocate natural resources in the name of its citizenry. But 'too often the law treats what are in reality indigenous peoples' lands as State lands … considered to be unencumbered with rights or … allocated to companies through a process that strips communities of the few rights that the government does recognise', allowing companies 90-year leaseholds (ibid.: 14). And Rist *et al.* (2009) note that oil palm development agreements with communities are often concluded without commitment by the companies, with recurring problems of land grabbing, with lack of transparency and appropriate consent procedures, exacerbated by the absence of clear land rights.

A recent UN report noted:

> Experience with existing and extensive oil palm plantations in other parts of Indonesia conclusively demonstrates that Indigenous peoples' property and other rights are disregarded, their right to consent is not respected, some are displaced, and they are left with no alternative but to become *de facto* bonded labourers gathering oil palm fruit for the companies that manage the plantations.
>
> (Quoted in Smolker *et al.* 2008: 30)

The agrofuels project in Indonesia (in both compositional and contextual terms) is expressed directly by the Plantation Office Head in Sanggau District:

> We believe that the oil palm estate has a good multiplier effect. The financial benefits from oil palm estates are by farmers on the estates, through wages for employment, as well as through the opportunities for the community to conduct business around the estate. These can contribute significantly to the development of the area. We are aware that the development of oil palm plantations can also impose high social and financial costs. Nonetheless, we still feel we are more fortunate compared with other districts [without oil palm]. *Due to the lack of financial support for [alternative] agricultural activities, particularly from the commercial banks, it is really hard to develop the agriculture sector in Sanggau District.* Therefore, the most feasible activities that can be conducted in Sanggau District are plantation activities especially oil palm estates.
>
> (Quoted in Colchester *et al.* 2006: 122, emphasis added)

While in contextual terms the financial attraction of agrofuels for an expanding global market outbids funds for agricultural development, in compositional terms plantations are valued for their multiplier effect, represented as 'financial benefits'. Such attention to the positive side of the ledger always underestimates the negative multiplier effects of industrial agriculture, in particular agrofuels. Ethnographic research on oil palm plantations in Kalimantan, Indonesia, confirms the combined negative social and ecological effect of agrofuel expansion:

> Forest and land availability have been greatly reduced, making it more difficult for the local communities to obtain NTFPs [Non-Timber Forest Products] and leading to a lack of farming lands. As there are not enough farming lands, farming has become more intensive. The same lands are used continuously, so that the soil does not have enough time to regain fertility. As there is not enough arable land, many people have given up rice farming and a linear regression can be seen in the diversity of crops cultivated in relation to the proximity of the plantation.... Availability of, and access to foods such as meat, vegetables and fruits has declined, so that more food has to be bought, leading to higher food expenses.
>
> (Orth 2007: 51)

Typically, *adat* communities diversify agricultural production with subsistence crops, rice, coffee, fruit trees, and *damar* trees, which yield a valuable resin. Testimony by affected villagers notes that companies offering benefits to participate in plantations encourage (or compel) land transfers, undermining customary agricultures – including the greater variety and therefore value of *damar* products over palm oil – and degrading the environment as 'changes in the vegetation cover have caused changes in species' distributions and have led to uncontrolled pests booming' – even though some smallholder oil palm producers improve their income and gain access to markets via new road systems (Colchester *et al.* 2006: 99). Social costs include alcohol abuse and breakdown in communality traditions via profit-seeking, leading to 'everything being measured only in economic terms' (Colchester *et al.* 2006: 100).

Malaysia signed the 2007 UN Declaration on the Rights of Indigenous Peoples, requiring states to 'consult and cooperate in good faith with the indigenous peoples concerned through their own representative institutions in order to obtain their free and informed consent prior to the approval of any project affecting their lands or territories…' (quoted in Colchester *et al.* 2007: 79), but the evidence suggests patterns of routine disregard for the integrity of communal lands, community decision-making, and compensating for loss or damage. Such patterns include unheralded occupation of territories, ignoring the fact that, as one resident observes, 'our livelihood is greatly dependent on the resources in our surroundings' (ibid:. 47). As reported in *Land is Life*:

> The people first found out about the oil palm scheme when workers started work on their lands, clearing the lands which included rubber trees and fruit trees belonging to the indigenous communities. As the oil-palm land clearing work continued, the rivers that supplied water to the people and the fish stock were affected. In addition to the crops, and polluted rivers, the people's burial ground and farm lands were also destroyed. People were then unable to hunt for the game which is an important element in their diet. There was no more rattan to harvest either, the raw material for handicrafts which had provided extra cash income to the communities. Jungle food sources, like vegetables, were also destroyed.
>
> (Colchester *et al.* 2007: 54)

Like Indonesia, Malaysia has a plural legal regime, including upholding custom under the constitution. Native courts are officially recognised in Sarawak, administering community affairs and local justice – and higher court judges do uphold native peoples' land claims. Nevertheless, the Sarawak government limits the exercise of native customary rights, refusing to reveal the location of lands actually subject to these rights and retaining the right to implement its policy of natural-resource-based development. Within this policy, native landowners must surrender their lands to the state for 60 years for development as joint ventures with private companies, despite the absence of clear principles regarding compensation to native landowners and reclamation of their lands when the leases expire (Colchester *et al.* 2007: 1–2). Colchester and his colleagues report that:

> As a direct result of its restricted interpretation of the extent of customary rights, companies are being given leases for oil palm development over supposedly 'vacant' and 'idle' State lands, which are, in fact, quite obviously inhabited, encumbered with customary rights and being actively used by local communities in their daily lives. The result is that most palm oil projects are contested by local communities.
>
> (Ibid.: 77–8)

The invisibilisation of cultural systems of social reproduction not based on private property is endemic to histories of colonisation. It justifies the appropriation of territory and resources for commercial and security purposes, whatever the social and environmental cost. Ironically, under pressure to promote 'green accumulation', states and corporations identify 'idle' land for expansion of commercial agrofuels but:

> growing evidence raises doubts about the concept of *idle* land. In many cases, lands perceived to be *idle, under-utilised, marginal* or *abandoned* by government and large private operators provide a vital basis for the livelihoods of poorer and vulnerable groups, including through crop farming, herding and gathering of wild products.
>
> (Cotula *et al.* 2008: 22–3)

The FAO has highlighted the marginalising effect of agrofuels on women in rural areas – its 2008 report notes that marginal lands provide important subsistence functions to rural peoples and are often farmed by women, who are denied access to property (Gaia Foundation *et al.* 2008: 4). In India, for example, *Jatropha* production targets 'waste lands' which sustain millions of people as 'commons' and pasturelands. In addition to pastoralists:

> refugees from development projects, displaced persons, jobless labourers and small farmers facing crop failure often rely on these lands as places where they can put their cattle during an emergency. If these lands are enclosed, the lifelines of many already disadvantaged people will be jeopardised.
>
> (GRAIN 2008a: 8)

The irony, of course, is that such displacement processes displace ways of life potentially more important to planetary sustainability.

In transitional moments within food regimes, resistance and construction of alternatives express contradictory relations, sometimes offering a glimpse of new ways of organising social reproduction, even if those ways fail to materialise fully in that particular time and place.

Brazil offers one such example – in a possible alternative way of organising energy production. Brazil's biofuel project is subdivided into two sectors: first, a relatively unregulated agro-industrial ethanol programme, centred in the São Paulo region, which has seen sugarcane expansion at the expense of dairy farming, orange groves and other staple crops; elsewhere agrofuel monoculture is 'uprooting local small-scale producers who become transformed into temporary labour often living in precarious conditions on the outskirts of local towns' (Wilkinson and Herrera 2008: 24). Lands assigned for agrarian resettlement projects are being accumulated via intense pressure on small farmers for sugarcane plantations worked by landless labourers under conditions of debt peonage. Once mined for ethanol, the lands revert to resettlement status, but in a degraded state (Smolker *et al.* 2008: 22).

Second, however, there is the government-regulated biodiesel programme, decentralised and designed to promote regional development and social inclusion. The National Petroleum Agency organises auctions through which firms, on acquiring a Social Seal provided by the Ministry for Agrarian Development, gain access to the biodiesel market. These firms must 'demonstrate that a given percentage of their raw material or crude oil has been contracted with family farms in agreement with the rural trade unions' (Smolker *et al.*, 2008: 8).

In a detailed report, examining the operation of this Biodiesel Programme by region, Wilkinson and Herrera observe that the economic viability of this programme is threatened by land access (particularly in the Northeast, where cattle ranching predominates) and low incomes from castor oil planting (pushing farmers to seek agro-ecological and niche farming), oil palm plantation preference in the north, and, in the centre-west, a preference for soybeans encourages an agribusiness takeover (2008: 60–1). The south, however, with its cooperative traditions, saw a recent surge of locally-focused cooperatives and associations experimenting with ethanol from sugarcane (and manioc and sorghum), biodiesel from *Jatropha*, and tung (tree crops) via a variety of intercropping systems both with tree crops and short-cycle food crops and joint processing and farming activities. Each producer is only allowed to plant two hectares of biofuel in order to ensure adequate food supply. Wilkinson and Herrera view this programme as a 'radical challenge to the dominant agribusiness model,' noting:

> Within this perspective the combined food and energy production systems are seen as strategies for increasing the autonomy of the less favoured family farm sector, an important feature of which includes the production of ethanol for local consumption. These projects are still at an early stage of development and so definitive conclusions cannot be drawn as to their feasibility. Nevertheless … they may well offer a complementary strategy for ethanol production in the family-farming context in other regions of the country. Regulatory adjustments permitting direct sales to the members of a producer cooperative thereby promoting decentralised distribution circuits increase the attractiveness of such a strategy.
>
> (2008: 57)

This programme, however marginal in the Brazilian landscape, models a decentralised, cooperative system integrating energy and food production as an alternative to monoculture and its displacement of foodstuffs. It also offers the possibility of substantiating the rural development claims of a biofuels programme, but on a foundation of cooperative production relations (as opposed to simple commodity production). Contrary to the initial solution of 'biofuelling poverty' proposed by Oxfam, which sought sustainability standards in the global biofuels market, this programme substitutes local markets as the goal and anchor of integrated rural development: the difference being that energy/food *sovereignty* retains and replenishes value, whereas the global market, even with principles, extracts and degrades value. Establishing such decentralised cooperative polycultures is the goal of the food sovereignty movement, and durability depends on farmer mobilisation to sustain such agricultures, even when supported by national states. The slogan of the food sovereignty movements – 'small farmers feed the world and cool the planet' – expresses such articulation between social justice and provisioning based on agro-ecological principles.

## Food regime ecology

Renaming biofuels 'agrofuels' not only reminds us of crop land competition and fuel displacing food,[13] but it also signals an ecological consequence whereby biofuel plantations displace biodiversity and, under the current agrofuels project, reproduce and deepen forms of greenhouse gas emission. As Campbell (2009) advocates, food regime analysis requires a political ecological perspective, founded in Marx's concept of the 'metabolic rift'. The concept refers to the separation of social production from its biological foundations and underlies the spatial separation of urban life from rural life as agriculture industrialises (cf. Foster 2000; Moore 2000). This, in turn, depends on manufacturing technologies, whose contribution to the metabolic rift involves expanding inputs of energy and natural resources, and industrial wastes – sometimes recycled today, but largely outside of natural cycles. Fossil fuel dependence is a fundamental consequence of this rift, contributing greatly to carbon emissions, and the associated agrofuels project. However, the metabolic rift is not only about a material transformation of production, with spatial consequences, it is also about an epistemological break (McMichael 2009c).

Following the separation of labour from its means of subsistence via the metabolic rift, productive relations, and social institutions, are increasingly embedded in the market, subordinated to value relations. The point is that *given* the metabolic rift, the ontological priority in social intercourse becomes capitalist value relations. Thus, the conversion of agriculture to a branch of industry privileges capital in its subordination of landed property in the name of value. But, in addition to a methodology that understands capital now as the dominant historical force (Marx 1973), the inversion is in the structure of thought as well, superimposing value relations on our understanding of the historical process. My point is that agrofuels symbolise this ontology, whereby meeting the unsustainable energy needs of a fossil-fuel dependent accumulation process is accomplished by subordinating agriculture to a non-food crop, to maintain value, if not food security. The agrofuels 'gold rush' reveals the one-dimensionality of value relations as embodied in capitalism and its structures of thought. That is, the metabolic rift is not only assuming greater significance in how we analyse the historical moment,[14] but also both its material *and* epistemic consequences need to be overcome. Restoring the social/natural metabolism to promote ecological sustainability will only materialise when we transcend the value calculus through which capital rules the world. Renaming biofuels 'agrofuels' is part of this discursive shift stemming from the crisis of the food regime.

In general, the constraints on the material, and discursive, world assert themselves in a proliferating literature on, and growing public recognition of, 'ecological feedback' – most notably in climate warming. In particular, the literature on biofuels and carbon markets includes a growing scientific challenge to attempts to measure and value emissions via a market-based metric. Analogously, the proliferating food sovereignty movement proposes restoring natural metabolism through social knowledges anchored in agro-ecological practices. Each movement embodies recognition of the inability of modern science and its industrial processes to interpret and manage natural cycles through market mechanisms.

In articulating the development paradigm's new market environmentalism, the EU Energy Commissioner stated in 2006 that, 'Biofuels are the only known substitute for fossil fuels in transport today. They contribute to our security of energy supply, reduce greenhouse gas emissions and create jobs in rural areas' (quoted in Gilbertson *et al.* 2007: 7). The EU, of course, has reconciled itself to new targets that can only be met by importing agrofuels from the global South. Accordingly, the UK Climate Change Minister claimed in 2007 that, 'the global community must as a matter of urgency work towards the development of internationally recognised standards for biomass grown to produce biofuels' (quoted in Gilbertson *et al.* 2007: 13). The subject of certification, of course, raises questions about how to standardise a sustainable biofuel metric. From a survey, Biofuelwatch claims that a 'majority of biofuel industry responses ... reject any mandatory safeguards.... Many responses suggest that not enough is known about life-cycle greenhouse gas emissions from biofuels, but nonetheless demand government support for rapid market expansion' (quoted in Gilbertson *et al.* 2007: 15).

Ignoring the precautionary principle in this way is doubly problematic, as lack of interest in a sustainable biofuels industry is dwarfed by the lack of concern for longer-term effects of greenhouse gas emissions. Just as carbon emissions from transport have hitherto been omitted from the globalisation ledger – discounting such negative 'externalities' and enabling a false economy – so this false economy is extended by proponents of an agrofuels project. The conversion of rainforests, peatlands, savannas, or grasslands to produce agrofuels in Brazil, Southeast Asia, and the US 'creates a "biofuel carbon debt" by releasing 17 to 420 times more $CO_2$ than the annual greenhouse gas (GHG) reductions these biofuels provide by displacing fossil fuel' (Fargione *et al.* 2008).

The problem of emissions is not simply that in most cases (other than perhaps sugarcane) agrofuels release more than they reduce in substituting for fossil fuel energy. The additional emissions produce side-effects, or 'externalities', which are now acknowledged in the scientific community. As the Transnational Institute (2007: 10) reports, 'Much of the evidence presented for agrofuels to reduce greenhouse gas emissions ignores the larger picture of "land use change"' (usually deforestation), soil erosion and nitrous oxide emissions.' Nobel Prize winner Paul Crutzen (2007) observes that biofuels raise rather than lower emissions, and from research with colleagues on nitrous oxide emissions from crop fertilisers, he concludes that:

> the replacement of fossil fuels by biofuels may not bring the intended climate cooling due to the accompanying emissions of $N_2O$ … depending on N content, the use of several agricultural crops for energy production can readily lead to $N_2O$ emissions large enough to cause climate warming instead of cooling by 'saved fossil $CO_2$'.
>
> (Crutzen *et al.* 2007)

Despite the acknowledgement of by-product emissions, there is also recognition that it is difficult to obtain invariant results from emission calculations. Biofuelwatch reports that the few calculations of agrofuel emissions from land use, deforestation and soil organic carbon loss have different methodologies and therefore substantial variation in their results (Gilbertson *et al.* 2007: 36). Servaas Storm (2009: 1020) notes, for example, that 'carbon savings' from offset projects are unmeasurable, because they are based on an unrealised counterfactual. For Larry Lohmann, offsets are a 'fictitious commodity', created by 'deducting what you hope happens from what you guess would have happened' (quoted in Storm 2009: 1020). As Storm notes, lack of verifiability leads to carbon imperialism, turning the South into a 'carbon dump' while sustaining Northern lifestyles. Joan Martinez-Alier (2009) reinforces this by noting that the Kyoto Protocol enabled the North to obtain property rights on carbon sinks in the South and the atmosphere in return for reduced emission targets.

While some argue for ecological restoration over land conversion for agrofuels as more likely to reduce carbon emissions, the point is that measuring emissions accurately for purposes of standardisation is impossible. Thus it was claimed by Berkeley scientists that, 'Including incommensurable quantities such as soil erosion and climate change into a single metric requires an arbitrary determination of their relative value' (quoted in Gilbertson *et al.* 2007: 37). And this is the case for attempts to calculate emissions along production chains, as well as life cycle analysis of emissions from the agrofuel complex. Gilbertson *et al.* conclude:

> Very few life-cycle greenhouse gas assessments are peer reviewed. There are currently no peer reviewed life-cycle greenhouse gas studies for biodiesel

> from palm oil, jatropha or soya, and peer review studies on sugar cane ethanol are limited to those looking at energy gains and fossil fuel displacement, rather than total greenhouse gas balances.
>
> (2007: 39)

Further, the Intergovernmental Panel on Climate Change (IPCC) admits that '$CO_2$ equivalences' are gross oversimplifications: 'the effects and lifetimes of different greenhouse gases in different parts of the atmosphere are so complex and multiple that any straightforward equation is impossible' (Lohmann 2008: 360). In spite of the focus on getting the calculations as accurate, or comprehensive, as possible, the overriding point is that this controversy over certification methodologies is a proxy for a more significant issue: namely, the cognitive dissonance in attempting to certify via an economic calculus quite incommensurate with an ecological calculus. The incommensurability lies in the difference between a virtual fractionation of carbon units as a standardising means of regulating a carbon-based economy, and the actual interactive complexity of carbon cycles, both natural and 'unnatural'.[15]

One clear form of such interactive complexity is illustrated in the concept of 'positive feedback', used by climate scientists to describe the self-acceleration of climate change. In a new IPCC summary in 2007, the panel notes that 'emission reductions … might be underestimated due to missing carbon cycle feedbacks' (quoted in Monbiot 2007). The likelihood of such feedback is why climate scientists argue that global temperatures should not be allowed to rise more than two degrees above pre-industrial levels – otherwise, by 2040 'living systems on the land will start to release more carbon dioxide than they absorb' (Monbiot 2006: 10). Under these conditions, plants shrivel and trees die, raising temperatures which, with decreasing rainfall (particularly in the tropics), kill more trees and plant life, the metabolism of soil microbes accelerates, releasing more carbon (already occurring in the UK: by the end of the twenty-first century 'the world's soils will eject the manmade carbon they have absorbed over the past 150 years' [Monbiot 2006: 10–11]), permafrost melt in the far north can release methane, and so on. However this phenomenon, otherwise known as the 'nemesis effect', plays out is testimony to the self-organising character of natural cycles.[16]

Climate change emergency policy is in effect a Canute-like attempt to reduce emissions, to stem warming, and thereby to head off 'Gaia's revenge' (Lovelock 2007). But it is likely to fail precisely because of the inability to subordinate ecological relations to a singular economic calculus. The discourse of sustainability has reached perhaps a high point in the recent IAASTD Report (2008), which is critical of industrial agriculture. Stating that 'business as usual is not an option', given the combination of climate, energy, water, and food crises, the IAASTD questions industrial agriculture and GM (genetically modified) food as the solution to the social and ecological crises associated with global agribusiness, on the grounds that markets fail to adequately value environmental and social harm. The report also questions the salience of a market-driven approach,[17] and its narrow focus on productivity, versus an integrative view of food, resource, and nutritional security, which underlines agriculture's multifunctional contribution to complex social reproduction issues. It advocates policies that 'promote sustainable agricultural practices (e.g. using market and other types of incentives to reward environmental services) [and] stimulate more technology innovation, such as agro-ecological approaches and organic farming to alleviate poverty and improve food security' (IAASTD 2009: 24). Further, the IAASTD recommends that monetary or other incentives for 'performance-based ecological services' recognise

> the importance of the multiple functions of agriculture and creates mechanisms to value and pay for the benefits of resource-conserving ecosystem services provided by sustainable agricultural practices, such as low-input and low-emission production, conservation tillage, watershed management, agroforestry practices, carbon sequestration, biological control and pollination, and conservation of agricultural biodiversity.
>
> (Ibid.: 24)

This report represents an attempt to straddle the boundary between market and non-market practices. The danger is of course that valuing nature and 'ecological services' performed by producers introduces a 'global values' language that abstracts from local particularities and practical knowledges. In fact, the concept of 'ecosystem services' is a proxy for a 'global ecology' discourse premised on market mitigation,[18] whereby compensation for services is an indirect form of consumption of the environment. Put another way, payment for ecosystem services 'relies on creating market mechanisms that attract investment from areas *requiring* ecosystem services – including maintenance of biodiversity – to areas *providing* these services, e.g., from urban to rural areas, and from the global "north" to the global "south"' (Sullivan 2008). A further danger is that through the economic calculus of scarcity, demand for environmental services increases their market value 'in ways that *out-compete* other forms and practices of value for the landscapes providing them' (ibid.). Thus a new industry of ecological accounting is born, which, through the development lens, establishes an offset industry (now formalised in the UN programme, REDD: Reducing Emissions from Deforestation and Degradation).[19]

A case in point is the Yasuní region in Amazonian Ecuador, where oil reserves are offset by environmental values, and there has been a political stand-off regarding whether or not President Correa can obtain carbon credits for preserving the forest. As Adam Ma'anit notes:

> The real danger is that once a dollar value has been assigned to something as arguably incalculable as a tree, a forest, or yes, even a human life, it allows the bean counters to start comparing costs and benefits. Economists can start to ask, when the price of oil hits $200 a barrel: Does the benefit of extracting a billion barrels of oil outweigh the cost of destroying the Yasuní National Park and the communities of people that live there?
>
> (2008: 19)

Bean-counting is a powerful discourse, as one commentator views it:

> The carbon dioxide emissions from extracting and burning the oil would be about 375 million tons, and emissions from deforestation would be 172 million a total of 547 millions tons. The World Bank has estimated the abatement cost for carbon dioxide at $14 to $20 per ton.... The cost to the world to abate these emissions will be between $1.7 billion and $2.4 billion for the extraction and burning, and $909 million for deforestation, for a total between $2.6 billion and $3.7 billion.
>
> Correa proposes that Ecuador issue bonds for the value of the carbon dioxide emissions avoided by preserving the forest. He promises to park the funds at a neutral bank and only spend them on social development and alternative-energy projects in Ecuador. If a future government of Ecuador decides to exploit the oil, they have to repay the bondholders plus interest.
>
> Preserving Yasuní is a rare win–win situation. The rich world (that created the climate problem) can help mitigate it in a relatively low-cost manner. Ecuador obtains the funds to help grow its relatively poor economy. Far from radical populism, this is economic efficiency at its finest.
>
> (Gallagher 2009)

The omission in this argument, beyond the subordination of ecology to a carbon market, is the inability to view this issue spatially and temporally. *Spatially*, a carbon market abstracts from the players' location, and here a seemingly healthy exchange obscures a continuing process of emitting greenhouse gases in or by the 'rich world' elsewhere. And *temporally*, mitigation of this sort does not, under present arrangements and practices, reduce the continuing flow of emissions from fossil-fuel use, which will continue to alter climates and compromise forests. Economic efficiency is a chimera insofar as it collapses the incommensurable into commensurable (and virtual) units of supply and demand.

Arguably, such dissonance provides the conditions for the scientific community, including social scientists of the food regime persuasion, to recognise the reductionism of assigning a market value to ecological processes and elaborate an ecologically relevant discourse which would begin to bridge the epistemic rift embodied in the market calculus. Henceforth, food regime analysis and its associated development and agrarian questions can no longer ignore 'ecological feedback'. The climate crisis, the intensification of 'biophysical override' via transgenic technologies (Weis 2007), and the biofuels 'revolution' – all expressions of the food regime, have made sure of this.

## Conclusion

This chapter has argued that 'agrofuels' represent the crisis of the current food regime insofar as they breach the implicit rules of the neoliberal world order, by which food security is to be guaranteed through corporate stewardship of the global market, as the most durable and efficient allocator of agricultural resources. While 'peak soil' is locked in an embrace with 'peak oil' via chemical agriculture, intensifying climate change, the resort to biofuels is an artificial solution. It is artificial in two senses: first, biofuels (first and second generation) are increasingly recognised as ineffectual in reducing greenhouse gas emissions; and second, biofuels displace food and food producers – revealing the falsity of corporate agriculture's claim to 'feed the world', while an emergent food/fuel complex offers fungible possibilities for profitable investments via alliances between agribusiness, energy, automobile and biotechnology companies, and states (McMichael 2009d). The bait-and-switch tactic, whereby neoliberal shortcomings are papered over with attempts at 'ecological modernisation' via 'internalising externalities' in the agrofuels project, is increasingly recognised as such. It represents the bankruptcy of a development paradigm invested in a market calculus, increasingly exposed by food riots, a burgeoning global food sovereignty movement, and alarming 'ecological feedback'.

## Note

The author thanks Kate Neville (for helpful feedback on an earlier version) and two anonymous reviewers.

1 Social movement critics rename biofuels *agrofuels* in recognition of their problematic environmental and social consequences, whether first- or second-generation. Cf. Corporate Europe Observatory (2007).

2

> US corn ethanol explains one-third of the rise in the world corn price according to the FAO, and 70 per cent according to the IMF. The World Bank estimates that the US policy is responsible for 65 percent of the surge in agricultural prices, and for … the former USDA Chief economist, it explains 60 percent of the price rise.
>
> (Berthelot 2008: 27)

3 Cf. Martinez-Alier (2002).

4 Just as Arrighi (1994) has argued, British and American hegemonies, backed with military/financial force, were founded in political-economic principles (e.g. freedom of trade, freedom of enterprise) adopted by rival states as relatively universal organising principles, so the WTO institutionalised a universally accepted organising principle (liberalisation), with military/financial/legal force standing behind adoption by member states, despite asymmetry of observance between North and South.

5 Note that ecological degradation characterised the imposition of tropical export agricultures by imperial powers (cf. Davis 2001).

6 This term comes from La Vía Campesina, an international coalition of peasant organisations.

7 Arguably, Henry Bernstein's (2008) plea to analyse the agrarian question today as a question of labour reproduces a classical, accumulation-centred episteme that is at odds with the reality of peasant *political mobilisation* as a new social class (class here because it is constituted as a political class via neoliberal capitalist process) – dedicated not to reproducing a traditional peasantry, but drawing on traditions (ecological knowledges, culture of the commons) of the 'peasant way' (as La Vía Campesina names it) to reconstitute smallholder agriculture around land rights, local markets, labour/knowledge cooperation, agro-ecological methods and 'agrarian citizenship' (Wittman 2009).

8 See GRAIN (2008b). Roughly 20 per cent of the global land grab is scheduled for agrofuel crops, which, alongside of projected export food crops, constitute a new investment frontier for food, financial, energy, and auto companies (Vidal 2009: 12).

9 E.g. the UK Gallagher Report (2008).
10 The Bank promotes land legislation to enable land sales to foreign investors.
11 Rist *et al.* (2009) note, for example, that oil palm production contributes over 63 per cent of smallholder household incomes in two locations in Sumatra, and that there is evidence of oil palm alleviating poverty.
12 For a development of this observation, see Araghi (2010).
13 The World Bank (2007) noted that the 'grain required to fill the tank of a sports utility vehicle with ethanol (240 kilograms of maize for 100 litres of ethanol) could feed one person for a year' (Policy Brief: 'Biofuels: The Promise and the Risks').
14 See for example, Foster (2000), Moore (2000), Clark and York (2005), McMichael (2008) and Wittman (2009).
15 See Lohmann (2006) for an extended discussion of this.
16 Analogously, agrofuels have distinct feedback effects through the mechanism of price as the value-form of capital accumulation. Thus certification schemes, focusing on 'sustainable' agrofuel production, are unable to address 'leakage' or displacement of production elsewhere. As TNI notes, 'Future certified palm oil, for example, might be produced from land deforested several years previously, while forest continues to be cleared for palm oil *for other markets*' (2007: 31, emphasis added).
17 The IAASTD emphasises that reinventing 'agriculture' requires experts in agricultural knowledge, science and technology to work with local farmers, and other professionals such as social and health scientists, governments and civil society.
18 For an extended treatment of this subject see McMichael (2009a).
19 Hari (2009: 16) notes that Greenpeace investigated an initial REDD-like model in Bolivia, where The Nature Conservancy, British Petroleum, Pacificorp, and American Electric Power in 1997 established a protected forest called the Noel Kempff Climate Action Project, preserving 3.9 million acres of tropical forest (to prevent release of 55 million tons of $CO_2$) allowing an equivalent release elsewhere from coal and oil operations. In addition, the money received for the offset was use to log a neighbouring forest.

## References


Araghi, F. 1995. Global depeasantization, 1945–1990. *The Sociological Quarterly*, 36(2), 337–68.

Araghi, F. 2000. The great global enclosure of our times. In F. Magdoff, F.H. Buttel, and J.B. Foster (eds) *Hungry for profit: the agribusiness threat to farmers, food, and the environment.* New York: Monthly Review Press, pp. 145–60.

Araghi, F. 2009. The invisible hand and the visible foot. Peasants, dispossession, and globalization. In A.H. Akram-Lodhi and C. Kay (eds) *Peasants and globalization. Political economy, rural transformation and the agrarian question.* London: Routledge, pp. 111–47.

Araghi, F. 2010. Accumulation by displacement. Global enclosures, food crisis, and the ecological contradictions of capitalism. *REVIEW*, 32(1), 113–46.

Arrighi, G. 1994. *The long twentieth century. Money, power and the origins of our times.* London: Verso.

Barndt, D. 2008. *Tangled routes. Women, work and globalization on the tomato trail.* Aurora, ON: Garamond Press.

Berthelot, J. 2008. The food crisis explosion: Root causes and how to regulate them. *Kurswechsel*, 3, 23–31.

Bernstein, H. 2008. Agrarian questions from transition to globalization. In A.H. Akram-Lodhi and C. Kay (eds) *Peasants and globalization. Political economy, rural transformation and the agrarian question.* London and New York: Routledge, 214–38.

Borger, J. 2008. UN chief calls for review of biofuels policy. *Guardian* online. Available at: www.theguardian.com/environment/2008/apr05/biofuels.food (Accessed 4 June 2008).

Borras, J., M. Edelman and C. Kay (eds) 2008. *Transnational agrarian movements confronting globalization.* Chichester: Wiley-Blackwell.

Campbell, H. 2009. The challenge of corporate environmentalism: social legitimacy, ecological feedbacks and the 'food from somewhere' regime? *Agriculture and Human Values*, 26, 309–19.

Clark, B. and R. York. 2005. Carbon metabolism: global capitalism, climate change, and the biospheric rift. *Theory & Society*, 34(5), 391–428.

Cleaver, H. 1977. Food, famine, and the international crisis. *Zerowork*, 2, 7–70.

Colchester, M., Marcus Colchester, Norman Jiwan, Andiko, Martua Sirait, Asep Yunan Firdaus, A. Surambo and Herbert Pane. 2006. *Promised land. Palm oil and land acquisition in Indonesia: Implications for local communities and indigenous peoples.* Moreton-in-Marsh, UK: Forest Peoples Programme and Perkumpulan Sawit Watch.

Colchester, M., Marcus Colchester, Wee Aik Pang, Wong Meng Chuo and Thomas Jalon. 2007. *Land is life. Land rights and oil palm development in Sarawak.* Moreton-in-Marsh, UK: Forest Peoples Programme and Perkumpulan Sawit Watch.

Corporate Europe Observatory (CEO). 2007. The EU's agrofuel folly: policy capture by corporate interests. June. Available at: www.corporateeurope.org (Accessed 4 June 2008).

Cotula, L., N. Dyer and S. Vermeulen. 2008. *Fuelling exclusion? The biofuels boom and poor people's access to land.* International Institute for Environment and Development (IIED) and Food and Agricultural Organization (FAO).

Crosby, A. W. 1986. *Ecological imperialism.* New York: Cambridge University Press.

Crutzen, P. 2007. Chemistry world, 21 September. Available at: www.rsc.org/chemistryworld/News/2007/September/21090701.asp (Accessed 4 June 2008).

Crutzen, P.J., A. Mosier, K.A. Smith, and W. Winiwarter 2007. $N_2O$ release from agrobiofuel production negates global warming reduction by replacing fossil fuels. *Atmospheric Chemical Physics Discussion*, 7, 11191–205.

Cutler, A.C. 2001. Critical reflections on the Westphalian assumptions of international law and organization: a crisis of legitimacy. *Review of International Studies*, 27(2), 133–50.

Davis, M. 2001. *Late Victorian holocausts. El Nino famines and the making of the third world.* London and New York: Verso.

Davis, M. 2006. *Planet of slums.* London: Verso.

Desmarais, A.A. 2007. *La Vıa Campesina. Globalization and the power of peasants.* Halifax: Fernwood Publishing.

Dixon, J. 2002. *The changing chicken: chooks, cooks and culinary culture.* Sydney: University of New South Wales Press.

Escobar, A. 2008. *Territories of difference. Place, movements, life, redes.* Durham, NC: Duke University Press.

Fargione, J., Joseph Fargione, Jason Hill, David Tilman, Stephen Polasky and Peter Hawthorne. 2008. Land clearing and the biofuel carbon debt. *Science*, 319(5967), 1235–8.

Food First. 2007. Colombian palm oil biodiesel plantations: A 'lose–lose' development strategy? *FoodFirst Backgrounder*, 13(4), 1–4.

Foster, J.B. 2000. *Marx's ecology. Materialism and nature.* New York: Monthly Review Press.

Friedmann, H. 1978. World market, state and family farm: social bases of household production in an era of wage-labour. *Comparative Studies in Society and History*, 20(4), 545–86.

Friedmann, H. 2000. What on earth is the modern world-system? Food-getting and territory in the modern era and beyond. *Journal of World-System Research*, 6(2), 480–515.

Friedmann, H. 2005. From colonialism to green capitalism: social movements and the emergence of food regimes. In F.H. Buttel and P. McMichael (eds) *New directions in the sociology of global development*, Volume 11. Oxford: Elsevier, 229–67.

Friedmann, H. and P. McMichael. 1989. Agriculture and the state system: the rise and fall of national agricultures, 1870 to the present. *Sociologia Ruralis*, 29(2), 93–117.

Gaia Foundation, Biofuelwatch, the African Biodiversity Network, Salva La Selva, Watch Indonesia and EcoNexus. 2008. Agrofuels and the myth of the marginal land. Briefing. Available at: www.cbd.int/doc/biofuel/Econexus%20Briefing%20Agrofuels-MarginalMyth.pdf (Accessed 13 August 2014).

Gallagher, E. 2008. *The Gallagher review of the indirect effects of biofuels production.* Renewable Fuels Agency, UK Government.

Gallagher, K.P. 2009. Paying to keep oil in the ground. *Guardian*, 7 August, 7.

Gilbertson, T., Tamra Gilbertson, Nina Holland, Stella Semino and Kevin Smith. 2007. *Paving the way for agrofuels. EU policy, sustainability criteria and climate calculations.* Amsterdam: Transnational Institute.

GRAIN. 2008a. Agrofuels in India, private unlimited. *Seedling*, April. Available at: www.grain.org/seedling/?id=543 (Accessed 20 January 2009).

GRAIN. 2008b. Seized. The 2008 land grab for food and financial security. Grain Briefing. Available at: www.grain.org/briefings/id?=212 (Accessed 21 January 2009).

Hari, J. 2009. The wrong kind of green. How conservation groups are bargaining away our future. *The Nation*, 22 March, 11–19.

Harvey, D. 2003. *The new imperialism.* Oxford: Oxford University Press.

International Assessment of Agricultural Knowledge, Science and Technology for Development (IAASTD). 2008. Executive summary of the synthesis report. Available at: www.agassessment.org/docs/SR_Exec_Sum_280508_English.pdf (Accessed 17 January 2009).

IAASTD. 2009. *Agriculture at a crossroads. Global summary for decision makers*. Washington, DC: Island Press.

Kautsky, K. 1988[1899]. *The agrarian question*, Volume 2. London: Zwan Publications.

La Vía Campesina. 2000. Bangalore Declaration, October 6. Available at: http://viacampesina.org/main_en/index.php?option=com_content&task=view&id=53&Itemid=28. (Accessed 3 November 2006).

Lohmann, L. 2006. Carbon trading. A critical conversation on climate change, privatization and power. *Development Dialogue*, 48, 1–358.

Lohmann, L. 2008. Carbon trading, climate justice and the production of ignorance: ten examples. *Development*, 51, 359–65.

Lohmann, L. 2009. Climate as investment. Corner House Briefings. Available at: www.cornerhouse.uk/resource/climate-investment (Accessed 8 January 2010).

Lovelock, J. 2007. *The revenge of Gaia: why the earth is fighting back – and how we can still save humanity.* London: Penguin.

Ma'anit, A. 2008. Costing the earth. *New Internationalist*, 413, 17–19.

McMichael, P. 1996. *Development and social change. A global perspective.* Thousand Oaks: Pine Forge Press.

McMichael, P. 1999. The global crisis of wage labour. *Studies in Political Economy*, 58, 11–40.

McMichael, P. 2005. Global development and the corporate food regime. In F.H. Buttel and P. McMichael (eds) *New directions in the sociology of global development.* Oxford: Elsevier Press, pp. 265–300.

McMichael, P. 2008. Agrofuels, food security, and the metabolic rift. *Kurswechsel*, 3, 14–22.

McMichael, P. 2009a. Contradictions in the global development project: geo-politics, global ecology and the 'development climate'. *Third World Quarterly*, 30(1), 251–66.

McMichael, P. 2009b. Food sovereignty, social reproduction and the agrarian question. In A.H. Akram-Lodhi and C. Kay (eds) *Peasants and globalization. Political economy, rural transformation and the agrarian question.* London: Routledge, 288–311.

McMichael, P. 2009c. A food regime genealogy. *The Journal of Peasant Studies*, 36(1), 139–69.

McMichael, P. 2009d. The agrofuels project at large. *Critical Sociology*, 35(6), 825–39.

Martinez-Alier, J. 2002. *The environmentalism of the poor. A study of economic conflicts and valuation.* Cheltenham: Edward Elgar.

Martinez-Alier, J. 2009. Socially sustainable economic de-growth. *Development and Change*, 40(6), 1099–119.

Marx, K. 1973. *Grundrisse.* New York: Penguin.

Mintz, S. 1986. *Sweetness and power. The place of sugar in modern history.* New York: Vintage.

Monbiot, G. 2006. *Heat. How to stop the planet burning.* London: Penguin.

Monbiot, G. 2007. This crisis demands a reappraisal of who we are and what progress means. *Guardian*, 4 December. Available at: www.guardian.co.uk/commentisfree/2007/dec/04/comment.politics (Accessed 8 January 2010).

Moore, J. 2000. Environmental crises and the metabolic rift in world-historical perspective. *Organization & Environment*, 13(2), 123–57.

Morgan, D. 1980. *Merchants of grain.* Harmondsworth: Penguin.

Orth, S.M. 2007. *Subsistence foods to export goods. The impact of an oil palm plantation on local food sovereignty, North Barito, Central Kalimantan, Indonesia.* Sawit Watch, Wageningen: Van Hall Larenstein.

Oxfam 2007. Bio-fuelling poverty. Why the ET renewable-fuel target may be disastrous for poor people. *Oxfam Briefing Note*, 1 November

Patel, R. 2007. *Stuffed and starved. Markets, power and the hidden battle over the world's food system.* London: Portobello Books.

Peine, E. 2009. The private state of agribusiness. PhD dissertation, Cornell University.

Perkins, J.H. 1997. *Geopolitics and the green revolution. Wheat, genes and the Cold War.* New York: Cambridge University Press.

Pritchard, B. 2009. The long hangover from the second food regime: a world-historical interpretation of the collapse of the WTO Doha Round. *Agriculture and Human Values*, 31, 297–307.

Pritchard, B. and D. Burch. 2003. *Agro-food globalization in perspective. International restructuring in the tomato processing industry.* Aldershot: Ashgate.

Rainforest Action Network. 2007. Getting real about biofuels. Available at: www.ran.org/content/fact-sheet-geeting-real-about-biofuels (Accessed 8 January 2010).

Rifkin, J. 1992. *Beyond beef. The rise and fall of the cattle culture.* New York: Penguin.

Rist, L., L. Feintrenie and P. Levang. 2009. The livelihood impacts of oil palm: smallholders in Indonesia, *Biodiversity and Conservation*, 19(4).

Sachs, W. 1993. *Global ecology.* London: Zed Press.

Sayer, D. 1987. *The violence of abstraction.* London: Blackwell.

Smolker, R., B. Tokar and A. Petermann. 2008. *The real cost of agrofuels: impacts on food, forests, peoples and the climate.* Global Forest Coalition & Global Justice Ecology Project.

Storm, S. 2009. Capitalism and climate change: can the invisible hand adjust the natural thermostat? *Development and Change*, 40(6), 1011–38.

Sullivan, S. 2008. Markets for biodiversity and ecosystems: reframing nature for capitalist expansion? Event organised by the International Institute for Environment and Development, and the IUCN Commission on Environmental, Economic, and Social Policy, Fourth World Conservation Congress, Barcelona.

Syal, R. 2009. Body Shop ethics under fire after Colombian peasant evictions. *Observer*, 13 September, 8.

Transnational Institute (TNI). 2007. Agrofuels. Towards a reality check in nine key areas. June.

Vidal, J. 2009. Food land grab 'puts world's poor at risk'. *Guardian Weekly*, 7 October, 25.

Walker, R. 2004. *The conquest of bread. A hundred and fifty years of agribusiness in California.* New York: New Press.

Weis, T. 2007. *The global food economy. The battle for the future of farming.* London: Zed Books.

Wilkinson, J. and S. Herrera. 2008. *Agrofuels in Brazil. What is the outlook for its farming sector?* Rio de Janeiro: Oxfam International.

Wittman, H. 2009. Reworking the metabolic rift: La Vía Campesina, agrarian citizenship and food sovereignty. *The Journal of Peasant Studies*, 36(4), 805–26.

World Bank. 2007. *World development report. Agriculture for development 2008.* Washington, DC: World Bank.

# 7 Agrofuels and land rights in Africa

*Emmanuel Sulle and Ruth Hall*

## Introduction

From the mid-2000s there has been a dramatic shift in the search for an expanded supply of agrofuels in response to rising demand in Western economies. Agrofuels, mainly biodiesel and ethanol, require substantial tracts of land to produce economically viable outputs. This has created conditions in which investors began prospecting to fill market gaps in Western countries following the latter's adoption of renewable energy policies. This has led to demands for large tracts of land to grow feedstock to produce agrofuels in the South, mostly Africa and Latin America, in turn generating debate among scholars, civil society organizations (CSOs) and activists about the potentials and threats associated with this new development (Oxfam 2008; Matondi *et al.* 2011; Fairhead *et al.* 2012; Locke and Henley 2013).

Since 2005, private companies have cleared and 'opened' large farmlands in Africa, while some have secured leases on land for biofuels but are yet to operationalize their projects on the ground and others are still negotiating for long-term leases. Initiatives to start new agrofuel plantations in Africa have slowed since 2008–2009 and, while a number of projects are operational or in the final stages of establishing their primary production and processing facilities, others have either collapsed or have abandoned their biofuel feedstocks and switched instead to the production of food crops. Nevertheless, existing and planned agrofuel projects threaten rural peoples' rights to access, own and use land and its associated resources, e.g. water and forests. Cases of agrofuel investments in Tanzania and Mozambique indicate that vulnerable groups in society, such as pastoralists and women, bear the massive risks of these investments.

Large-scale agrofuel investments in plantations of sugar cane, oil palm, *jatropha curcas* (henceforth 'jatropha')[1] or soy can be found in many African countries, most notably Sierra Leone, Ghana, Ethiopia, Tanzania, Mozambique and Zambia (Land Matrix 2013). In each of these countries, competition over land between external investors and local people has produced conflict and controversy as tensions arise over land rights, notably the question of who has the authority to transact unregistered land, including 'community', 'customary' or 'public' land. Most starkly, tensions have arisen over land uses and the decisions made – often by state authorities – to promote agrofuels rather than food production, leading to disputes over the 'food versus fuel' contradictions in public policy.

While the move to use agrofuels as renewable content in the fuel supplies of developed countries has been guided by clear policy targets, in many African countries these initiatives usually started without a guiding policy, strategies or an institutional and legal framework in place – compounding the existing problems of weak land tenure rights in law and policy. These governance gaps and policy vacuums prompted fears among scholars, land rights activists and CSOs that African countries were inadequately prepared to ensure that the expected benefits of agrofuels accrue to the nation as a whole, and specifically to those directly affected (Hultman *et al.* 2012).

In this chapter we draw on our research on agrofuels and land rights in Africa (Sulle and Nelson 2009; Hultman *et al.* 2012; and Hall 2011a, 2011b), and specifically on field-based research over the past two years on the expansion of sugar cane in Mozambique and Tanzania, as well as several jatropha projects in those countries, many of which have meanwhile ceased production or switched to non-agrofuel crops. We summarize existing knowledge about the character, scale and distribution of such deals in Africa, and through a critical examination of what is known, develop a broader framing for understanding agrofuels expansion there. The aim is to provide a detailed analysis of the impacts of agrofuel expansion on the land rights and livelihoods of the poor in specific settings, and to locate these within the wider debates concerning agrofuel expansion in Africa. Special attention is paid to vulnerable groups such as women and pastoralists – these being the most affected by the initial development of agrofuel in Africa.

This chapter is divided into five sections. The next section explores the drivers of the expansion of agrofuels into African farmland, the current status of agrofuel projects, and clashing views over the impacts of agrofuels on local people, their land rights, land uses and production. It is followed by a discussion of early agrofuel projects in operation in both Tanzania and Mozambique, and the politics surrounding their emergence and, in some cases, their demise. We then examine the ecological and socio-economic impacts of agrofuel expansion and vulnerable groups' land rights, and conclude by pointing to what we argue are six distinct debates concerning agrofuels in Africa.

## Expansion of agrofuels in Africa – drivers, the current state of affairs and clashing views

The main influx of agrofuel investors into African farmland took place between 2005 and 2009 (Hultman *et al.* 2012; Sulle and Nelson 2009; Songela and Maclean 2008), as mostly foreign investors, often with little experience with specific feedstocks, entered into the industry. Apart from the cases of South Africa, Malawi and Zimbabwe, which started agrofuel production in order to to address their fuel crises some decades ago, the agenda of agrofuel development in Africa has been externally driven.

### *Drivers*

The principal drivers of this influx include Western policies, like the US Energy Independence and Security Act of 2007, which set the requirement of producing at least nine billion gallons of ethanol in 2008 and up to 36 billion gallons of agrofuel a year by 2022, and the European Union (EU)'s target of a 10 per cent share of agrofuels in total fuels by 2020 (Magdoff 2008: 40). While the EU is under pressure to drop its target of 10 per cent to 5 per cent renewable content in its fuelstocks, the US has retained its targets and in addition it has implemented a direct subsidy programme for corn producers. The rise in US demand for ethanol has been conditioned not merely by ecological concerns but by geopolitics – as a response to growing oil consumption and surging security concerns over the nation's reliance on sources of energy from the unstable Middle East (Statoil 2011; Hermele 2012). Financial speculation is considered to be one of the reasons why big companies are looking for areas to generate cheap returns and agriculture is considered to be one of the rapidly growing sectors, given the growing world population (Merian Research and CRBM 2010). Financial institutions such as social security and hedge funds financed a number of the failed attempts to secure land in Tanzania (Oakland Institute 2011).

Given these incentives and opportunities, driven by policies as well as global market conditions, Western companies responded by attempting to secure land to produce agrofuels around the world. For instance, companies from the UK, Sweden, Germany and the Netherlands opened up agrofuel projects in Tanzania and more than seven other African countries (Land Matrix 2013). Most of these companies had export markets in view and the domestic use of the agrofuels produced was barely a priority (for specific cases in Tanzania see Songela and Maclean 2008; for Mozambique, Burgess 2012 and Borras *et al.* 2011a). Further drivers of agrofuel development in Africa include growing incentives to generate environmentally friendly sources of energy to address climate change; the view that agrofuels can support the goals of rural development and poverty reduction through employment creation; and concerns for energy security, as African countries prioritize domestically produced agrofuels to feed into their grids and supply transportation energy (Borras *et al.* 2011b).

In many developing countries, agrofuel projects were promoted on the premise that they would be pro-poor and pro-(rural)development. This was done without clear evidence on the ground until the time that many of these projects, particularly the introduction of jatropha in rural China and India, proved unproductive (Ariza-Montobbio *et al.* 2010). Equally, Kant and Wu (2011) note extraordinary failures of jatropha projects in Tanzania, Kenya, China and India. The introduction of jatropha was backed by state agencies in Kenya and Tanzania despite the absence of a clear understanding of soil, feedstocks' productivity, labour relations or the technological needs of producing biodiesel from jatropha seeds (see Hunsberger, Chapter 8 in this volume). In many respects, the drive towards these agrofuels was shaped by political exigencies rather than technical assessments.

Furthermore, whether or not they are seen as part of a strategy to mobilize popular support ahead of elections, as in Mozambique in 2004 (Schut *et al.* 2010) and Zimbabwe in 2008 (Matondi *et al.* 2011), agrofuels and/or their significance within African states cannot be understood primarily as an externally driven imposition, but rather as the intersection of local and exogenous forces, and therefore a feature of politics at multiple scales. Together with financial interests among business and political elites, and a need to demonstrate progress towards national self-sufficiency in energy (ibid.), these 'endogenous drivers' combined with 'exogenous drivers' to produce a sudden agrofuels 'boom' in several African countries.

### *Current state of affairs*

While some countries have encouraged certain feedstocks for agrofuel development, others have imposed restrictions on the use of certain feedstocks. For example, government officials in Tanzania promoted the use of non-edible jatropha, hoping that this would reduce competition over the use of food crops for the production of agrofuels (E. Mfugale, personal communication March 2009). In contrast, South Africa issued a restriction on the cultivation of jatropha on the grounds that the crop has substantial environmental impacts. South Africa also restricted the use of maize for ethanol production due to food security concerns (GEXSI 2008).

Despite the early influx of investors searching for African farmland to develop agrofuel feedstocks, very few projects have materialized. A recent study by the Overseas Development Institute (Locke and Henley 2013) shows projects currently engaged in agrofuel production in only four African countries (Ethiopia, Mozambique, Tanzania and Zambia), which contrasts starkly with the total 'land grab' figures reported and widely cited, such as Oxfam's (2011) estimate of 227 million ha transacted since 2001 and the World Bank's (Deininger and Byerlee 2011) figure of 56 million ha under negotiation between 2009 and 2010. It also contradicts databases such as the Land Matrix (2013), which gave a figure of approximately 34,285,661 ha of concluded deals in 2013 and, in December 2013, reported some 201 large-scale land deals for agrofuels in Africa, a further 339 deals for food crops and 108 deals for non-food agricultural purposes. Other land deals are for tourism, multiple uses and non-specified. The recent ODI study identified the total area under agrofuel cultivation in four African countries as being much smaller than the amount of land authorized for development. The difference between areas authorized and areas actually cultivated[2] is stark in several countries in southern Africa (see Table 7.1). In Zambia for instance, only 4,000 ha are being cultivated out of a total of 654,000 ha authorized for agrofuel production.

Though we are able to secure reliable and updated information for only a small selection of African countries, looking at the present stages of agrofuel projects shows clearly that many projects, especially jatropha, are still at the planning stage. In addition, and contrary to media-driven perceptions, among those

*Table 7.1* Areas under agrofuel cultivation

| *Country* | *Hectares cultivated* | *Hectares authorized* |
|---|---|---|
| Ethiopia | 15,000 | 576,000 |
| Mozambique | 6,000 | 209,000 |
| Tanzania | 2,000 | 66,000 |
| Zambia | 4,000 | 654,000 |

Source: Locke and Henley (2013: 7).

projects that are operational, jatropha does *not* stand out as the dominant crop, except in Ethiopia, Tanzania and Zambia (Anseeuw *et al.* 2012).

The fact that it has mainly been investors from outside the African continent who have initiated agrofuel projects warrants questions concerning markets, production, processing and transportation. While investors from Europe may have markets back home in mind, the costs of processing and transporting such outputs are high, host countries prefer domestic energy supplies, and these factors have forced companies to switch their attention to creating local markets.

So where do agrofuels produced in Africa go? Information is patchy, and no comprehensive picture is possible at present. Tanzania exports small quantities of agrofuels to Finland (URT 2012). Biodiesel from Tanzania and India have been used by Air New Zealand, which since 2008 has powered its aircraft with a 50–50 blend of biodiesel from jatropha and standard A1 jet fuel (ENS 2008, cited in Sulle and Nelson 2009). Whether it has sustained this is, however, unclear as most of the jatropha projects in both Tanzania and India have not yet yielded much, while some have been closed. More broadly, the goal of many investors – to supply EU markets – has not been realized.

## *Clashing views over agrofuels in Africa*

The proponents of agrofuels argue that their production will mitigate climate change, create jobs, facilitate rural development, improve access to social services and provide infrastructure, and thus reduce poverty (Deininger and Byerlee 2011; Murphy *et al.* 2011). In support of agrofuel production, the International Federation of Agricultural Producers (IFAP) – an organization formed by business-oriented farmers – argues that agrofuels are a 'new market opportunity' for farmers that can help to diversify risk while also mitigating climate change and promoting rural development (cited in Borras *et al.* 2010, 585). These proponents of agrofuels argue that, when governments have adequate policy and regulatory frameworks in place, a win–win situation is achievable (Burgess 2012: 6) and that the risks associated with the displacement of people and the allocation of 'idle', 'unused' or 'marginal' land for agrofuel production can also mitigate competition over the use of resources (Burgess 2012; Murphy *et al.* 2011).

Critics argue that investors' interests are to make profit, and that they therefore target the most fertile land to produce agrofuels – and land with good access to water supply and close to infrastructure and urban centres – rather than marginal lands (Oxfam 2013). These are the same areas prized by local farmers, and so arguments about under-utilization appear misplaced. Cotula and Vermeulen (2010), using a number of case studies in sub-Saharan Africa, found that investors have targeted, and been allocated, land with access to water and markets, and often with sufficient rainfall. For instance, in Gaza Province in Mozambique, investors were allocated fertile land situated close to a dam built by the government, which was capable of supplying irrigation water and generating electricity to enable commercial operations. This was precisely the land used by local farmers, and it was very important to them (Burgess 2012). Equally, in Tanzania, agrofuel investors have concentrated on the coastal regions with plenty of rainfall and easy access to the ports of Dar es Salaam, Lindi and Mtwara.

## Agrofuels and land rights – insights from Tanzania and Mozambique

Perceptions of the merits and demerits of agrofuel expansion have emerged within national political discourses, and are shaped by competing notions of what constitutes 'development' and competing routes to achieving it. Cases from Tanzania and Mozambique illustrate how responses to such projects, and assessments of the ecological and economic agendas driving them, have been conditioned by the broader politico-economic interests at local and national levels.

### *Experience from Tanzania*

President Jakaya Kikwete of Tanzania has been an ardent supporter of agrofuel investments in his country, framing them as pro-development. On taking office after the 2005 general election, Kikwete instructed the Ministry of Energy and Minerals to develop a strategy to promote agrofuels, especially bioethanol, and reiterated this call in his subsequent monthly televised addresses to the nation. On his first official visit to the Rufiji delta in southern Tanzania, he encouraged villagers to welcome the Swedish company (SEKAB Tanzania Ltd), which was planning to develop ethanol from sugar in the area. The President's interest seems to have grown after visits to bioethanol plants in both Sweden and the USA (Kamanga 2008).

Tanzania's land law makes the stakes high for communities on the land targeted by investors. In order for investors to secure rights to 'village land', these rights have to be converted into 'general land' under national state ownership (a de facto expropriation), after which the Tanzania Investment Centre (TIC) becomes the custodian with the mandate to conduct transactions of this land with investors. While Tanzanian villagers aim to benefit from investments, they have to forego their land in perpetuity, because according to Tanzania's Land Act

No. 4 of 1999 and Village Land Act No. 5 of 1999 any land transferred from ‘village land’ to ‘general land’ ceases to be village land and so, regardless of how a specific investment fares, a village may not access this land and the resources on it in the future.

In Tanzania, four major biofuel companies started operations, some of which are still operational, others having collapsed (Sulle and Nelson 2013). The Swedish company EcoEnergy Tanzania Ltd.[3] holds 22,500 ha in Bagamoyo District (the home district of President Kikwete) on a 99-year lease. It acquired this as a sub-lease from another Swedish company, the above-mentioned SEKAB Tanzania Ltd., which had acquired the land in 2008 to produce ethanol for the European market (ibid.) but which collapsed in 2009. The acquired land is the former, non-operational, Zanzibar People’s Ranch (RAZABA), which the TIC offered to foreign investors with the apparent aim of establishing a flagship project in the President’s district. According to Songela and Maclean (2008), there were no established villages on the ranch, yet 14 households and some peasants were compensated for being displaced, as informal occupation and land use had followed the closure of the ranch. In 2013, EcoEnergy began developing 7,800 ha of sugar cane for the production of sugar (and later ethanol), but for local markets (Locher and Sulle 2013).

Sun-Biofuels Tanzania Ltd, a UK-based company, obtained 8,211 ha of land out of the total of 81,000 ha it requested from 12 villages, across five wards,[4] in Kisarawe District, Coastal Region (Kamanga 2008; Sulle and Nelson 2009). At the time of Sulle’s recent field visit, the company had only managed to develop a jatropha plantation on 2,000 ha of its allocated land. Field research found that villagers were involved in the negotiation process for the land acquired by Sun-Biofuels, but they were not certain how much land was allocated to the company, for how long and for what amount of money. The government evaluator determining the compensation was Ardhi University, also a client of the company. The company paid a token amount of TZS260 million (US$208,000)[5] in compensation to members of the 12 villages which had trees on their land, and promised a further TZS540 million (US$348,387) at a later stage, but as of December 2013, it had paid this second sum to the District Council (DC). The DC is set to distribute the received amount to affected villages in terms of financial support to villagers’ initiated social projects (R. Geta, personal communication, 27 December 2013). The compensation was based on the area the company planted with trees, rather than on the value of the land and natural resource use foregone by the local communities (Kamanga 2008, Sulle and Nelson 2009). These uses included the cultivation of cash and food crops (mainly cassava and legumes), and forest-based economic activities such as the commercial production of charcoal, and the harvesting of traditional medicines, mushrooms, fuelwood and building materials. Due to the slow development of its land in 2011, the company sold its shares to a new investor, 30 Degree East, leaving communities that had been affected by the loss of access to common property resources without recourse to the original investor, which did not fulfil its promises of hospitals, schools, water infrastructure and village offices.

In Tanzania, much of the 'idle' land allocated for agrofuel feedstock cultivation is in direct or indirect use, even if this use is seasonal, cyclical or part of diversified land-use patterns. Pastoralists frequently use these lands for seasonal grazing, and the same lands are sources of firewood, charcoal, traditional medicines, building materials and wild fruits. In assessing the value of natural forests in Tanzania's Kisarawe District, as part of an assessment for the calculation of compensation to communities affected by a jatropha investment, the World Bank (2008) established that the forested natural environment provided 100 per cent of traditional medicine, 75 per cent of building materials and 90 per cent of the energy needs of these communities. Threats posed to local people by the jatropha project, were resource exclusion, exacerbated land-related conflicts, marginalization and corruption (Cotula andVermeulen 2011; Hall 2011a; Oxfam 2008).

### *Experience from Mozambique*

Mozambique has been approached by more than 20 large-scale agrofuel investors wanting to acquire more than 1,000 ha each, which, by 2012, had collectively acquired a total area of 589,268 ha (Burgess 2012). The projects approved and currently underway are to develop plantations for jatropha, sugar cane and palm oil, as well as other agrofuel feedstocks such as soy, sweet sorghum, cassava and sunflower. Following a government moratorium on the allocation of land for large-scale investments in 2009, new approaches employed by investors in the agrofuel sector have been to seek additional lands or to propose outgrower schemes to source the raw materials for their processing plants (ibid.). To date, South African and Brazilian companies are the leading investors in Mozambique's sugar industry, while other investors originate from the UK, Portugal, Italy and the USA. Most investors are targeting the export markets, with only 10 per cent of the produce aimed at the domestic market (ibid.).

Mozambique has one of the highest registration rates for large-scale agricultural investments in Africa. By mid 2008, the country had pending requests to acquire more than 12 million ha, most of them related to agrofuels (Peters 2009). The Brazil–EU partnership, 'Sustainable Development of Bioenergy' was aimed at assisting the Mozambican government to develop a large sustainable agrofuels industry (see Franco *et al.* 2010). This led to the publication of the Biofuels Policy Strategy in 2009 which, apart from the government's decision to halt further agrofuel projects, has yet to address the critical problem of land rights. Mozambique's Land Law of 1997 provides statutory recognition and protection of informal land rights acquired through custom and use, and enables holders of such rights to be confirmed through the allocation of a DUAT,[6] a heritable and transferrable certificate of ownership for a period of 50 years. Yet by 2011, only 12 per cent of communal land had been demarcated (Deininger andByerlee 2011). As a result, despite statutory recognition, community lands are susceptible to appropriation by the state for allocation on leasehold to new investors (Norfolk and Tanner 2007). In addition, Nhantumbo and Salomão (2010) found that legal provisions such as the requirements with regard to consultations and compensation are not properly enforced due to the stark asymmetry of political and economic power between (usually poor) rural communities and wealthy international (and domestic) investors.

As indicated in a number of studies (Borras *et al.* 2011a; Burgess 2012; Nhamtumbo and Salomao 2010), Mozambique is among the African countries that registered a comparatively large number of biofuel investors. Massingir Agro-Industrial (MAI), one of the major agrofuel investors in Mozambique, aims to develop a sugar plantation in Massingir District, Gaza Province, situated north of Maputo. Between 2007 and 2009, the Mozambican government allocated 30,000 ha in the same area, under a 50-year renewable lease, to a company called ProCana, a subsidiary of the London-based Central African Mining and Exploration Company (CAMEC), to produce ethanol-based plastics for South Africa (Borras *et al.* 2011a). The allocated land is adjacent to the Oliphants River and thus fertile and irrigable in all seasons. In November 2010, the Mozambican government revoked ProCana's contract, ostensibly because it had not initiated its planned developments and therefore had not complied with the terms of its lease – yet the investors, local government authorities and the local community are unclear or disagreed regarding the real reasons for the lease being cancelled (interview with stakeholders in Massingir, April 2013).

In October 2011, the Mozambican government allocated almost exactly the same land – 31,000 ha – to MAI, a consortium of the South African sugar company Transvaal Suiker Beperk (TSB, with a 51 per cent shareholding) and a consortium of Mozambican businessmen, known as SIAL (with a 49 per cent shareholding). Like ProCana before it, MAI has a 50-year DUAT; it is expected to invest US$740 million over this period (Allafrica 2012) and ultimately plans to plant 37,000 ha of sugar cane, to be processed into sugar, ethanol, molasses and animal feed and used for the generation of electricity; 80 per cent of this produce will be exported to Europe (Hall and Paradza 2012). In terms of the deal, MAI will assist the local community to develop food gardens on 1,000 ha of land. However, as of January 2014, the company had yet to start planting sugar cane and establishing a processing factory, and was still at the preparatory stage (S. Mabasso, personal communication, January 2014).

In 2007, the German Company Elaion was allocated 1,000 ha in Sofala Province, where the local population already faced food insecurity due to vagaries of the weather (floods and droughts) and high food prices; the deal reduced communities' access to natural resources on which they depended to support their livelihoods, thus accentuating their vulnerability (Nhamtumbo and Salomão 2010). In Mozambique, as in many other countries in Africa, rural communities derive their livelihoods not only from growing crops and keeping livestock, but also from forest products such firewood, charcoal, timber beams, twine, honey, grass, bamboo and forest foods. Loss of access to such common property resources is seldom adequately accounted for in large-scale land deals, in which negotiations generally focus only on the loss of residential, cropping and grazing land. In Mozambique, the emergence of agrofuel projects that plan large-scale plantations takes place in a context where the majority (nearly 70 per cent) of the population lives in rural areas and 80 per cent of them are engaged in small-scale farming (FAO 2012; Burgess 2012).

## Agrofuel investments and land rights

The experience of agrofuel expansion in Tanzania and Mozambique, as outlined above, draws attention to processes of displacement of the rural poor; to the commodification of land and labour; and to the suppression of alternative modes of production and consumption. In both countries, it is evident that governments have played a significant role in facilitating the acquisitions of large tracts of land for agrofuel investments. For instance, in Tanzania, under the current Village Land Act No. 5 of 1999, the capacity of village authorities to allocate land to a foreign investor is limited. When an investor in collaboration with government officials from higher authorities (district or central government) has identified an area of village land for investment, the Commissioner of Lands transfers this land from the Village Land category into General Land. The same land will then be handed over to the TIC, which offers the Derivative Right to a foreign investor and itself (TIC) remains the custodian of the land. Should it remain unaddressed, this legal practice, in which communities lose their land because of the investment, is likely to cause more poverty rather than development among the poor majority (Sulle and Nelson 2013).

In all cases, the main problem associated with large-scale land acquisitions for agrofuel investments is the lack of state recognition of rural communities' rights to land under customary ownership. This lack of recognition has significant impacts on vulnerable groups, particularly women and pastoralists.

Likewise, in Mozambique the former communal land offered to PROCANA was communal land. Yet, after the PROCANA deal collapsed, this land was not returned to the community; rather, district officials accompanied and introduced the new investor to the communities in Massingir District. Communities were thus convinced that a superior investor had come and that they would receive the promised benefits.

### *Agrofuels and women's land rights*

UN conventions recognize women's equal rights to access, own, control and inherit land as well as adequate housing and property (UN-HABITAT 2002). It has also been argued that well-executed, large-scale acquisitions of land for investment purposes can provide opportunities to both women and men (Behrman *et al.* 2012). Yet women in Africa are systematically denied these rights (Benschop 2004). The expansion of agrofuels in Africa is another challenge that threatens to diminish women's access to land in rural areas. Women play a significant role in rural household production and reproduction. In sub-Saharan Africa, women contribute about 60–80 per cent of the labour for food production for both household consumption and sales.[7] However, the commercial crops typically targeted in agrofuel investments are farmed mainly by men, i.e. women are likely to be marginalized in the production and ownership of these crops (Arndt *et al.* 2011; see White *et al.*, Chapter 4 in this volume).

Worldwide, communal land is owned under customary tenure by at least 75 per cent of the poor; yet, the customary ownership of land by local communities is not properly recognized by the state (Alden Wily 2011). In many African countries, even in states where land laws are considered progressive, women remain marginalized. This is because, in Africa, in decision-making processes regarding investment deals, women's voices are heard less due to the fact that most of the negotiations are undertaken by government officials, local chiefs and elites, positions that are dominated by men (Tandon and Wegerif 2013). This situation favours local elites as the beneficiaries of agrofuel investments, while marginalizing vulnerable groups and the poor majority.

### *Ecological and socio-economic impacts*

One of the problems associated with the expansion of large-scale agrofuel plantations is the dramatic changes in agroecology associated with the introduction of monoculture farming, the introduction of invasive species, a decline in biodiversity and sometimes irreversible damage to the local environment (Schoneveld *et al.* 2011; Songela and Maclean 2008; Gordon-Maclean *et al.* 2008). Mounting criticism of the negative environmental impacts of agrofuels expansion – in addition to the socio-economic impacts – prompted an intergovernmental intervention in the southern African region to support national governments to develop coherent agrofuel strategies. A Southern African Development Community (SADC) Biofuels Task Force, created in 2011, was mandated to enlist scientific expertise in order to 'advance the uses of biofuels in the region' (SADC Biofuels Taskforce 2011: 1) and engaged in building scenarios of land-use change to identify the potential costs and benefits of agrofuel expansion (see Table 7.2).

While environmental impacts are clearly identified, socio-economic impacts are largely listed under the 'unknown' including, notably, the impact of these land- use changes on 'poverty alleviation'. This comprehensively called into question the fundamental assumptions underpinning the SADC Framework for Sustainable Biofuel Use and Production, approved by the SADC Energy Ministers on 29 April 2010, the stated purpose of which was 'to accelerate their initiatives in developing biofuels as a source of alternative and cheap environmentally friendly fuel but also for rural development and poverty reduction' (SADC 2010).

## Conclusions

Agrofuels in Africa have been promoted on the grounds that they will improve domestic energy security, address climate change through the reduction of greenhouse gases, create jobs and enhance rural development. Amid the need to develop new alternative sources of energy, Africa – regarded as a resource-rich

*Table 7.2* Trade-offs linked with different land use change scenarios

| *Land use change scenario* | *Benefits* | *Costs* | *Unknown* |
|---|---|---|---|
| Mature forest to biofuel | • Economic returns from agrofuels (formal employment, smallholder income)<br>• Revenue for central/local government<br>• Able to access more productive lands<br>• Timber rents/revenue during plantation establishment<br>• No indirect land use change | • Loss of carbon<br>• Loss of biodiversity/habitat<br>• Loss of NTFPs<br>• Water impacts<br>• Loss of access to EU market | • Poverty alleviation |
| Mixed landscapes with shifting agriculture to biofuel | • Revenue for central/local government<br>• Risk to smallholders under unproven business models/ speculation | • Loss of cropland<br>• Loss of economic value of common property (grazing land, NTFPs, charcoal)<br>• Displacement (if industrial plantation) | • Poverty alleviation<br>• Displaces pressures from wood fuel |
| Permanent cropland to biofuel | • Risk to smallholders under unproven business models/ speculation<br>• No direct land use change | • Loss of cropland<br>• Displacement (if industrial plantation)<br>• High indirect land use effects due to displaced land uses | • Poverty alleviation<br>• Displaces pressures from wood fuel |

Source: SADC biofuel taskforce (2011) (edited by authors).

continent with cheap land and water – has become the centrepiece for the corporate strategies of many Western companies. African government and private sector elites in search of a political base and patronage networks have embarked on promoting agrofuels to rural areas through investment agencies.

The reality of agrofuels in Africa is complex, and we are witnessing the emergence of a new scepticism within governments and growing caution among investors, while community responses continue to emphasize the dangers of a trade-off between food and fuel. Concrete policy and advocacy initiatives are needed to safeguard the land rights of rural Africans, especially women and pastoralists, whose land rights are most insecure. Given that a number of African states are in the process of re-drafting their constitutions or reforming their land policies and laws, it is crucial that land rights are fully recognized and protected in these key documents and that adequate institutional mechanisms for their implementation are created. Several positive developments are underway. In Tanzania, the Tanzania Gender Network Programme (TGNP) has been in the forefront of educating communities about the impacts of large-scale agrofuel investments on women. TGNP has also been encouraging members of communities to present their views on land rights and resources ownership to the National Constitutional Review Commission, set to formulate the country's new constitution by the end of 2014. In Zambia, the national NGO, Zambia Land Alliance, has been in the forefront to ensure that women's land rights are protected by the country's new constitution by making various submissions to the constitutional review team (Zambia Land Alliance 2004).

The debate over agrofuels in Africa can best be understood as several quite distinct sets of questions, which are often coalesced into pro- and anti-biofuel arguments. First, is a set of questions regarding the *land rights, resource access* and *livelihoods* of rural people affected by large-scale agrofuel investments, including development-induced dispossession. Second, are salient concerns regarding *social differentiation between 'winners' and 'losers'*, as certain locals stand to benefit from new investments through employment, rents and changes in local economies, while others, especially women, who lose land and common pool natural resources, are often excluded from new circuits of accumulation. Third, is a set of politico-economic questions centring on how agrofuels feature in *national political discourse and party politics*, especially as agrofuels – as a route to energy self-sufficiency – have been seen to signify a source of political autonomy, including autonomy from Western influence and from volatile international oil markets. Fourth, are questions regarding the *destination markets* for agrofuel feedstocks, and the extent to which these are to contribute towards realizing policy targets in developed countries or realizing energy security in African countries. Fifth, are technical questions regarding *profitable feedstocks and the feasible scale of production* – for example, whether even large areas planted with feedstocks can make any meaningful contribution to national fuel supplies. Sixth, are questions regarding the *agroecological impacts* of agrofuel production, especially in the form of large monoculture plantations, and the political ecology of the arge-scale loss of biodiversity and clearcutting of forests in the name of green energy.

In relation to all the above, mounting evidence shows that the euphoric era of large-scale agrofuel investments in Africa is over.

## Notes

1 Jatropha has long been planted in Africa and Asia as a protective hedge around homesteads, gardens and fields, since it is poisonous and not browsed by animals (Kempf 2007).
2 In this chapter, the cultivation of agrofuels refers to the growing of feedstocks such as jatropha, soy and sugar cane, mainly for the purposes of producing ethanol or biodiesel, depending on the type of feedstock used.
3 For details, see www.ecoenergy.co.tz/ (accessed: 3 February 2014).
4 In Tanzania, a ward is the administrative unit above a village. Two or more villages form a ward.
5 The exchange rates used here are for 2008 and 2013, namely US$1 = TZS1,250 and TZS1,550 respectively.
6 DUAT is an abbreviation of the Portuguese words: *Direito de uso e aproveitamento da terra*, meaning '*the right to use and benefit from the land*'.
7 See, www.fao.org/sd/fsdirect/fbdirect/fsp001.htm (accessed: 3 February 2014).

## References

Alden Wily, Liz (2011) '"The law is to blame": The vulnerable status of common property rights in sub-Saharan Africa', *Development and Change* 42(3), 733–757.

Allafrica (2012) 'Mozambique: Guebuza visits sugar project in Massingir' (http://allafrica.com/stories/201211120298.html, accessed: 26 January 2014).

Anseeuw Ward, Mathieu Boche, Thomas Breu, Markus Giger, Jann Lay, Peter Messerli and Kerstin Nolte (2012) *Transnational land deals for agriculture in the Global South. Analytical report based on the Land Matrix database.* Bern, Montpellier, Hamburg: Centre for Development and Environment (CDE), Centre de Coopèration Internationale en Recherche Agronomique (CIRAD), German Institute of Global Areas Studies (GIGA) (http://landportal.info/landmatrix/media/img/analytical-report.pdf, accessed: 24 May 2013).

Ariza-Montobbio, Pere, Lele Sharachchandra, Giorgos Kallis and Joan Martinez-Alier (2010) 'The political ecology of Jatropha plantations for biodiesel in Tamil Nadu, India', *The Journal of Peasant Studies* 37(4), 875–897.

Arndt, Channing, Rui Benfica and James Thurlow (2011) 'Gender implications of biofuels expansion in Africa: The case of Mozambique', *World Development* 39(9), 1649–1662.

Behrman Julia, Ruth Meinzen-Dick and Agnes Quisumbing (2012) 'The gender implications of large-scale land deals', *The Journal of Peasant Studies* 39(1), 49–79.

Benschop, Marjolein (2004) *Women's rights to land and property.* Nairobi: UN-HABITAT, Commission on Sustainable Development.

Borras Jr., Saturnino M., Philip McMichael and Ian Scoones (2010) 'The politics of biofuels, land and agrarian change: Editors' introduction', *The Journal of Peasant Studies* 37(4), 575–592.

Borras Jr., Saturnino M., David Fig and Sofia Monsalve Suárez (2011a) 'The politics of agrofuels and mega-land and water deals: Insights from the ProCana case, Mozambique', *Review of African Political Economy* 38(128), 215–234.

Borras Jr., Saturnino M., Ruth Hall, Ian Scoones, Ben White and Wendy Wolford (2011b) 'Towards a better understanding of global land grabbing: An editorial introduction', *The Journal of Peasant Studies* 38(2), 209–216.

Boudreaux, Karol (2013) 'Addressing land rights can make social change possible', *Guardian* (www.theguardian.com/global-development-professionals-network/2013/may/23/land-rights-partnerships-social-change, accessed: 17 June 2013).

Burgess, Clare (2012 'Large scale biofuel projects in Mozambique: A solution to poverty?', thesis presented to the School of Social and Political Sciences in partial fulfilment of the requirements for the degree of Master of Development Studies in the field of Development Studies in the School of Social and Political Sciences, University of Melbourne, Melbourne.

Cotula, Lorenzo and Sonja Vermeulen (2010) 'Over the heads of local people: Consultation, consent, and recompense in large-scale land deals for biofuels projects in Africa', *The Journal of Peasant Studies* 37(4), 899–916.

Cotula, Lorenzo and Sonja Vermeulen (2011) 'Contexts and procedures for farmland acquisitions in Africa: What outcomes for local people?', *Development* 54(1), 40–48(9).

Deininger, Klaus and Derek Byerlee (2011) *Rising Global Interest in Farmland: Can It Yield Sustainable and Equitable Benefits?* Washington, DC: World Bank.

Fairhead, James, Melisa Leach and Ian Scoones (eds) (2012) 'Green grabbing: a new appropriation of nature?', *The Journal of Peasant Studies* 39(2), 237–261.

FAO (United Nations Food and Agriculture Organisation) (2012) *Trends and impacts of foreign investment in developing country agriculture: Evidence from case studies.* Rome: FAO.

Franco, Jennifer, Les Levidow, David Fig, Lucia Goldfarb, Mireille Hönicke and Maria Mendonca (2010) 'Assumptions in the European Union biofuels policy: Frictions with experiences in Germany, Brazil and Mozambique', *The Journal of Peasant Studies* 37(4), 661–698.

GEXSI (2008) *Global market study on jatropha project inventory: Africa*. Prepared for the World Wide Fund for Nature (WWF). London/Berlin: GEXSI LLP (www.jatropha-alliance.org/fileadmin/documents/GEXSI_Jatropha-Project-Inventory_AFRICA.pdf, accessed: 28 September 2013).

Gordon-Maclean, Andrew, James Laizer, Paul Harrison and Riziki Shemdoe (2008) *Biofuel industry study, Tanzania.* Tanzania and Sweden: World Wide Fund for Nature (WWF).

Hall, Ruth (2011a) 'Land grabbing in southern Africa: The many faces of the investor rush', *Review of African Political Economy* 38(128), 193–214.

Hall, Ruth (2011b) *Land grabbing in Africa and the new politics of food security.* Policy Brief 041. Brighton: Future Agricultures Consortium (http://r4d.dfid.gov.uk/PDF/Outputs/futureagriculture/FAC_Policy_Brief_No41.pdf, accessed: 3 December 2013).

Hall, Ruth and Gaynor Paradza (2012) 'Foxes guarding the hen-house: the fragmentation of 'the state' in negotiations over land deals in Congo and Mozambique', paper presented at the International Conference on Global Land Grabbing II, 17–19 October 2012, organized by the Land Deals Politics Initiative (LDPI) and hosted by the Department of Development Sociology at Cornell University, Ithaca, NY.

Hermele, Kenneth (2012) *Land matters: Agrofuels, unequal exchange, and appropriation of ecological space.* Lund University, Lund: Media-Tryck.

Hultman Nathan, Emmanuel Sulle, Christopher Ramig and Seth Sykora-Bodie (2012)

'Biofuels investments in Tanzania: Policy options and sustainable business models', *The Journal of Environment Development* 21(3), 339–361.

Kamanga, Khoti (2008) *The agrofuel industry in Tanzania: A critical enquiry into challenges and opportunities*. Dar es Salaam: Land Rights Research and Resources Institute, Joint Oxfam Livelihood Initiative for Tanzania.

Kant, Promode and Shairong Wu (2011) 'The extraordinary collapse of Jatropha as a global biofuel', *Environmental Science & Technology* 45, 7114–7115.

Kempf, Mathias (2007) *Jatropha production in semi-arid areas of Tanzania. A feasibility study*. Dodoma: Rural Livelihood Development Company (www.tnrf.org/files/E-INFO-RLDC_Jatropha_Production_in_Semi-Arid_Areas_of_Tanzania_2007.pdf, accessed: 26 January 2014).

Land Matrix (Land Matrix Global Observatory) (2013) The online public database on land deals. International Land Coalition (ILC), Centre de Coopération Internationale en Recherche Agronomique pour le Développement (CIRAD), Centre for Development and Environment (CDE), German Institute for Global and Area Studies (GIGA) and Deutsche Gesellschaft für Internationale Zusammenarbeit (GIZ) (www.landmatrix.org/en/, accessed: 4 February 2014).

Locher, Martina and Emmanuel Sulle (2013) 'Foreign land deals in Tanzania: An update and a critical view on the challenges of data (re)production', LDPI Working Paper 31 (www.plaas.org.za/sites/default/files/publications-pdf/LDPI31Locher%26Sulle_0.pdf, accessed: 7 June 2013).

Locke, Anna and Giles, Henley (2013) *Scoping report on biofuels projects in five developing countries*. London: Overseas Development Institute.

Magdoff, Fred (2008) 'The political economy and ecology of agrofuels', *Monthly Review*, July–August 2008, 34–50.

Matondi, Prosper B., Kjell Havnevik and Atakilte Beyene (eds) (2011) *Biofuels, land grabbing and food security in Africa*. London: Zed Books.

Merian Research and CRBM (2010) *The vultures of land grabbing: The involvement of European financial companies in large-scale land acquisition abroad*. Brussels: Campaign to Regulate Finance for Development.

Murphy, Richard, Jeremy Woods, Mairi Black and Marcelle McManus (2011) 'Global developments in the competition for land from biofuels', *Food Policy* 36(1), S52–S61.

Nhantumbo, Isilda and Alda Salomão (2010) *Biofuels, land access and rural livelihoods in Mozambique*. London: International Institute for Environment and Development (IIED).

Norfolk Simon and Christopher Tanner (2007) *Improving tenure security for the rural poor: Mozambique country case study*. Rome: Food and Agricultural Organization of the United Nations.

Oakland Institute (2011) *Understanding land investment deals in Africa: Agrisol Energy and Pharos Global Agriculture Fund's Land Deal in Tanzania*. Land Deal Brief, June 2011, Oakland: The Oakland Institute (www.oaklandinstitute.org/land-deal-brief-agrisol-energy-and-pharos-global-agriculture-fund%E2%80%99s-land-deal-tanzania, accessed: 30 September 2013).

Oxfam (2008) *Another Inconvenient Truth: How biofuel policies are deepening poverty and accelerating climate change*. Oxford: Oxfam Briefing Paper 114.

Oxfam (2011) *Land and Power: The Growing Scandal Surrounding the New Wave of Investments in Land*. Oxford: Oxfam Briefing Paper 151.

Oxfam (2013) Pan African Land Hearing, event held in Johannesburg, 13–15 August 2013.

Peters, Floor (2009) 'Socio-economic impact study of biofuel plantation on farm households in Mozambique', Policy Brief 1–53. Netherlands Ministry of Foreign Affairs Government and Wageningen University.

SADC (Southern African Development Community) (2010) *SADC Framework for Sustainable Biofuels.* Gaborone: SADC, approved by the SADC Energy Ministers Meeting on 29 April 2010.

SADC Biofuels Taskforce (2011) 'The SADC CSIR Workshop on Biofuels and Science/Policy Exchange'. Protea Hotel OR Tambo, Kempton Park, Johannesburg, 29 August 2001.092011.

Schoneveld, George, Laura German and Eric Nutakor (2011) 'Land-based investments for rural development? A grounded analysis of the local impacts of biofuel feedstock plantations in Ghana', *Ecology and Society* 16(4), 10 (http://dx.doi.org/10.5751/ES-04424–160410, accessed: 3 February 2014).

Schut Marc, Maja Slingerland and Anna Locke (2010) 'Biofuel developments in Mozambique: update and analysis of policy, potential and reality', *Energy Policy* 38, 5151–5165.

Songela, Francis and Andrew Maclean (2008) *Scoping exercise (Situation Analysis) on the biofuels industry within and outside Tanzania.* Dar es Salaam: WWF Tanzania Programme Office, Energy for Sustainable Development.

Statoil (2011) 'Energy perspectives: Long-term macro and market outlook' (www.statoil.com/en/NewsAndMedia/News/Downloads/Statoil%20%20Energy%20Perspectives.pdf, accessed: 11 October 2013).

Sulle, Emmanuel and Fred Nelson (2009) *Biofuels, land access and rural livelihoods in Tanzania.* London: IIED.

Sulle, Emmanuel and Fred Nelson (2013) 'Biofuels investment and community land tenure in Tanzania: The case of Bioshape, Kilwa District', Future Agricultures Consortium Working Paper 73. Brighton.

Tandon, Nidhi and Marc Wegerif (2013) 'Securing rights and livelihoods for rural women in the context of corporate land investments: Learning from experiences in Africa', paper presented at the Annual World Bank Conference on Land and Poverty, Washington, DC, 8–11 April 2013.

UN-HABITAT (2002) 'UN Special Rapporteur on Adequate Housing: Study on Women and Adequate Housing', April E/CN.4/2003/55, p. 9 (www.unhchr.ch/housing, accessed: 23 September 2013).

URT (United Republic of Tanzania) (1999) *Land Act No. 4.* Dar es Salaam: Government printer.

URT (1999) *Village Land Act No. 5.* Dar es Salaam: Government printer.

URT (2012) Ministry of Natural Resources and Tourism (MNRT) Export Data, Dar Es Salaam: Ministry of Natural Resources and Tourism.

World Bank (2008) *Putting Tanzania's Hidden Economy to Work: Reform, Management and Protection of its Natural Resource Sector.* Washington, DC: The World Bank.

Zambia Land Alliance (2004) *Land is Life, Submissions to Constitutional Review Commission.* Lusaka: Zambia Land Alliance (www.mokoro.co.uk/files/13/file/lria/submissions_to_const_review_comm_zla.pdf, accessed: 26 January 2014).

# 8 The discursive and material flexibility of *Jatropha curcas*

*Carol Hunsberger*

## Introduction

Major agrofuel feedstocks such as corn, sugarcane, oil palm and soy fit Borras *et al.*'s (2012) definition of 'flex crops': they can be used for multiple, interchangeable purposes such as food, feed, fuel and commercial or industrial products. These authors suggest that the material flexibility of these crops makes them particularly attractive to investors because it allows them to decide what to produce and sell based on price signals, thus diversifying their investment while dealing with a single crop. Indeed, production of these four crops has increased dramatically in recent years. Soy production alone nearly doubled – from 50.4 million hectares to 98.8 million hectares – in the 30 years leading up to 2009, with half of this growth occurring after the year 2000 (FAO 2012).

Like soy, oil palm and sugar cane, the oilseed shrub *Jatropha curcas* (hereafter jatropha) has been promoted as a source of fuel and has spread rapidly in the global South (GEXSI 2008). However, it does not fit the flex crop pattern. Jatropha cannot be consumed as food and can only be made into livestock feed if it is first detoxified. Aside from a few places where jatropha soap and fertilizer are sold (see e.g. Favretto *et al.* 2012), markets for jatropha products other than energy have not emerged. Thus, jatropha has relatively little flexibility in a material sense.

Despite this low material flexibility – and I argue, at times because of it – jatropha has been the subject of multiple discourses that actors have used to encourage its spread. In this chapter, I use the idea of 'discursive flexibility' to examine how claims about jatropha's potential to achieve multiple goals have shaped its recent history. By discursive flexibility I mean the ability to strategically switch between or activate multiple discourses in pursuit of an objective. I ask how the fluidity and coexistence of discourses about jatropha have influenced the uptake – and sometimes rejection – of jatropha as an energy crop.

This task fits with one of the core interests of political ecology: untangling and critically examining the relationship between material and discursive dimensions of human–environment interactions. Here, I interpret discourse broadly as a narrative that includes language as well as the assumptions and representations through which language is translated into social meaning (Grillo 1997), and take the position that discourse and materiality are mutually influential. Precedents for this approach can be found in Jessop's (2004) analysis of how 'economic imaginaries' help institutions define meaning and develop strategies for particular activities; Hay and Rosamond's (2002) assertion that 'discursive constructions' can shape actors' understandings of what is possible, with tangible consequences; and Tsing's (2005) work on the role that discourse ('myth' and 'spectacle') play in building momentum for new enterprises. By probing the interactions between discourses attached to jatropha and their relationship to its materiality, I hope to contribute to a better understanding of jatropha's trajectory as well as to push the concept of flex crops in a new direction.

The chapter proceeds as follows. First, it reviews jatropha's material uses and explores how its low material flexibility has played into debates about agrofuels. Then, drawing on semi-structured interviews conducted in Kenya in 2009, it traces how actors' stated expectations for jatropha were reflected in multiple representations of the crop. Three examples of 'discursive flexibility in action' are then presented. The final sections synthesize the discussion on discursive flexibility and make a case for its importance to flex crop research.

## Jatropha and material flexibility

### *Material uses*

Liquid fuel has been the focus of most recent jatropha initiatives. When jatropha seeds are crushed, the resulting straight vegetable oil (SVO) can be burned in modified lamps and used in some diesel engines. Because of its high viscosity, further processing is needed to turn jatropha SVO into biodiesel, which can be blended with fossil fuel or used in a wider range of engines. Airplane test flights were conducted in New Zealand, the United States and Brazil from 2008 to 2010 running partly on jatropha-based fuel. Using jatropha oil to generate electricity is the goal of some projects, which seek to enable grain milling, cold storage, crop processing or non-agricultural economic opportunities. Lighting, transport, electricity and stationary engines may be subject to different incentives and link to different value chains but, arguably, they can be viewed as a bundle of pathways that all represent 'jatropha for energy' via liquid fuel.

Jatropha oil can also be made into soap. In the 1830s, jatropha seeds grown in Cape Verde were exported to Portugal and France for this purpose (Heller 1996). More recently, farmers in Mali who produce 'white' soap for commercial sale, made from jatropha oil, reportedly earned up to US$94 per year (Favretto *et al.* 2012). However, this type of soap cannot be made as a by-product of energy production, since it uses the same part of the plant. 'Black' soap, made from residues, can reduce household expenses by substituting for purchased soap, but it does not have a market in Mali (ibid.).

After jatropha seeds have been pressed to extract oil, the leftover seed cake can be used as fertilizer or compressed into briquettes for cooking fuel. The seed cake can also be fermented to produce biogas; residue from this process reportedly retains enough nutrients to also act as a fertilizer (Jongschaap *et al.* 2007). The plant's fruit coats have been tested as fuel for special stoves designed to burn sawdust. In Tanzania and Kenya, these fruit coats were found to be suitable for some purposes (fast-cooking foods like vegetables) but not others (slow-cooking foods like dried beans), largely because of the stove's design. Accessing fruit coats to use as fuel was a limiting factor in the study areas (Grimsby and Borgenvik 2013).

Jatropha acts as a living fence, since it can be planted densely from cuttings, helps reduce soil erosion and is not grazed by animals. It has a variety of medicinal uses: the oil has been used to treat skin diseases and relieve pain; the leaves processed and used as an antiseptic after childbirth as well as to treat coughs; and the sap used to treat wounds due to its antimicrobial and blood-clotting properties (Heller 1996). Orwa *et al.* (2009) add that jatropha seeds are used as a purgative and its branches as a chewing stick, while parts of the plant have also been used to treat fevers, baldness, jaundice, syphilis and haemorrhoids. Preparations from the seeds have been deployed as a rat poison, while the plant's tannins have been used for inks and dyes (ibid.)

### *How flexible is jatropha?*

Jatropha cannot easily 'flex' in the sense of producers or investors deciding which of several products to make from it. In theory, jatropha oil could be used to produce soap if not energy – but in practice, large-scale soap production is not likely to emerge as an alternative value chain. Jatropha oil could be directed toward various (energy) applications, but each energy pathway has its own infrastructure requirements, making it less likely that SVO and biodiesel would be interchanged. Some suggest that marketing by-products is essential for jatropha energy production to be economically viable without subsidies (Pipal 2012). However, diverting other parts of the plant would remove nutrients, particularly nitrogen, from the soil (Jongschaap *et al.* 2007). Rather than adding flexibility, it seems that the strategy of selling by-products could actually reduce it: if a venture's success relies on not one, but two or more dependable value chains operating simultaneously, then switching between products becomes even less likely.

Regarding multi-functionality, the literature suggests two barriers to simultaneously using different parts of the plant: first, a high level of technological or financial resources may be needed to make full use of by-products; and second, there appear to be incompatibilities of scale between producing large quantities of energy from jatropha and maintaining the cultural or regulating functions of the plant that represent additional (largely non-economic) value (van der Horst *et al.* 2014).

Recent studies further illustrate jatropha's low flexibility and its implications. Seed yields and economic returns from jatropha production have been lower than expected (GTZ 2009; Iiyama *et al.* 2013); dispossession or land tenure disputes have marred several projects (Schoneveld *et al.* 2011; Smalley and Corbera 2012); markets for jatropha seeds have failed to develop (Hunsberger 2010;

McCarthy *et al.* 2012); and, in general, poverty alleviation benefits that some hoped would result from growing jatropha have not materialized (Ariza-Montobbio *et al.* 2010). Grimsby *et al.* (2012) report that without subsidies, jatropha production in Tanzania depends on having a labour force willing to work for US$0.90 per day – a rate so low that farmers reportedly allowed poor people and children to harvest their seeds for free. Baka (2014) finds that in Tamil Nadu, India, elderly women were the only ones harvesting jatropha due to low wages and the short harvest period; younger workers said they would look for opportunities in other regions if jatropha cultivation replaced *Prosopis juliflora*, which sustained their current livelihoods. Numerous jatropha initiatives have been discontinued because of unfavourable economic prospects (Messemaker 2008; Schoneveld 2011).

While some projects have reported modest gains for some smallholders (Favretto *et al.* 2012), increased income for plantation workers (Schoneveld *et al.* 2011), or opportunities (albeit low-paid and seasonal) for elderly women (Baka 2014), overall this body of work suggests that farmers who entered into jatropha production have faced substantial risks (see also German *et al.* 2011). Without markets for other jatropha products, producers have found few ways to benefit where energy markets and infrastructure have failed to develop.

### *Inflexibility as an asset*

Despite these risks to farmers, low material flexibility helped to drive interest in jatropha. In 2007–2009, historic peaks in global food prices coincided with a period of backlash against agrofuels, from which jatropha emerged relatively unscathed. Much criticism of agrofuels at the time focused on the theme of 'food versus fuel'. In 2007, Jean Ziegler, then United Nations Special Rapporteur on the Right to Food, declared agrofuel production to be a 'crime against humanity' because of its alleged contribution to hunger (Ferrett 2007). La Via Campesina (2009) asserted that agrofuel production would divert food crops for fuel, influence which crops were planted and compete with food production for irrigation water. Agrofuel demand also reportedly drew down grain reserves, affecting the available supply of crops such as maize (Ogg 2009), while many critiqued agrofuels' impacts on food prices (Brown 2007). These arguments emphasize that agrofuels can affect three of the four elements of food security as defined by the Food and Agriculture Organization of the United Nations (2008): physical availability of food, physical and economic access to food and stability of these dimensions over time.

Jatropha appeared to circumvent these critiques in two ways: (1) as an inedible crop, it could not cause food to be directly converted into fuel; and (2) due to its reputation for growing in dry and nutrient-poor conditions, it was expected not to compete for land and water with food production. Further, some claimed that growing jatropha would improve rural food security by (3) providing a micro-climate for food crops where conditions would otherwise be too harsh; (4) generating income that could be used to purchase food; and (5) providing a local energy source to improve food processing or storage (Brittaine and Lutaladio 2010). Together these arguments helped jatropha assume 'sustainable' status.

The perception that jatropha was more ethical than other energy crops found its way into policy documents. Kenya's draft biodiesel strategy of 2008 stated that in light of food security and poverty concerns, jatropha 'stands out among others as the primary non-food biodiesel crop that will be promoted for development in the bio-diesel industry' (Government of Kenya 2008: 11). India's policy similarly encouraged non-edible oilseed cultivation on 'wastelands', implicitly favouring jatropha (Baka 2011). As policymakers have increasingly shown awareness and concern about the potential negative impacts of agrofuel production on environments and livelihoods (Hunsberger *et al.*, 2014; Bailis and Baka 2011), jatropha's 'sustainable' potential continues to be reflected in policy documents. For example, as long as jatropha is not produced on 'high-carbon land' it will probably not violate the sustainability criteria of the 2009 EU Renewable Fuel Directive, even if it is unlikely to achieve a favourable carbon balance on 'degraded' land (Romijn 2011).

Jatropha's low material flexibility positioned it as a 'sustainable' energy crop and accelerated its spread even as it placed risk on farmers and reduced options for investors. Exploring how jatropha has been discursively represented can help interpret this apparent contradiction.

## Jatropha and discursive flexibility

Actors associated with jatropha have shown remarkable flexibility in producing and maintaining discourses about the crop. Media coverage illustrates a wide range of coexisting views. Positive news reports have called it a 'wonder shrub' (Mutua 2007), 'trophy tree' (Obala 2010), 'saviour' (Cheboi 2008) and 'resource of dreams' (Thomas 2009). Negative media coverage has used equally strong language, such as 'blunder crop' (Lane 2009) and 'biofuel gone bad' (Time 2009), stressing that jatropha has fallen short of expectations.

Academic literature has questioned claims about jatropha's ability to produce abundant oil in harsh growing conditions. Researchers have expressed concerns over the financial, social and ecological risks of a large-scale experiment with jatropha, doubts that such a system would benefit farmers, scepticism that enough is known about jatropha's agronomy, and disbelief that jatropha can produce commercial quantities of oil (Achten *et al.* 2007; Brittaine and Lutaladio 2010; Jongschaap *et al.* 2007). The persistence of optimistic messages in the media despite increasingly critical research findings suggests that neither of these perspectives has become dominant over time.

The following sections summarize discourses and representations of jatropha offered by actors in Kenya, drawing on semi-structured interviews conducted from February to July 2009 in Nairobi and Coast Province. The interviews used here include three government, nine NGO, five private sector, three donor and four research representatives, and were conducted as part of a larger project comparing the claims of key jatropha actors with the experiences of smallholder farmers who were growing jatropha in two rural areas in Kenya (Hunsberger 2012). To maintain anonymity, interviewees are cited using codes that correspond to their affiliations, recognizing that some fit in more than one category.[1]

### *Motivations and expected benefits*

When asked what motivated them to work with jatropha and what they hoped it would achieve, interviewees identified potential environmental, economic, livelihood and food security benefits.

Actors in Kenya linked jatropha to a two-pronged environmental agenda. Climate change mitigation featured strongly, with several using terms such as 'clean,' 'renewable' and 'sustainable' to describe energy from jatropha. Many climate-related comments were based on fuel from jatropha releasing less carbon than fossil fuels when burned, though some also referred to the plant's ability to store carbon as it grows. Thus, actors hoped jatropha-as-energy would lower emissions while jatropha-as-plant would act as a carbon sink. A second set of environmental motivations was related to conservation. Interviewees stated that jatropha could increase forest cover, prevent soil erosion and rehabilitate 'marginal' or degraded lands. For example, a government representative stated: 'We thought that jatropha being a crop that does well in marginal areas, would help in rehabilitating some of those areas. It would play a double purpose of rehabilitating those areas as well as producing energy' (GO1).

Actors gave economic reasons for promoting jatropha that included direct gains from producing and selling energy as well as spin-off effects of making cheaper or more reliable energy available to other economic sectors. In other words, jatropha production was seen as both a productive activity in its own right and an enabler of activities in other areas. At a national level, actors expected that jatropha could displace imports of fossil-fuel diesel and kerosene, promoting a more favourable trade balance (GO1, GO2, RS2). Some mentioned the possibility of exporting to countries with favourable biodiesel policies (NGO2, NGO4), but most focused on the goal of a domestic market for jatropha products. If jatropha-based biodiesel production became established in Kenya, actors expected it to benefit the agriculture, transport and industrial sectors. Interviewees believed that jatropha would have a more stable price than fossil fuel, an attractive feature given the turbulence of world oil prices in 2008 (and again in 2011).

Actors talked of producing straight vegetable oil (SVO) as a way to support individual farmers, as well as to reach broader rural economic development objectives. Some thought that farmers could generate income by selling seeds, either through contracts with processors or informal agreements with neighbours who wanted seeds to plant. However, most thought that farmers and rural communities would benefit more from using SVO for household lighting and to support activities such as grain milling than from simply selling seeds. Several spoke of the money households could save by producing their own fuel for lighting instead of buying kerosene. Others said they wanted to see farmers increase food production by using SVO to run irrigation pumps, or increase their income by using grain mills or cold storage to add value to other crops. Actors also spoke of using SVO to generate electricity, enabling rural people to take up income-generating activities that were not previously possible.

Interviewees identified education and health benefits as livelihood improvements that they believed would accompany the use of SVO for lighting or cooking. Some described how children in families that could not regularly afford kerosene would have greater opportunity to study in the evenings if their households grew jatropha for lamp oil. Some pointed to anticipated health benefits from using jatropha oil in the home, claiming that it produces less smoke than kerosene, firewood or charcoal, and would cause fewer lung problems from indoor air pollution (GO1, DN1).

Food security was another important theme. Several actors stated that jatropha could provide energy without threatening the availability of food, using arguments introduced earlier: that it would not divert food supplies to produce fuel or take up arable land. A government representative (GO2) summarized:

> We didn't want a feedstock that would be competing with human beings for food, so we didn't want to go into staple foods like maize, to use them for biodiesel production. We also didn't want a feedstock that would compete with resources for food crop production, for example land … the high potential areas for food production, the areas that have enough rainfall, have fertile soils… Because if you plant jatropha and then it turns out to be more economical in terms of what the farmer would get by selling or producing, they might opt to produce jatropha and leave maize. That would be very risky for this country.

Others expected jatropha to enhance agricultural productivity in two ways: first, by protecting crops from being grazed or trampled by animals when grown as a fence around existing fields, (NGO7). Second, when grown in very hot, dry environments, jatropha's root system and shade could create a micro-climate enabling sesame or beans to grow between the rows (NGO4). Some used the argument that jatropha could improve households' access to food by increasing their purchasing power, either by providing income or reducing the expense of buying kerosene.

### *Multiple discourses*

To summarize, interviewees in Kenya envisioned a wide range of positive outcomes from jatropha at different scales. At a household level, some espoused providing fuel for lighting, reducing fuelwood and charcoal use, and increasing incomes. At a village level, they envisioned producing electricity for small businesses in areas not served by the national grid. Regionally, some endorsed producing jatropha commercially and selling energy to large companies or power stations. At a national level, they mentioned increasing the country's

GDP, reducing spending on foreign exchange and buffering against world oil price fluctuations. An additional goal was to increase food security through increased rural purchasing power, the expansion of agriculture into semi-arid lands and reduced damage to crops by animals. Several also spoke about international objectives related to reducing carbon emissions to combat climate change.

These perspectives reveal two main discourses that differ greatly in their scale and form. On the one hand, the desire to achieve import substitution, commercial production or major reductions in carbon emissions motivated a model of national economic development through large-scale plantations, centrally coordinated agrofuel production or international partnerships. This approach, based on delocalized markets, production and value chains, is more likely to be of interest to corporate, private investors who see potential for an economically robust sector to emerge, particularly if oil prices rise again in future. Theoretically, it fits with a (neo)liberal perspective in which private enterprise provides an engine for economic growth, which in turn stimulates 'trickle-down' benefits for the general population.

On the other hand, objectives related to poverty reduction and agricultural development through value addition motivated an approach that includes local processing to power smaller-scale applications such as decentralized electricity production, tractors and irrigation pumps. This is rural development in a broader sense, where there is room to consider the health and cost-saving benefits of replacing kerosene with jatropha for household lighting as well as crop protection on individual farms by using jatropha fences to guard against wildlife and livestock. This scenario, advocated for biofuels in general by Milder *et al.* (2008) and for jatropha specifically by Achten *et al.* (2010), is perhaps better suited to NGO-run projects that have few economies of scale, can use donor money to absorb start-up costs and are less profit-oriented than large commercial operations. This approach is broadly consistent with a capabilities perspective, which prioritizes expanding the range of options available to individuals and households so they can make choices that are beneficial and meaningful to them.

Thus, broadly competing discourses of jatropha-led development in Kenya were construed and articulated by those responsible for promoting and regulating such a process. One of these discourses advocates large-scale production as a cash crop to meet national objectives, while the other promotes small-scale cultivation to support household or farmer group objectives. That jatropha can be seen as an instrument of both market-led national economic growth and community-led rural development helps explain why it has appealed to actors with widely varying priorities. This 'duality of discourse' (Hilhorst 2001) creates an opportunity for actors to strategically choose which discourses to activate in which situations, maintaining a creative flexibility to link discourses together or to shift between them.

## Discursive flexibility in action

Examples from Kenya illustrate three apparent strategies for mobilizing discourses in flexible ways.

### *Strategy 1: shifting representations of jatropha's purpose*

Actors who changed how they presented jatropha's purpose over time provide one example of discursive flexibility. One approach for growing jatropha in Kenya has been to use it as a living support for vanilla vines. Two of the NGO representatives interviewed initially promoted jatropha for this purpose, before interest in jatropha for biofuel became widespread. One (NGO4) described how a visiting trade delegation prompted the realization that vanilla was being grown as a profitable cash crop in Uganda but not in western Kenya, which is only a short distance away and shares similar ecological conditions. The interviewee looked for a tree species that could act as a scaffold for vanilla, settling on jatropha.

A private sector interviewee (PS1) reported observing that where farmers had tried to grow vanilla and jatropha together, the vanilla vines were often dead or stunted, while the jatropha was growing reasonably well. He encouraged these farmers to change their strategy and keep the jatropha, saying:

> When I went to help these farmers sort out their vanilla-jatropha issue, and already they had planted it, put a lot of money, I told them instead of cutting, *change your perception* that you will get oil from this plant.
>
> (emphasis added)

An NGO representative (NGO2) described a similar shift in focus, stating that vanilla and jatropha could be decoupled and jatropha pursued for its own sake.

This shift in strategy may appear to be an example of material flexibility, but I argue that it is not. Changing the stated reason to grow jatropha – from a support plant to a source of fuel – without an existing market or the processing capacity to produce energy appears to be a way of keeping interest focused on the crop itself rather than on achieving successful results. It can be interpreted as an attempt to stem losses or deflect attention away from failure rather than as a choice that producers could make between viable options.

### *Strategy 2: portraying different approaches as co-dependent*

Actors who presented large- and small-scale jatropha production as mutually dependent illustrate another form of discursive flexibility. Several interviewees stated that small- and large-scale approaches to growing jatropha would stimulate and mutually reinforce each other. Some described how large investments could build processing capacity, raise interest among potential buyers and encourage more farmers to plant, thus building momentum that could, in turn, enable smaller projects. One researcher (RS4) stated that large projects, if successful, could provide examples to motivate and support small initiatives. Going further, an NGO interviewee (NGO3) said this when asked what would be the best approach to growing jatropha in Kenya:

> I would say the hybrid. Whoever is able to plant more, let him plant because until we get a critical mass we cannot talk about jatropha as a business. When we get sizable quantity, that's the time you will find probably investors bringing crushing machines, investing in an industry of transesterification and all that. But until we get to that point we cannot. At the same time, the villager there who can get a few kilos, extract oil, she can use them for local lighting, for the stove and all that. So to me, a hybrid would work very well. The commercial would assist, would be a link to the small-scale.

These comments imply that it was not only possible, but necessary that small- and large-scale production should occur side by side. Two distinct approaches – articulated as commercial production led by the corporate, private sector and livelihood-oriented production for local use – were seen as not only compatible, but interdependent. Bundling together approaches in this way creates the impression that jatropha would *necessarily* achieve a wide range of goals simultaneously, a theme to which I shall return in the discussion section.

### *Strategy 3: blurring discourses*

In contrast to local production for local use, some actors promoted an outgrower role for small-scale producers. Typically, an outgrower arrangement involves small-scale farmers producing a crop on their own land under contract for a company that also operates a nucleus plantation. This means inserting small-scale producers into a commercial value chain where they would sell jatropha seeds into a centralized system for cash. Several interviewees drew parallels between a potential jatropha industry and existing systems of coffee, tea, sugar and wheat production, presenting these as examples of how large and small farms could successfully coexist.

The outgrower production model appears to offer a different kind of 'win–win' situation than the parallel advance of commercial and local production. Those who espoused the outgrower approach described it as able to achieve multiple benefits. In terms of ecology, small plots would be preferable to large, monoculture plantations, while in terms of land tenure, farmers would keep the title to their own land – an arrangement that might appeal to smallholders, while also sparing investors from controversial negotiations over land (Pipal 2012). An outgrower system could (in theory) also achieve the economic objectives associated with large-scale, centralized production. In this way, some promoted a nucleus-outgrower production model as compatible with many goals of both large- and small-scale approaches.

For outgrowers, cultivating jatropha under contract would mean selling seeds for cash rather than generating energy to meet local needs. A donor (DN1) expressed reservations about the likelihood that an outgrower approach would benefit small-scale farmers:

> No one wants to set up a plantation where he'll have to spend so much money and receive so little yield. So what they'll opt to do is have outgrowers grow it for them, and then they'll source it from the outgrowers, and then put punitive measures to make sure they receive the maximum out of that … How long will it take for a farmer to pay off his debt? It's a very risky thing.

Actors who promoted the outgrower model portrayed jatropha as good for small farmers and good for the country, while glossing over risks that small-scale farmers would likely face. The discursive flexibility here lies in portraying one production model (the outgrower model) as achieving the aspirations of both large-scale and small-scale discourses simultaneously without acknowledging that many of the unique benefits of small-scale production for local use would no longer occur. Fears that the outgrower model would place the risks of experimenting with jatropha on small-scale farmers have been borne out by recent experiences in Zambia (German *et al.* 2011). Promoting outgrowing as an ideal approach blends large- and small-scale discourses in a way that goes beyond the 'mutually dependent' position described above. Instead of two coexisting approaches that would each achieve a discrete set of goals, this position merges them into one that could supposedly do the same. Conflating multiple goals in this way arguably makes the spread of jatropha harder to contest.

## Discussion

Much of jatropha's discursive flexibility comes from the idea that it can achieve different sets of goals depending which production model is used. The examples discussed above show actors in Kenya equating large-scale jatropha production with national and international goals, small-scale production for local use with household or community goals and outgrower systems with 'the best of both worlds'. The idea that different outcomes (and benefits for different groups of people) can be achieved by choosing between several possible production models may give the appearance of material flexibility, but is more accurately understood as discursive flexibility. Unlike the decision of which end-product to make, which maintains investors' flexibility until the last possible moment, the choice of production model must be made early on and involves considerable lock-in; it is not easy to switch among large-scale plantations, small-scale production for local use and nucleus-outgrower arrangements, once they are underway.

In the second strategy above, what I have called discursive flexibility takes the form of promoting jatropha as though it can achieve multiple objectives simultaneously through 'mutually supporting' commercial and local production strategies. This seems to be a case of discursive *plurality* and its strategic deployment. Tsing (2005) points out that 'global' projects are most effectively constructed when they roll together objectives at different scales. For jatropha, the bundling of 'global' discourses of climate change, national objectives on energy security and poverty reduction, and local interest in value addition and rural economic opportunities has produced 'thick collaboration' between actors and across scales (Hunsberger 2013).

Perhaps a conceptual parallel can be drawn between multi-functionality in the material and discursive realms. Producing and marketing products from several parts of the jatropha plant at once may seem to fit the flex crop ideal of 'diversification ... within a single crop sector' (Borras *et al.* 2012: 851) – but at a commercial scale, such multi-functionality might not add robustness because it would depend on two or more functioning value chains rather than one. Similarly, bundling together the benefits of mutually dependent small- and large-scale production sounds appealingly as though jatropha activities would be more stable and benefit more people than if a single strategy were pursued alone – but the corollary to having two strategies 'supporting' each other is that both are threatened if either one experiences problems. If this is the case, then maintaining positive discourses becomes crucially important to those with an interest in promoting jatropha, even if this task becomes detached from delivering the benefits that the discourses promise.

## Conclusions

This chapter has examined what the concept of discursive flexibility can contribute to understanding: (1) the political ecology of jatropha; and (2) flex crops. Regarding the first, I have argued that jatropha has low material flexibility and that this is a unique feature among agrofuel crops. Because few marketable products can be made from jatropha and markets for these products are not well developed, producers cannot easily shift from one value chain to another (e.g. from biodiesel to soap). Producers face higher risks as a result. For other agrofuel crops, energy production adds another layer to a political economy that already includes food, feed, commercial or industrial products.

Paradoxically, jatropha's low material flexibility helped propel it by positioning it as a 'sustainable' energy source. I argue that in terms of its ability to attract donor, investor and government support, jatropha's low material flexibility has largely been compensated by high discursive flexibility. This is because actors have been able to portray jatropha as capable of simultaneously achieving national or 'global' goals related to economic growth and climate change, and 'community' goals related to livelihoods and local ecologies. Changing the stated reason for growing jatropha in the absence of a relevant value chain, presenting large plantations and small-scale production for local use as mutually dependent and conflating various goals through an outgrower model, provide three examples of discursive flexibility.

Regarding the second objective, this analysis makes a case for expanding the focus of 'flex crop' research to include attention to discursive flexibility – particularly how strategies to deploy multiple discourses or shifts between them are used to justify, compensate for, shield or attack a crop, as well to channel benefits in specific directions. Key questions include: who has the power to mould and exercise discursive flexibility, and in what ways? Who stands to benefit (and not benefit) from resulting decisions and outcomes? While this chapter has focused on legitimating, even synergistic discourses, the interplay of clashing discourses is equally salient and offers a tempting context in which to explore power.

This analysis has reinforced the political ecology perspective that, since discourses are highly influential in defining human–environment relationships and are never neutral, they and their tangible consequences deserve to be problematized. As Hall *et al.* argue, 'struggles over resources are, simultaneously, struggles over meaning' (2012: 166). While materialist explanations of crop flexibility are important to understanding the dynamics of agrofuels, they can be further enhanced by adding an analytical layer that probes the influences of shifting and interacting discourses.

## Note

1 DN=donor; GO=government; NGO=NGO; PS= private sector; RS=researcher.

## References

Achten, W.M.J., E. Mathijs, L. Verchot, V.P. Singh, R. Aerts and B. Muys (2007) 'Jatropha biodiesel fueling sustainability?', *Biofuels Bioproducts & Biorefining-Biofpr* 1, 283–91.

Achten, W.M.J., W.H. Maes, R. Aerts, L. Verchot, A. Trabucco, E. Mathijs, V.P. Singh and B. Muys (2010) 'Jatropha: from global hype to local opportunity', *Journal of Arid Environments* 74(1), 164–165.

Ariza-Montobbio, Pere, Sharachchandra Lele, Giorgos Kallis and Joan Martinez-Alier (2010) 'The political ecology of Jatropha plantations for biodiesel in Tamil Nadu, India', *The Journal of Peasant Studies* 37(4), 875–897.

Bailis, Robert and Jennifer Baka (2011) 'Constructing sustainable biofuels: governance of the emerging biofuel economy', *Annals of the Association of American Geographers* 101(4), 827–838.

Baka, Jennifer (2011) 'Biofuels and wasteland grabbing: how India's biofuel policy is facilitating land grabs in Tamil Nadu, India', International Conference on Global Land Grabbing. Sussex.

Baka, Jennifer (2013) 'What wastelands? A critique of biofuel policy discourse in South India', *Geoforum* 54, 315–323.

Borras, Saturnino M., Jennifer C. Franco, Sergio Gómez, Cristóbal Kay and Max Spoor (2012) 'Land grabbing in Latin America and the Caribbean', *The Journal of Peasant Studies* 39(3–4), 845–872.

Brittaine, Richard and NeBambi Lutaladio (2010) 'Jatropha: a smallholder energy crop: The potential for pro-poor development', in *Integrated Crop Management*. Rome: Food and Agriculture Organization of the United Nations.

Brown, Lester (2007) 'Massive diversion of US grain to fuel cars is raising world food prices' (www.earth-policy.org/plan_b_updates/2007/update65, accessed: 2 February 2014).

Cheboi, S. (2008) '"Evil tree" turns out a saviour as fuel prices bite', *Daily Nation*, 16 February.

FAO (2008) 'An introduction to the basic concepts of food security', in *Food security information for action practical guides*. Rome: EC-FAO Food Security Programme.

FAO (2012) *FAO statistics*. Rome: Food and Agriculture Organization of the United Nations.

Favretto, Nicola, Lindsay C. Stringer and Andrew J. Dougill (2012) 'Cultivating clean energy in Mali: policy analysis and livelihood impacts of Jatropha curcas', Leeds/London: Centre for Climate Change Economics and Policy, Working Paper 84, Sustainability Research Institute Paper 28.

Ferrett, Grant (2007) 'Biofuels "crime against humanity"' (http://news.bbc.co.uk/2/hi/7065061.stm, accessed: 11 November 2013).

German, Laura, George C. Schoneveld and Davison Gumbo (2011) 'The local social and environmental impacts of smallholder-based biofuel investments in Zambia', *Ecology and Society* 16(4).

GEXSI (2008) *Global market study on Jatropha: Final Report*. London/Berlin: prepared for WWF.

Government of Kenya (2008) *Strategy for the development of the bio-diesel industry in Kenya 2008–2012*. Nairobi: Ministry of Energy: Renewable Energy Department.

Grillo, R.D. (1997) 'Discourses of development: The view from anthropology', in Grillo, R.D. and R.L. Stirrat (eds) *Discourses of development: anthropological perspectives*, Oxford and New York: Berg, 1–33.

Grimsby, Lars Kåre, Jens Bernt Aune and Fred Håkon Johnsen (2012) 'Human energy requirements in Jatropha oil production for rural electrification in Tanzania', *Energy for Sustainable Development* 16(3), 297–302.

Grimsby, Lars Kåre and Erik Johans Langfeldt Borgenvik (2013) 'Fuelling sawdust stoves with jatropha fruit coats', *Sustainable Energy Technologies and Assessments* 2, 12–18.

GTZ (2009) *Jatropha reality check: a field assessment of the agronomic and economic viability of Jatropha and other oilseed crops in Kenya*. Nairobi: GTZ.

Hall, Derek, Philip Hirsch and Tania M. Li (2012) *Powers of exclusion: land dilemmas in Southeast Asia*. Singapore and Honolulu: National University of Singapore Press and University of Hawaii Press.

Hay, I. and B. Rosamond (2002) 'Globalization, European integration and the discursive construction of economic imperatives', *Journal of European Public Policy* 9(2), 147–167.

Heller, Joachim (1996) *Physic nut. Jatropha curcas L. Promoting the conservation and use of underutilized and neglected crops*. Gatersleben/Rome: Institute of Plant Genetics and Crop Plant Research/International Plant Genetic Resources Institute.

Hilhorst, Dorothea (2001) 'Village experts and development discourse: 'progress' in a Philippine Igorot village', *Human Organization* 60(4), 401–411.

Hunsberger, Carol (2010) 'The politics of Jatropha-based biofuels in Kenya: convergence and divergence among NGOs, donors, government officials and farmers', *The Journal of Peasant Studies* 37(4), 939–962.

Hunsberger, Carol (2012) *Great aims, small gains: Jatropha-based biofuels and competing discourses of development in Kenya*. Ottawa: Carleton University, Department of Geography and Environmental Studies.

Hunsberger, Carol (2014) 'Jatropha as a biofuel crop and the economy of appearances: experiences from Kenya', *Review of African Political Economy* 41(140), 216–231.

Hunsberger, Carol, Simon Bolwig, Esteve Corbera and Felix Creutzig (2014) 'Livelihood impacts of biofuel crop production: implications for governance', *Geoforum* 54, 248–260: 10.1016/j.geoforum.2013.09.022.

Iiyama, Miyuki, David Newman, Cristel Munster, Meshack Nyabenge, Gudeta W. Sileshi, Violet Moraa, James Onchieku, Jeremias Gasper Mowo and Ramni Jamnadass (2013) 'Productivity of Jatropha curcas under smallholder farm conditions in Kenya', *Agroforestry Systems* 87(4), 729–746.

Jessop, Bob (2004) 'Critical semiotic analysis and cultural political economy', *Critical Discourse Studies* 1(2), 159–174.

Jongschaap, R.E.E., W.J. Corré, P.S. Bindraban and W.A. Brandenburg (2007) *Claims and facts on Jatropha curcas L.: Global Jatropha curcas evaluation, breeding and propagation programme*. Wageningen: Plant Research International.

La Vía Campesina (2009) 'Industrial agrofuels fuel hunger and poverty', in *The Via Campesina Notebooks*. Jakarta: La Vía Campesina.

Lane, Jim (2009) 'The blunder crop' (http://archive.is/Pfk6K, accessed: 13 November 2013).

McCarthy, John F., Jacqueline A.C. Vel and Suraya Afiff (2012) 'Trajectories of land acquisition and enclosure: development schemes, virtual land grabs, and green acquisitions in Indonesia's Outer Islands', *The Journal of Peasant Studies* 39(2), 521–549.

Messemaker, Lode (2008) 'The green myth? Assessment of the Jatropha value chain and its potential for pro-poor biofuel development in Northern Tanzania', International Development Studies, Faculty of Geosciences. Utrecht, Utrecht University.

Milder, Jeffrey C., Jeffrey A. McNeely, Seth A. Shames and Sara J. Scherr (2008) 'Biofuels and ecoagriculture: can bioenergy production enhance landscape-scale ecosystem conservation and rural livelihoods?', *International Journal of Agricultural Sustainability* 6(2), 105–121.

Mutua, Kitavi (2007) 'The wonder shrub', *The East African*. Nairobi.

Obala, Roselyn (2010) 'Trophy tree to unlock Kenya's full potential', *The Standard*, Nairobi.

Ogg, Clayton W. (2009) 'Avoiding more biofuel surprises: the fuel, food and forest tradeoffs', *Journal of Development and Agricultural Economics* 1(1), 12–17.

Orwa, C., A. Mutua, R. Kindt, R. Jamnadass and A. Simons (2009) *Agroforestree Database: a tree reference and selection guide, Version 4.0* . Nairobi: ICRAF.

Pipal (2012) *Economic viability of growing Jatropha curcas as a sustainable biofuel feedstock in East Africa: Final project report to DEG*. Nairobi: Pipal Ltd.

Romijn, Henny A. (2011) 'Land clearing and greenhouse gas emissions from Jatropha biofuels on African Miombo woodlands', *Energy Policy* 39(10), 5751–5762.

Schoneveld, George C. (2011) 'The anatomy of large-scale farmland acquisitions in Sub-Saharan Africa', Working Paper No. 85. Bogor, Indonesia: Center for International Forestry Research.

Schoneveld, George C., Laura A. German and Eric Nutakor (2011) 'Land-based investments for rural development? A grounded analysis of the local impacts of biofuel feedstock plantations in Ghana', *Ecology and Society* 16(4).

Smalley, Rebecca and Esteve Corbera (2012) 'Large-scale land deals from the inside out: findings from Kenya's Tana Delta', *The Journal of Peasant Studies* 39(3–4), 1039–1075.

Thomas, Torsten (2009) 'The resource of dreams', *Sun and Wind Energy* (5), 226–231.

*Time Magazine* (2009) 'Biofuel gone bad: Burma's atrophying jatropha', *Time Magazine*, 13 March.

Tsing, Anna (2005) *Friction: an ethnography of global connection*. Princeton: Princeton University Press.

van der Horst, D., T. Chibwe and S. Vermeylen (2014) 'The hedgification of maizescapes: scalability and multi-functionality of Jatropha curcas live fences in a smallholder farming landscape in Zambia', *Ecology and Society* 19(2), 48.

# 9 Socio-environmental conflicts and agrofuel crops

## The case of oil palm expansion in Colombia

*Victoria Marin-Burgos*

### Introduction

The aim of this chapter is to offer some empirical evidence on the complex relations among the material, socio-economic and political factors that shape the socio-environmental conflicts connected with the expansion of agrofuel crops – in particular, oil palm.

The contemporary expansion of crops that can be transformed into agrofuels in the global South, and of oil palm in particular, has been the source of much concern among civil society organizations and scholars. This expansion has often been connected with a policy-driven demand for agrofuels in the global North, especially in the European Union (EU) (Bailey 2008; Friends of the Earth 2008; Monbiot 2005; Pye 2010; Rice 2010; Wittington 2009), and is seen as a source of negative impacts in the producing countries that may result, or has already resulted, in socio-environmental conflicts (Colchester *et al.* 2006; Borras *et al.* 2010; Friends of the Earth *et al.* 2008; Mingorance 2006; Sheil *et al.* 2009; The Rainforest Foundation UK 2013; Wakker 2005; White and Dasgupta 2010). Socio-environmental conflicts are conflicts over access to natural resources and the burdens of environmental impacts, and over the protection of the material and cultural basis of people's livelihoods. These conflicts are rooted in differences in values and inequalities in power and wealth among human groups (Bryant 1998; Escobar 2006; Martínez-Alier 2002; Muradian *et al.* 2012). I argue that socio-environmental conflicts connected with the expansion of agrofuel crops involve complex interrelations among varied material, socio-economic and political factors. Agrofuel production and demand may be embedded in these relations, but may not be the only, or the main, driver of the conflicts.

An analysis of the territorialization processes connected with the expansion of agrofuel crops in specific socio-spatial contexts helps to shed light on the multiplicity of interrelated factors involved in the socio-environmental conflicts associated with this expansion. This enables us to identify the role of agrofuels in these conflicts. Colombia is a relevant case, as it is currently the fifth-largest producer of palm oil in the world and the largest in the Americas. The agribusiness of oil palm cultivation and palm oil production in Colombia started to consolidate in the 1960s.[1] However, the expansion of oil palm cultivation has been relatively large and accelerating since 2002, when Alvaro Uribe Vélez' government decided to give strong support to the use and production of agrofuels. This accelerated increase in oil palm cultivation has been accompanied by conflicts over access to land due to the attempts of palm oil producers or oil palm growers to expand oil palm plantations.

In this chapter, I explore the territorialization process in the case of the 2000–2010 expansion of the palm oil frontier in Colombia, drawing upon: (1) one case of conflict over access to land involving a local association of peasants (ASOPRODAGRO PC) and a palm oil company (Indupalma Ltda.) in the San Alberto municipality, Cesar province;[2] and (2) government support for the palm oil agroindustry during that period.

The analysis is based on data collected from various primary and secondary sources: (1) documents and secondary information produced by the palm oil agroindustry, the government, researchers and journalists reporting the history of government support for the palm oil agroindustry, in particular between 2000 and 2010; and (2) interviews, group meetings and observations obtained from fieldwork carried out between August and December 2011 in Colombia. The fieldwork was mainly based in Bogotá, the capital of Colombia, in order to interview government officials, representatives of the palm oil agroindustry and key informants. The fieldwork also involved several field visits to the San Alberto municipality in September 2011. During the field visits, I carried out group meetings (group meeting 1 on 4 September 2011; and group meeting 2 on 8 September 2011) and interviews with members of ASOPRODAGRO PC. The participants in the group meetings and interviews represented 18 of the 64 families that are part of the association. The field visits also resulted in direct observations, interviews, and one group meeting with key informants (group meeting 3), by means of which I collected information on the history of the expansion of oil palm cultivation in the San Alberto municipality, the perceived effects of this expansion in the local population and more information on the conflict between the members of ASOPRODAGRO PC and Indupalma.

This chapter is structured as follows. The next section presents the conceptual framework that forms the basis of the analysis of the case. In the third section, I describe how the palm oil frontier has expanded in Colombia between 2000 and 2010. In the fourth section, I analyse the territorialization of place in the light of the socio-environmental conflict between ASOPRODAGRO PC's members and Indupalma. In the fifth section, I analyse the political dimension of the territorialization process that facilitated the 2000–2010 expansion of oil palm cultivation. The chapter concludes with lessons learned from the role of agrofuels in the socio-environmental conflicts connected with the expansion of oil palm cultivation in Colombia.

## Conceptual framework: commodity frontiers, socio-environmental conflicts and territorialization

The increasing cultivation of agrofuel crops, such as oil palm, can be categorized as the expansion of commodity frontiers, resulting in a crop boom.

Moore has defined the expansion of commodity frontiers as the process of 'production and distribution of specific commodities, and of primary goods in particular, that have restructured geographic spaces in such a way as to require further expansion' (Moore 2000: 410). The expansion of commodity frontiers entails profound socio-ecological transformations at the sites of expansion (ibid.). Specific local geographies are socially and physically modified as the frontier expands through different mechanisms of resource access and control. As local people defend access to the resources and territories that constitute the basis of their livelihoods, such an expansion often results in socio-environmental conflicts (Martínez-Alier 2002; Muradian *et al.* 2012).

Because they are flex crops, agrofuel crops have expanded their frontiers rapidly since 2000, when policies to stimulate agrofuel production and consumption were adopted in different countries. Borras *et al.* define flex crops as 'crops that have multiple uses (food, feed, fuel, industrial material) that can be easily and flexibly inter-changed' (Borras *et al.* 2012, 851). The accelerated expansion of agrofuel crop frontiers, in particular those of oil palm, is well captured by Hall's concept of crop booms:

> crop booms may be defined as taking place when there is a rapid increase in a given area in the amount of land devoted to a given crop as a monocrop or near monocrop, and when that crop involves investment decisions that span multiple growing seasons (usually because it is a tree crop that takes some years to grow to maturity and begin producing).
>
> (Hall 2011: 840)

In the case of crop booms, land control is inherent to the frontier expansion. Peluso and Lund (2011: 668) define land control as 'the practices that fix or consolidate forms of access, claiming, and exclusion for some time'. As land control practices affect local people's access to the territories that sustain their livelihoods, agrofuel crop booms may elicit socio-environmental conflicts. Socio-environmental conflicts are one of the core themes in the field of political ecology (Martínez-Alier 2002; Leff 2012), as they result from both the socio-environmental transformation of places and asymmetrical power relations. Thus, for a better understanding of socio-environmental conflicts connected with agrofuel crop booms, the concept of power needs to be articulated together with the notion of territorialization. I understand territorialization as the control of people and resources by controlling territory and land in order to exercise political power or through the exercise of political power (Peluso and Lund 2011). This notion of territorialization is rooted in Sack's definition of human territoriality, i.e. 'the attempt to affect, influence, or control actions and interactions (of people, things, and relationships) by asserting and attempting to enforce control over a geographic area' (Sack 1983: 55). Sack's conceptualization of territoriality as a strategy of control is limited to the physical form of territorialization as control over a geographic area. In order to articulate land control practices with the power relations involved in socio-environmental conflicts, this notion of territorialization has to be broadened to include not only the control of places, but also the ability to influence structural and relational mechanisms of access to resources through political action. Drawing upon Ribot and Peluso's theory of access (Ribot and Peluso 2003), I argue that control over access to resources can be enabled by influencing the political space where mechanisms of access are moulded, i.e. the social and political practices within which territorialization is deployed. It is not only the territory as such that counts, but also the exercise of political power through or for territorialization (Cox 2003). According to Ribot and Peluso, capital, knowledge and technology, labour and access to the market are structural mechanisms of access to resources. Who is able to control access to resources through these mechanisms, and how, depends on political and economic circumstances (Ribot and Peluso 2003: 153). Therefore, influencing policy-making and implementation might be a strategy for land control, thus constituting a means of territorialization. The notion of territorialization that encompasses both physical and political dimensions is used for the empirical analysis contained in this chapter.

## Expansion of the palm oil frontier, agrofuels and socio-environmental conflicts in Colombia

The palm oil agroindustry started to consolidate in Colombia in the 1960s. Mostly national entrepreneurs started to cultivate and produce palm oil to supply the domestic market with the support of the national government (Aguilera 2002; Ospina Bozzi 2007; Vargas Tovar 2002). From the 1960s to the end of the 1980s, government support and the protection of domestic production by policies to control imports of oils and fats facilitated the establishment and growth of the national palm oil sector. From the 1960s until the end of the 1980s, oil palm cultivation and palm oil production consolidated as a flourishing agroindustry controlled by national entrepreneurs and oriented toward the domestic market for traditional uses (food, oleochemicals, soaps and animal feed). During the 1990s, the saturation of the traditional domestic market, the liberalization of the market and the lack of appropriate government support resulted in the stagnation of oil palm cultivation and palm oil production (Aguilera 2002; García Reyes 2011; Mesa Dishington 2009). However, as of 2000, Colombia became part of the global trend of the accelerated expansion of palm oil frontiers. The palm oil boom in Colombia between 2000 and 2010 meant a substantial increase in palm oil production for exports and for the biodiesel domestic and export market, and a 'crop boom' (Hall 2011), i.e. the accelerated expansion of oil palm cultivation.

Given the flex-crop nature of palm oil – that allows its transformation not only into final goods for the food, oleochemical, cosmetics and animal feed markets, but also into biodiesel – expansion of the palm oil frontier in Colombia between 2000 and 2010 was highly influenced by the global agrofuels boom that started at the beginning of the twenty-first century (see Borras *et al.* 2010; FAO 2008; Makkar 2012; and White and Dasgupta 2010 on the global agrofuels boom). However, the characteristics of the palm oil commodity chain, as well as the politics of the palm oil business in the Colombian palm oil political economy, shaped the ways in which the global agrofuels boom unfolded in Colombia. A commodity chain is understood here as 'sets of interorganizational networks clustered around one commodity or product, linking households, enterprises, and states to one another within the world-economy' (Gereffi *et al.* 1994: 2). The Colombian palm oil commodity chain is largely domestically oriented (national entrepreneurs producing as long as possible for the domestic market) and highly integrated, i.e. high levels of integration of oil palm cultivation and palm oil production, with industrial processing activities (Observatorio Agrocadenas 2006: 60; see also Table 9.1).

Moreover the commodity chain has to deal with commercialization constraints, as the traditional domestic market is saturated and there is a lack of international competitiveness (Mesa Dishington 2009). For the domestically-oriented and highly integrated Colombian palm oil sector, the agrofuels boom emerged as a twofold business opportunity to cope with the commercialization and competitiveness challenges that jeopardized the palm oil producers' capital accumulation.

First, since the Colombian President, Alvaro Uribe Vélez (2002–2010), embraced the global enthusiasm for agrofuels, his government promoted the creation of a national biodiesel market that meant the enlargement of the domestic market for palm oil (ibid.). A blending mandate and tax exemptions were adopted during his reign of government to stimulate demand.

Second, the use of vegetable oils for biodiesel production in other countries – especially in the EU – increased the demand for these commodities in the international market, thus creating market opportunities for palm oil exports (FAO 2008; Makkar 2012; MVO 2009a, 2009b and 2011).

Thus, the Colombian palm oil sector adopted a two-pronged strategy to seize the opportunities that the agrofuel/palm oil boom represented for the development of the agribusiness: (1) to support the increase in sales to the domestic market by establishing a national biodiesel market and industry; and (2) to take advantage of the available opportunities in the international market during the time required to establish the domestic biodiesel market/industry, and as long as those opportunities were profitable. This strategy is well illustrated by the words of the executive president of the National Federation of Oil Palm Growers, Fedepalma:

> Colombia has high production costs…; and we know that we will achieve competitiveness by making the work more efficient and productive, and by finding niche markets in which we can sell our products at prices that are profitable in relation to these costs. For example, with biodiesel we have the possibility of expanding our participation in the local market, which has the best prices…
>
> As long as the global market is large enough for the most competitive producers to place their products without meeting the existing demand, the market will have to pay the prices of producers like us, who produce at higher prices. Otherwise it will run the risk of being undersupplied.
>
> (own translation from Mesa Dishington 2009: 173)

*Table 9.1* Commodity chain integration of seven major palm oil producers in Colombia

| *Business/Family group (all of them are national)* | *Oil palm cultivation* | | | *Palm oil extraction and commercialization* | | | | | | *Palm oil processing for biodiesel* | |
|---|---|---|---|---|---|---|---|---|---|---|---|
| | *Own plantations* | *Productive alliances* | *Supply agreements with independent growers* | *Extraction* | | | *Commercialization* | | | *Own firm* | *With other firms' ownership* |
| | | | | *Own plant* | *Plant with other firms* | *Plant with productive alliances* | *Own firm* | *With other firms* | *With productive alliances* | | |
| Murgas-Davila family Oleoflores business group | ✓ | ✓ | ✓ | ✓ | ✓ | ✓ | ✓ | ✓ | ✓ | ✓ | |
| Haime-Gutt family Indupalma/Grasco business group | ✓ | ✓ | ✓ | ✓ | | | ✓ | | | | |
| Eder family Manuelita business group | ✓ | | ✓ | ✓ | | | ✓ | | | ✓ | |
| Dávila Abondano family Daabon business group | ✓ | ✓ | | ✓ | | | ✓ | | | | ✓ |
| Lacouture Pinedo family | ✓ | ✓ | | ✓ | | | ✓ | ✓ | | | |
| CasaLuker business group | ✓ | | ✓ | ✓ | ✓ | | ✓ | ✓ | | | ✓ |
| Espinosa family business group | ✓ | | | ✓ | ✓ | | ✓ | ✓ | | | ✓ |

Source: author's own construction.

This strategy was strongly supported by the government. A series of government measures enabled palm oil producers to establish a domestic biodiesel industry and increase oil palm cultivation and palm oil production as detailed in the fifth section below. The government resorted to existing policy tools that supported private investment, and adopted new legal instruments to stimulate agrofuel consumption and production. However, much of the government support stimulates oil palm cultivation and palm oil production regardless of its actual final use and destination (biodiesel, other oleochemical uses, food production, domestic market or exports). As a result, there was a substantial increase in exports starting in 2003 – especially to the EU – until sales to the new policy-driven domestic biodiesel market started to rise in 2008 (Fedepalma 2009, 2010, 2011).

The law that established a general biodiesel blending mandate was adopted in 2004 (Law 939, 2004), but the specific blending targets started to be gradually implemented as of 2008 (García-Romero and Calderón-Etter 2012). In the meantime, the palm oil agroindustry started to increase palm oil production and build biodiesel refineries. Palm oil production rose rapidly from 2003 until bud rot disease – an infection that often results in the death of the palms – affected about 40,000 hectares of oil palm, leading to a decrease in palm oil production in 2010 (Cabra Martínez 2010).[3] Likewise, six biodiesel refineries were built and put into operation between 2005 and 2010 (one is already out of operation) (Fedebiocombustibles 2011). The national biodiesel industry started gradually to absorb a larger share of the palm oil sales and production as of 2008. By 2010, it already absorbed about 40 per cent (Fedepalma 2011). By 2012, the share of the palm oil sales to the domestic market absorbed by the biodiesel industry was 56 per cent, while the share in the total sales (including exports) was 45 per cent (Fedepalma 2013). The five operating biodiesel refineries are totally or partially owned by national palm oil producers that are part of wealthy business groups. Some of these palm oil producers are vertically integrated with both biodiesel production and other processing businesses that also absorb palm oil produced by the same businesses groups, as shown in Table 9.1.

Thus, by 2010 a national biodiesel industry controlled by traditional domestic palm oil producers supplied a policy-driven national biodiesel market, and absorbed a significant portion of domestic palm oil production.

The increasing palm oil production for exports, and the domestic biodiesel market, resulted in an accelerated expansion of oil palm cultivation. The area cultivated almost tripled between 2000 and 2010, increasing from 157,000 hectares in 2000 to 404,000 hectares in 2010 (Fedepalma 2005 and 2011). By the end of 2012, there were 452,000 hectares under cultivation (Fedepalma 2013). The land control practices and socio-environmental transformations connected with this expansion have resulted in socio-environmental conflicts with local people who do not wish to participate in oil palm cultivation, and whose livelihoods depend on the land reclaimed for plantations, as illustrated by the case presented in the following section.

## Territorialization and socio-environmental conflicts

### *The geography of the contemporary expansion of the palm oil frontier in Colombia*

The geography of palm oil expansion is determined by the physical requirements of oil palm cultivation and palm oil production and the way in which production is organized. Oil palm cultivation and palm oil production in Colombia are organized in what the agroindustry calls 'palm oil nuclei'. A palm oil nucleus comprises an extraction plant and its supply base (oil palm plantations that supply it with oil palm fresh fruit bunches – FFB). The potential size of the supply base of a palm oil nucleus depends on the installed processing capacity of the plant (tons of FFB per hour) and the cultivation productivity (tons of FFB harvested per hectare). The larger the processing capacity and lower the productivity, the larger the supply base. The locations of the palm oil nuclei and the sites of expansion of cultivation, as well as the form this expansion takes, are determined by two requirements of the crop. First, the oil palm requires certain edaphoclimatic conditions to grow – i.e. specific soil qualities, altitude, topography, temperature, humidity, precipitation, solar radiation and hydric balance – that determine the physical location and potential expansion of palm oil nuclei. Second, palm oil has to be extracted from the FFB within 12 hours of being harvested to ensure the quality of the oil (Cenipalma and Fedepalma 2009). Therefore, palm oil extraction plants and oil palm plantations that supply them need to be located close to each other (ibid.). As a consequence, the expansion of cultivation necessarily takes place around palm oil extraction plants, forming enclaves in geographic regions with the proper edaphoclimatic conditions for the palms to grow.

The amount of land used to cultivate oil palm almost tripled between 2000 and 2010 in Colombia (Fedepalma 2005 and 2011). There are four geographical regions of oil palm cultivation and palm oil production: the northern zone, the central zone, the eastern zone and the south-western zone. Although the expansion of the frontier took place within these nature-bounded regional limits, the vastness of the regions allowed the occurrence of the crop boom represented by the remarkable expansion in hectares between 2000 and 2010. This expansion took place in the municipalities where oil palm had traditionally been grown until the end of the 1990s and it spread across new municipalities, thus shaping the new frontier. The number of municipalities with oil palm plantations more than doubled during the 2000–2010 period, from 47 in 1999 to 106 in 2010. New municipalities are either contiguous to one another, forming new enclaves, or contiguous to former municipalities, enlarging old enclaves.

As the palm oil agroindustry is essentially land-based, the physical territorialization of places suitable for oil palm cultivation around palm oil extraction plants is inherent to the expansion of the palm oil frontier. This means that palm oil producers/growers attempt to control access to land by delimiting and asserting control over geographic areas suitable for oil palm cultivation and close to palm oil extraction plants. This physical dimension of territorialization for the expansion of oil palm cultivation involves both land control practices and socio-environmental transformations. Physical territorialization limits access to lands and territories by local people who do not wish to participate in oil palm cultivation, and whose livelihoods are dependent on those lands and territories, thus leading to resource access struggles that take the form of socio-environmental conflicts, as illustrated by the case presented below.

### *Palm oil expansion and socio-environmental conflicts: the case of* **ASOPRODAGRO PC** *v.* **INDUPALMA** *in San Alberto*

ASOPRODAGRO PC is a peasant association consisting of 64 peasant families struggling against Indupalma Ltda – a large oil palm grower and palm oil producer company – in order to maintain access to plots of land they have used for a long period of time to sustain their livelihoods. The plots add up to about 250 hectares, but they are scattered over a large territory that comprises several estates in which an Indupalma's oil palm plantation is located, in the *Puerto Carreño* village in the municipality of *San Alberto*, south of *Cesar* province. The land tenure rights to the plots are contested, and the dispute started to escalate in 2002, when Indupalma started to reclaim the plots to incorporate them into one of its plantations (group meetings and interviews with ASOPRODAGRO PC members during field visits to San Alberto municipality in September 2011).

The municipality of San Alberto is not a new frontier of palm oil expansion. On the contrary, it is one of the municipalities in which oil palm cultivation started in Colombia in the 1960s. There are about 8,500 hectares of oil palm planted and a palm oil extraction plant, which are all controlled by Indupalma. San Alberto is a palm oil enclave in which the local economy is highly dependent on Indupalma's activities. According to the documents pertaining to San Alberto's history, before the 1960s there were few settlers in the area in which San Alberto is located today. The population started to grow with the establishment of Indupalma in the area, as it required a workforce for the plantations and the palm oil extraction plant (Fundesvic 2011; Murcia Navarro 2010). At present, about three-quarters of the people living in the urban zone of San Alberto (70 per cent of the population) belong to families linked to the palm oil agribusiness (San Alberto's Mayor's office, no date available).

ASOPRODAGRO PC's members are farmers who have long inhabited and used the contested lands for agricultural activities that are the basis of their livelihoods (group meetings 1 and 2, and interviews with ASOPRODAGRO PC members during field visits to the San Alberto municipality in September 2011). Most of them are without formal titles.[4] Some plots have been occupied by the same family for nearly 40 years. Others have been acquired by the current landholders from previous ones through different informal transfer arrangements such as oral negotiation or sale-letters (group meetings 1 and 2 with ASOPRODAGRO PC members).

The size of the plots varies from family to family, but they are all smallholdings ranging from 0.5 to 7 hectares. The ASOPRODAGRO PC members practise small-scale farming and the raising of livestock for subsistence and sell some of the produce in the local market. Their plots are a sort of micro-landscape that combines the family house with a mosaic of crops and livestock (observations during field visits to San Alberto municipality in September 2011). The plots are enclosed by thousands of hectares of palms, perfectly aligned and protected, with symbols meant to keep people other than Indupalma's personnel away. Such symbols include 'poisoned area' warnings and a security gate at the beginning of the public road that connects both the plantation and the plots with the main road towards San Alberto town (observations during field visits to San Alberto municipality in September 2011). The members of ASOPRODAGRO PC recount that since the 1960s, Indupalma has expanded the plantations by occupying public lands and buying from the original settlers the landholdings in public lands they had also occupied. As the plantation expanded, the original settlers of that territory were enclosed and constrained to sell, and only few resisted.

In the mid 1980s, Indupalma entered into an economic crisis that precipitated the company to the edge of closure. The closure was avoided when the company embarked on a restructuring process at the beginning of the 1990s (Fadul Ortíz 2001; Prieto Méndez 2008). Some of the initial measures taken to overcome the crisis were reducing personnel and stopping the harvest and maintenance of about 3,000 hectares of palm (Prieto Méndez 2008). As a consequence, there was no expansion of Indupalma's palm oil frontier during the 1990s, and hence the pressure of the company on the settlers located close to the plantations lessened during that period (group meeting 1 with ASOPRODAGRO PC members). Once the company started to recover from the crisis in 2000, it entered into a process of business renovation and development. Although it did not enter into the business of biodiesel production, it started to renovate and enlarge its own plantations, and expanded the palm oil frontier in neighbouring municipalities by the incorporation of new growers through contract farming arrangements (Indupalma 2011; Fadul Ortíz 2001; Murcia Navarro 2010; Prieto Méndez 2008). ASOPRODAGRO PC members recall that it was precisely as of 2002 that Indupalma's attempts to gain control over their plots intensified. Once the plantations started to be renovated, ASOPRODAGRO PC members started to come into conflict with Indupalma because some of their livestock were grazing in areas of the renewed plantations. However, Indupalma's territorialization strategy was not limited to fencing livestock out of the plantation. The company sought the control of ASOPRODAGRO PC members' plots to incorporate them into the plantation. Indupalma's means of territorialization were varied: the establishment of a security gate as mentioned above, retention of ASOPRODAGRO PC members' livestock, attempts at physical eviction (by Indupalma security personnel or with intervention by the police) and prohibiting Indupalma's personnel from buying products from ASOPRODAGRO PC members. The clashes between ASOPRODAGRO PC members and Indupalma's security personnel were frequent and escalated to the point that violent episodes occurred, such as the destruction of crops and physical attacks. Ultimately, both ASOPRODAGRO PC's members and Indupalma resorted to legal procedures from 2007 onwards. While some members of ASOPRODAGRO PC began titling processes, Indupalma filed 14 lawsuits claiming ownership of the lands. The lower court dismissed Indupalma's lawsuits in 2010. By the time of the last communication with ASOPRODAGRO PC members (June 2012), the titling processes that they had started were still pending.

### *Land tenure disputes or socio-environmental conflicts?*

At first sight, the struggles of ASOPRODAGRO PC's members against palm oil producers/growers are land tenure disputes rooted in the complex history of land occupation. A story in which land concentration, appropriation of public lands and irregular transfer of land tenure rights overlap with legitimate but unrecognized occupation by peasants, attempted eviction and institutional weakness. Land tenure disputes in Colombia are embedded in the historical agrarian structure. Inequitable distribution of access to land expressed in high levels of land concentration is a characteristic of Colombian agrarian history (Fajardo 2009). There has never been an agrarian reform and land titles are seldom seen. Small farmers gained access in the second half of the twentieth century through colonizing practices which have never been formally recognized.

However, territorialization for oil palm cultivation is not only founded on land ownership claims. The exclusionary character of the palm oil landscape underlies the territorialization processes. An industrial oil palm plantation is not compatible with the agricultural practices of peasants. In contrast to the mosaic of agricultural products and activities found on a traditional peasant farm, an oil palm plantation is a monoculture that cannot coexist with people, animals or other crops that are not functional to the development of the plantation. Therefore, local people who refuse to devote their lands to oil palm cultivation in the area of influence of a plantation have to be excluded. The same applies to animals and crops that may affect the development of the palms. Moreover, territorialization occurs at all scales (Sack 1983: 56). As palm nuclei expand, taking the form of enclaves, territorialization advances from the estate to the geographical region, thus producing considerable ecological transformations that affect the material basis of local people's livelihoods. The following quotes from members of ASOPRODAGRO PC illustrate the exclusionary character of industrial oil palm cultivation and how it transforms territories and ecologies:

> Palms live alone, I mean, any tree standing where the palms are has to be cut – native trees – so the palms can grow. Thus, I have concluded that these crops and their owners are alike, they want to extinguish everything that dwells where they are.
>
> (Group meeting 1 with members of ASOPRODAGRO PC)

> This area was rich in fish. If you threw the net there you could catch 10 or 15 fish as many times as you threw it. Now there are only water channels where those plantations are. They canalised water streams and destroyed the native forest. Only water channels were left; it is like making a water channel in the middle of a road.
>
> (Group meeting 1 with members of ASOPRODAGRO PC)

Thus, the struggles of the ASOPRODAGRO PC members over access to land connected with the expansion of the palm oil frontier are not only about land tenure rights. They are struggles to protect the places and ecologies that constitute the material and cultural basis of livelihoods, i.e. they are socio-environmental conflicts (Martinez-Alier 2002; Muradian *et al.* 2012; Escobar 2006). This case shows that the expansion of agrofuel crop frontiers in contexts of agrarian struggle may transform agrarian conflicts into socio-environmental conflicts. This corresponds with what some scholars have identified as the 'greening of the agrarian question' (Gerber and Veuthey 2010). The 'greening of the agrarian question' opens up new perspectives for empirical analysis at the intersection between political ecology and agrarian studies.

## Policy interventions as a means of territorialization

The territorialization of place connected with the contemporary expansion of the palm oil frontier in Colombia was strongly facilitated by the territorialization of political space, i.e. influencing policy-making and policy implementation to control access to resources for the development of the palm oil agroindustry.

Strong government support enabled the Colombian palm oil agroindustry to seize the business opportunities of the global agrofuel/palm oil boom. Oil palm cultivation and palm oil production have been promoted and supported by different governments since the 1960s (Aguilera 2002; García Reyes 2011). However, government support was especially strengthened during the two successive governments of Álvaro Uribe Vélez from 2002 to 2010. A wide range of instruments such as tax, price, market and agricultural incentives were adopted and implemented to support the establishment of a domestic biodiesel industry and increase oil palm cultivation and palm oil production. This government support for the development of the palm oil agroindustry, and the consequential expansion of the palm oil frontier, needs to be understood in the context of the national armed conflict and the fight against drugs in Colombia. The agrarian question was the seed for the formation of guerrilla groups in the 1960s, resulting in a violent country-wide armed conflict that evolved and became more complex as the agrarian issues merged with the incursion of the narco-business into rural areas, the emergence of illegal paramilitary forces, the participation of both guerrilla and paramilitary groups in the narco-economy and the consequent entanglement of the armed conflict with the government war against drugs (Molano Bravo 2004; Reyes Posada 2009). It is in this rural setting that the contemporary expansion of oil palm cultivation is played out. In fact, Uribe Vélez' administration's support for the palm oil agroindustry was based on the idea that oil palm cultivation could replace coca cultivation as an alternative for rural development.

There is a reciprocal constitutive relation between land access and oil palm cultivation. Controlling or gaining access to land is necessary for oil palm cultivation. At the same time, palm oil producers and oil palm growers are able to benefit from land by controlling oil palm cultivation. Therefore, structural mechanisms of access that enable the cultivation of oil palm and ensure market access for the products resulting from different processes along the commodity chain operate as means to gain, control or maintain land access. Oil palm cultivation requires access to capital, technology, knowledge and labour. Moreover, the ability to benefit from devoting land to this use and/or from controlling oil palm cultivation depends on access to markets for both the raw materials produced in the plantation (the FFB), and the palm oil extracted from them. The government support that enabled the development of the palm oil industry, and the production of biodiesel between 2002 and 2010, translated into several of these structural mechanisms of access, as shown in Table 9.2.

Most of these incentives were available not only for oil palm cultivation, and palm oil or biodiesel production, but also for other agricultural and industrial activities. However, it was the combination of them and the way they were allocated that enabled oil palm growers/palm oil producers to gain access to land for the accelerated expansion of the palm oil frontier.

The measures to stimulate the creation of a national biodiesel market enlarged the domestic demand for palm oil, benefiting mostly those palm oil producers that entered into the business of biodiesel production through vertical integration. However, as Table 9.2 shows, most of the instruments adopted and implemented by the government in favour of the palm oil agroindustry supported the agricultural process – oil palm cultivation – and palm oil production and commercialization regardless of the actual final use (biodiesel, other oleochemical uses, food production). Therefore, government support enabled palm oil producers and oil palm growers to expand oil palm cultivation even if they were not biodiesel producers, as in the case of Indupalma. Indupalma has no investments in biodiesel production. Nonetheless, the expansion of Indupalma's palm oil frontier was facilitated by the instruments to support agroindustrial activities (cultivation and palm oil production). The expansion of oil palm cultivation through 'productive alliances' was facilitated by agricultural credit and subsidies (Indupalma 2011). Furthermore, Indupalma's new oil palm extraction plant *Oro Rojo* was declared an agroindustrial tax-free zone (MCIT 2013).

The special support for the palm oil agroindustry during the Álvaro Uribe Vélez government was rooted in the ability of the agroindustry to influence policy-making and implementation. Influence on policy-making and implementation as a means of territorialization was mainly exercised through the National Federation of Oil Palm Growers – Fedepalma. Fedepalma organizes the palm oil producers and oil palm growers that represent the major share of the country's oil palm cultivation and palm oil production. Furthermore, because almost all

*Table 9.2* Instruments of government support for the palm oil agroindustry along the commodity chain and the structural mechanisms of access they represent.

| *Position in the commodity chain* | *Access mechanism* | | |
|---|---|---|---|
| | *Access to capital* | *Access to technology and knowledge* | *Access to markets* |
| Agroindustry crop production | • Income tax exemption products from oil palm plantations (10 years)<br>• Agricultural soft credit (plantation and maintenance)<br>• ICR subsidy 1999–2007<br>• ICR subsidy AIS-ICR 2007–2010<br>• Subsidy interest rate soft credit (AIS-LEC)<br>• Subsidy productive alliances<br>• Guarantee agricultural credit | • Co-funding research, development and innovation (fondo concursal ciencia y tecnología)<br>• Co-funding irrigation and drainage (subsidy)<br>• Soft credit technical assistance<br>• Subsidy technical assistance (ICR technical assistance) | |
| Agro-industry crop processing (crude palm oil extraction) | • Agro-industrial tax-free zones<br>• Agricultural soft credit | | • Subsidy to cover the fee of ERC.<br>• Price regulation (Price Stabilization Fund) |
| Industrial transformation | • Agroindustrial tax-free zones<br>• Public investment in biodiesel refineries | • Public funding research biodiesel<br>• Plan Biocom Colciencias (palm oil biodiesel) | • Price regulation to guarantee income to biodiesel producers |
| Demand | | | • Obligatory biodiesel blending targets<br>• VAT and global tax exemption for biodiesel |

Sources: author's own construction.

Acronyms

AIS = Agro Ingreso Seguro Programme (governmental programme aimed at protecting the income of agricultural producers and improving the competitiveness of agricultural activities).
ICR = Rural Capitalisation Incentive (ICR subsidy).
LEC = Special line of credit with subsidised interest rate.
ERC = Subsidy for Exchange-rate Risk Coverage.
VAT = Value added tax.

the biodiesel refineries are partially or totally owned by palm oil producers, Fedepalma is a member of the Biofuels National Federation (Fedebiocombustibles), and the executive president of Fedepalma and some palm oil producers are members of Fedebiocombustibles' board of directors. Fedepalma has had substantial influence on the adoption of government incentives and policies that have supported the development of both the palm oil agroindustry and the agrofuels sector.

Among Fedepalma's achievements is the establishment of the Price Stabilization Fund in 1996 and the applicability of the Rural Capitalization Incentive (ICR subsidy) to the planting and maintaining of late-maturing crops in 1999. Mesa Dishington (2009) provides a detailed historical account of Fedepalma's efforts to influence policy-making and its accomplishments. However, it was during Álvaro Uribe Vélez' government that Fedepalma achieved unprecedented government support for the palm oil agroindustry. Fedepalma contributed to the formulation of the law that contains both the general biodiesel mandate and the ten-year tax exemption for the income derived from oil palm cultivation (Law 939, 2004). It also contributed to the formulation of the policy to promote agrofuel production in Colombia adopted in 2008 (CONPES 2008) and provided the guidelines for what became the public policy to support the development of the palm oil sector in 2007 (CONPES 2007). It was also thanks to Fedepalma that special lines of government agricultural credit and subsidies were redefined to make them suitable for financing the establishment and maintenance of plantations with late-maturing crops. Fedepalma also influenced the design and implementation of a programme through which the government allocated funds for subsidizing palm oil producers' acquisition of exchange-rate risk coverage instruments (Mesa Dishington 2009).

## Conclusion

Local socio-environmental conflicts connected with the expansion of energy crops – such as oil palm – in the global South are often considered to be a direct consequence of the emergence of a policy-driven agrofuel market on the global scale. In particular, they are often connected to increased agrofuel production and demand in the global North. Drawing on the contemporary expansion of the palm oil frontier in Colombia and the case of a land access conflict connected with it, this chapter shows that such socio-environmental conflicts can only be understood in the light of complex interrelations among material, socio-economic and political factors operating at different levels. Processes of the territorialization of both place and political space shed light on those interrelations. In the case of Colombia, a policy-driven domestic agrofuel market was the vehicle through which the national palm oil sector materialized a territorialization project for capital accumulation.

The national agrofuels boom and the resulting expansion of the palm oil frontier that underlaid socio-environmental conflicts were manifestations of territorialization processes dependent on local geographies, the material characteristics of the crop and socio-economic and political processes. Thus, looking at agrofuels as direct causes of socio-environmental conflicts overlooks the complex interrelation of the multiple factors involved in the expansion of the frontiers at the basis of such conflicts.

## Notes

1 I use the term oil palm to refer to the tree and differentiate it from the oil extracted from its fruits, i.e. the palm oil. The difference is relevant because not all oil palm growers are palm oil producers. However, when I refer to the palm oil industry or agroindustry, it includes both palm growers and palm oil producers.

2 ASOPRODAGRO PC means Asociación de Productores Agropecuarios y Agrícolas del Corregimiento de Puerto Carreño (Association of agriculture and livestock producers from *Puerto Carreño* village).

3 Although the disease started in 2006 in the south-western region of oil palm cultivation, it spread to other regions in 2009. The accumulated loss of oil palm hectares reflected in palm oil production decreased in 2010. By 2010, the number of hectares affected by the disease amounted to 40,000.

4 In group meeting 1 with ASOPRODAGRO PC members, they stated that only five families had formal titles.

## References

Aguilera, María M. (2002) *Palma Africana en la Costa Caribe: Un semillero de empresas solidarias*, Documentos de Trabajo Sobre Economía Regional 31, Cartagena, Colombia: Banco de la República.

Bailey, Robert (2008) *Another inconvenient truth: how biofuel policies are deepening poverty and accelerating climate change.* Oxford, UK: Oxfam International.

Borras Jr., Saturnino M., Philip McMichael and Ian Scoones (2010) 'The politics of biofuels, land and agrarian change: editors' introduction', *The Journal of Peasant Studies* 37(4), 57–592.

Borras Jr., Saturnino M., Jennifer C. Franco, Sergio Gómez, Cristóbal Kay, and Max Spoor (2012) 'Land grabbing in Latin America and the Caribbean', *The Journal of Peasant Studies* 39(3–4), 845–872.

Bryant, Raymond L. (1998) 'Power, knowledge and political ecology in the third world: a review', *Physical Geography* 22(1), 79–94.

Cabra Martínez, Jorge A. (2010) 'Avances en el Manejo Sanitario de la Palma de aceite en Colombia: organización, tecnología, aplicación, gestión de recursos', power point presentation at the meeting of the National Federation of Oil Palm Growers-Fedepalma held in Cali-Colombia on June 8–10, 2011 (http://portal.fedepalma.org//congreso/2011/presentaciones.htm, accessed: 23 August 2013).

Cenipalma-Centro Corporación de Investigaciónde la Palma de Aceite and Fedepalma-Federación Nacional de Cultivadores de Palma de Aceite (2009) *Agenda Prospectiva de Investigación y Desarrollo Tecnológico para la Cadena Productiva de Palma de Aceite en Colombia con Énfasis en Oleína Roja.* Bogotá, Colombia: Cenipalma, Fedepalma.

Colchester, Marcus, Norman Jiwan, Andiko, Martua Sirait, Asep Yunan Firdaus, A. Surambo and Herbert Pane (2006) *Promised land: palm oil and land acquisition in Indonesia: implications for local communities and indigenous peoples.* Moreton-in-Marsh, UK, and Bogor, Indonesia: Forest Peoples Programme, SawitWatch, HuMA and ICRAF.

CONPES (Consejo Nacional de Política Económica y Social) (2007) *CONPES 3477 – Estrategia para el Desarrollo Competitivo del Sector Palmero Colombiano*. Bogotá, Colombia: Departamento Nacional de Planeación.

CONPES (Consejo Nacional de Política Económica y Social) (2008) *CONPES 3510 – Lineamientos de Política Para Promover la Producción Sostenible de Biocombustibles en Colombia*. Bogotá, Colombia: Departamento Nacional de Planeación.

Cox, Kevin (2003) 'Political geography and the territorial', *Political Geography* 22(6), 607–610.

Escobar, Arturo (2006) 'Difference and conflict in the struggle over natural resources: a political ecology framework', *Development* 49(3), 6–13.

Fadul Ortíz, Miguel (2001) *Informe Alianzas por la paz: el modelo Indupalma*. Bogotá, Colombia: Indupalma.

Fajardo, Darío A. (2009) *Territorios de la agricultura colombiana*. Bogotá, Colombia: Departamento de Publicaciones Universidad Externado de Colombia.

FAO (2008) *The state of food and agriculture 2008. Biofuels: prospects, risks and opportunities*. Rome: Food and Agriculture Organization.

Fedebiocombustibles (Federación Nacional de Biocombustibles de Colombia) (2011) 'Cifras Informativas del Sector Biocombustibles – Biodiesel de Palma de Aceite' (www.fedebiocombustibles.com/files/Cifras%20Informativas%20del%20Sector%20 Biocombustibles%20-%20BIODIESEL(13).pdf, accessed: 4 July 2013).

Fedepalma (Federación Nacional de Cultivadores de Palma de Aceite) (2001) 'Geografía Palmera' (http://portal.fedepalma.org/GeoPalma/index.htm., last access: 4 July 2013).

Fedepalma (Federación Nacional de Cultivadores de Palma de Aceite) (2005) *Statistical yearbook 2005*. Bogotá, Colombia: Fedepalma.

Fedepalma (Federación Nacional de Cultivadores de Palma de Aceite (2009) *Statistical yearbook 2009*. Bogotá, Colombia: Fedepalma.

Fedepalma (Federación Nacional de Cultivadores de Palma de Aceite) (2010) *Statistical yearbook 2010*. Bogotá, Colombia: Fedepalma.

Fedepalma (Federación Nacional de Cultivadores de Palma de Aceite) (2011) *Statistical yearbook 2011*. Bogotá, Colombia: Fedepalma.

Fedepalma (Federación Nacional de Cultivadores de Palma de Aceite) (2013) *Boletín Económico Marzo de 2013*. Bogotá, Colombia: Fedepalma (http://web.fedepalma.org/sites/default/files/files/Balance%202012%204Q%281%29.pdf, accessed: 27 November 2013).

Friends of the Earth (2008) *Fuelling destruction in Latin America, the real price of the drive for agrofuels*. Amsterdam, The Netherlands: Friends of the Earth International.

Friends of the Earth, LifeMosaic and Sawit Watch (2008) *Losing ground: the human rights impacts of oil palm plantation expansion in Indonesia*. London, Edinburgh and Bogor: Friends of the Earth, LifeMosaic and Sawit Watch.

Fundesvic (2011) *Las Familias Trabajadoras de la Palma Contamos nuestra Historia. Memoria de las víctimas del Sur del Cesar. Cartilla No. 1: "…y empezó nuestra sueño"*. Bucaramanga, Colombia: Fundesvic, Asociación Minga, Sintraproaceites.

García Reyes, Paola (2011) 'La Paz Perdida. Territorios colectivos, palma africana y conflicto armado en el Pacífico colombiano', Masters thesis, México: FLACSO.

García-Romero, Helena and Laura Calderón-Etter (2012) *Evaluación de la política de Biocombustibles en Colombia*. Bogotá, Colombia: Fedesarrollo (www.fedesarrollo.org.co/wp-content/uploads/2011/08/Evaluaci%C3%B3n-de-la-pol%C3%ADtica-de-Biocombustibles-en-Colombia.pdf, accessed: 20 May 2013).

Gerber, Julien F. and Sandra Veuthey (2010) 'Plantations, resistance and the greening of the agrarian question in coastal Ecuador', *Journal of Agrarian Change* 10(4), 455–481.

Gereffi, Gary, Miguel Korzeniewicz and Roberto P. Korzeniewicz (1994): 'Introduction: global commodity Chains', in Gereffi, Gary and Miguel Kozeniewicz (eds) *Commodity chains and global capitalism*. Westport, CT: Greenwood Press, 1–14.

Hall, Derek (2011) 'Land grabs, land control, and Southeast Asian crop booms', *The Journal of Peasant Studies* 38(4), 837–857.

Indupalma (2011) *Informe de Sostenibilidad 2010–2011*. Bogotá, Colombia: Indupalma (www.unglobalcompact.org/system/attachments/15386/original/Informe_Sostenibilidad_INDUPALMA_webpage.pdf?1337024883, accessed: 12 September 2013).

Leff, Enrique (2012) 'Political ecology: a Latin American perspective', in UNESCO-EOLSS Joint Committee (eds) *Encyclopedia of Life Support Systems (EOLSS) – Encyclopedia of Social Sciences and Humanities – Culture, Civilization and Human Society*. Oxford, UK: UNESCO, Eolss Publishers.

Makkar, Harinder P.S. (ed.) (2012) *Biofuel co-products as livestock feed – Opportunities and challenges*. Rome: Food and Agriculture Organization.

Martínez-Alier, Joan (2002) *The Environmentalism of the Poor*. North Hampton, NH: Edward Elgar Publishing.

MCIT (Ministerio de Comercio, Industria y Turismo) (2013) *Informe de Zonas Francas Aprobadas*. Bogotá, Colombia: MCIT (www.mincomercio.gov.co/minindustria/publicaciones.php?id=168, accessed: 31 May 2013).

Mesa Dishington, Jens (2009) *Lo gremial, pilar del desarrollo palmero*. Bogotá, Colombia: Fedepalma.

Mingorance, Fidel (2006) *The flow of palm oil Colombia – Belgium/Europe A study from a human rights perspective*. Bogotá, Colombia: HREV (Human Rights Everywhere) for Coordination Belge pours la Colombie (http://cbc.collectifs.net/doc/informe_en_v3–1.pdf, accessed: 18 April 2011).

Molano Bravo, Alfredo (2004) 'Coca, land and corruption', in Baud, Michiel and Donny Meertens (eds) *Colombia from the Inside: Perspectives on Drugs, War and Peace*. Amsterdam: CEDLA, 65–78.

Monbiot, George (2005) 'Worse Than Fossil Fuel', *Guardian* (www.monbiot.com/2005/12/06/worse-than-fossil-fuel/, accessed: 21 August 2013).

Moore, Jason W. (2000) 'Sugar and the expansion of the early modern world-economy: commodity frontiers, ecological transformation, and industrialization', *Rev. Fernand Braudel Cent*. 23, 409–433.

Muradian, Roldan, Mariana Walter, Joan Martinez-Alier (2012) 'Hegemonic transitions and global shifts in social metabolism: implications for resource-rich countries. Introduction to the special section', *Global Environmental Change* 22(3), 559–567.

Murcia Navarro, Andrés (2010) *Informe de Consultoría*. Bogotá, Colombia: Indupalma.

MVO (Product Board for Margarine, Fats and Oils) (2009a) *Fact sheet Rapeseed 2008*. Rijswijk, The Netherlands: MVO.

MVO (Product Board for Margarine, Fats and Oils) (2009b): *Market Analysis Oils and Fats for Fuel 2009*. Rijswijk, The Netherlands: MVO.

MVO (Product Board for Margarine, Fats and Oils) (2011): *Fact sheet Palm Oil 2010*. Rijswijk. The Netherlands: MVO.

Observatorio Agrocadenas (2006) *Agroindustria y competitividad: Estructura y dinámica en Colombia 1992–2005*. Bogotá, Colombia: Ministerio de Agricultura y Desarrollo Rural – IICA.

Ospina Bozzi, Martha L. (2007) *The Faces of Oil Palm*. Bogotá, Colombia: Fedepalma.

Peluso, Nancy L. and Christian Lund (2011) 'New frontiers of land control: introduction', *The Journal of Peasant Studies* 38(4), 667–681.

Prieto Méndez, Andrés (2008) *Informe el Modelo de Gestión de Indupalma.* Bogotá, Colombia: Indupalma.

Pye, Oliver (2010) 'The biofuel connection – transnational activism and the palm oil boom', *The Journal of Peasant Studies* 37(4), 851–874.

Reyes Posada, Alejandro (2009) *Guerreros y Campesinos, el despojo de la tierra en Colombia.* Bogotá, Colombia: Grupo Editorial Norma.

Ribot, Jesse C. and Nancy L. Peluso (2003) 'A theory of access', *Rural Sociology* 66(2), 153–181.

Rice, Tim (2010) *Meals per gallon: The impact of industrial biofuels on people and global hunger.* London: ActionAid UK.

Sack, Robert D. (1983) 'Human territoriality: a theory', *Annals of the Association of American Geographers* 73(1), 55–74.

San Alberto's Mayor's office (no date available) 'San Alberto – Cesar, Puerta de Oro del Sur del Cesar' (http://sites.google.com/site/datosdesanalbertocesar/, last access 12 September 2013).

Sheil, Douglas, Anne Casson, Erik Meijaard, Meine van Nordwijk, Joanne Gaskell, Jacqui Sunderland-Groves, Karah Wertz and Markku Kanninen (2009) 'The impacts and opportunities of oil palm in Southeast Asia: what do we know and what do we need to know?. Occasional paper no. 51', Bogor, Indonesia: CIFOR.

The Rainforest Foundation UK (2013) *The Seeds of Destruction – Expansion of Industrial Oil Palm in the Congo Basin*, London: The Rainforest Foundation UK.

Vargas Tovar, Ernesto (2002) 'Forum: the 1960s and 1970s oil palm: from farm to enterprises', *Palmas* 23(3), 86–92.

Wakker, Eric (2005) *Greasy palms: the social and ecological impacts of large-scale oil palm plantation development in Southeast Asia.* England, Wales and Northern Ireland: Friends of the Earth.

White, Ben and Anirban Dasgupta (2010) 'Agrofuels capitalism: a view from political economy', *The Journal of Peasant Studies* 37(4), 593–607.

Wittington, Eliot (2009) *Growing Pains: The Possibilities and Problems with Biofuels.* Christian Aid, UK.

# 10 Green grabbing – the case of palm oil expansion in so-called degraded areas in the eastern Brazilian Amazon

*Maria Backhouse*

Agrofuels have increasingly come under criticism in recent years. This criticism has not yet led to the cessation of state support, however. Instead, political decision-makers are presently discussing strategies to limit the potentially negative effects of agrofuel production. One strategy is to plant the crops for agrofuels in so-called degraded areas in order to prevent a shortage of foodstuffs or the deforestation of primary forests via direct or indirect changes in land use and competition for land. The World Bank estimates the area of 'currently uncultivated, nonforested land that would be ecologically suitable for rainfed cultivation in areas with less than 25 persons/square kilometer ($km^2$)' at approx. 445 million hectares worldwide (Deininger *et al.* 2011: 77). Even Brazil, one of the world's largest producers and consumers of agrofuels, is following this strategy. Since 2010, the Brazilian government has supported the production of palm oil in the Amazon basin, among other things for biodiesel production. The main cultivation areas are to be areas which are already deforested, preferably so-called degraded grazing land in the Amazonian state of Pará (EMBRAPA and MAPA 2010).

From the critical perspective of the political ecology of agrofuels, however, there can be no generally valid definition of so-called degraded areas. Areas which development strategists evaluated as degradated can, in fact, be the basis for the livelihoods of peasants or traditional local communities. (Blaikie and Brookfield 1987; Nalepa and Bauer 2012). The agroindustrial change in the use of so-called degraded areas can therefore strengthen the processes of appropriation and displacement of peasants in favour of agribusinesses which have been taking place for decades (Borras and Franco 2012; Nalepa and Bauer 2012). In this chapter, I argue on the basis of a case study[1] on the expansion of palm oil in the Brazilian state of Pará, that palm oil production in so-called degraded areas is a form of *green grabbing*.

The term *green grabbing* was first used by John Vidal (2008) and later discussed in a special issue of *The Journal for Peasant Studies* as a new capitalist form of the appropriation of nature (Fairhead *et al.* 2012). According to the authors, the term 'appropriation' means 'the transfer of ownership, use rights and control over resources that were once publicly or privately owned – or not even the subject of ownership – from the poor (or everyone including the poor) into the hands of the powerful' (ibid.: 238). The expression 'green grabbing' is used for the worldwide processes of appropriation and valuation of natural resources (such as the trade in $CO_2$ certificates) for environmental ends (ibid.). Green grabbing differs from simple land grabbing in that it is initiated by (national or transnational) environmental or climate protection measures – for example via the agroindustrial opening of allegedly degraded land. Environmental and climate policy objectives thus not only serve as 'green' legitimation strategies for land grabbing, but can themselves lead to processes of displacement or the loss of control over land access and land use as a result of specific political stimuli, for example support for the production of agrofuels. Green grabbing is also characterized by the fact that it is linked to new alliances of actors among the private sector, the state and NGOs, as well as by specific legitimation strategies and practices (ibid.: 239). Green grabbing involves not only the material process of appropriation, however, but also a specific 'discursive framing' (ibid.: 241). The creation and commodification of natural resources such as $CO_2$ and 'biofuels' and the production of agreement to this market-based form of environmental protection across all the political camps can only be understood with this additional focus on the scientific, political and everyday production of knowledge and new truths.

The thesis that palm oil expansion in Brazil represents green grabbing will be substantiated in this chapter in the following way: first, I will define green grabbing as the expression of continuous primitive accumulation in order to diagnose more precisely the analytical substance of green grabbing. I argue somewhat differently to Fairhead *et al.*, according to whom green grabbing can be an expression of primitive accumulation but is not necessarily such (Fairhead *et al.* 2012: 238). Continuous primitive accumulation describes the establishment or restructuring of capitalist relationships of ownership and production for the appropriation of surplus value or expanded reproduction from a Marxist perspective. On the basis of a re-interpretation of primitive accumulation, I shall identify the material, political and discursive dimensions of the flexible analysis concept of green grabbing. The analysis of the expansion of the oil palm plantations in northern Pará as a process of green grabbing follows in the third section. I proceed here from the history of the oil palm in Pará and then analyse the material, political and discursive dimensions of the current agroindustrial expansion. Taking the example of the powerful narrative of the degraded Amazon areas together with the state supported expansion of palm oil in Brazil, I will show how a natural resource is produced and legitimized for its agroindustrial development, and how societal agreement to this is produced and resistance made difficult. Finally, the results will be summarized.

## Analysis of green grabbing

I agree with the diagnosis of critical political ecology that the present socio-ecological symptoms of crisis such as climate change, the energy crisis and the agricultural crisis are expressions of a multiple, capitalist crisis of society

(Demirović *et al.* 2011). Primitive accumulation – understood as the establishment or restructuring of capitalist relationships of ownership and production – *can* be a strategy for dealing with this by developing new fields of accumulation (e.g. arable land) or creating them (e.g. emissions trade), by means of dispossession, enclosures or privatization. In this way the socio-ecological crisis offers capital the opportunity to place superfluous capital from other areas (e.g. the finance sector) in these fields (Zeller 2010: 103). Thus, I understand green grabbing as a central form of this crisis management.

I understand primitive accumulation (Marx 1962 [1867]: 741–791) not as a unique historical phase during the emergence of capitalism, but as an 'inherent and continuous element of modern societies' (De Angelis 2001: 3). Massimo de Angelis' interpretation of primitive accumulation as a continuous element of capitalism provides two important impulses for the more precise development of the analytical concept of green grabbing. He places the analytical focus on the process of the *separation* of the producers and means of production (De Angelis 2001: 7). In doing so, he makes an important differentiation: in contrast to the processes of separation within simple capital accumulation, in the case of primitive accumulation we are talking about the *original* creation of capitalist relationships of production and ownership or their extensive restructuring within capitalism (De Angelis 2001: 8f.) This original or repeated separation is set in motion by *extra-economic* means such as state intervention, a renewal of the legislative framework or direct violence. It is decisive that it is not the methods of separation such as robbery or enclosure that are characteristic of continuous primitive accumulation, but the creation or restructuring of capitalist social *relations*.

De Angelis, by describing the processes of separation as contested and in their historical course as contingent, also makes clear the political dimensions of the separation, or of continuous primitive accumulation (De Angelis 2001: 16). Applied to green grabbing, this means that whether in a particular region the agroindustrial production of agrofuel crops prevails against other forms of land use, also depends on the actors, their alliances and conflicts. This opens up a field of social conflicts: with this analytical focus on the actors and their resistance, the process of separation also has a subversive, emancipatory potential.

The means of separation are not necessarily based on violence. If the restructuring processes are convincingly legitimized across the social classes, i.e. have become hegemonial in a Gramscian sense, green grabbing can also take place ostensibly peacefully (cf. Kelly 2011: 692). In order to be able to grasp the discursive dimension of green grabbing, I supplement the ideology-critical dimension of Marx' concept of primitive accumulation by Hall's ideology-theoretical deliberations. The ideology-critical dimension in Marx is indicated, first, in his criticism of bourgeois economics (Hymer 1971; Perelman 1983; Kalmring 2013), in which he destroys the foundations of the bourgeois self-conception (liberty, equality, fraternity) with his descriptions of the violent development of capitalism. Second, Marx describes primitive accumulation as a disciplining process through which the industrial proletariat is created (Negt and Kluge 1983).

Following Hall's ideology-critical deliberations, we can take these two moments together and ask with what narratives and practices dominant ideologies are produced and green grabbing in so-called degraded areas is legitimized as 'right' and 'without alternative' even among the subjugated classes (Hall 1981, 1988). Hall understands ideologies not as false consciousness or intended manipulation on the part of the ruling classes, but in the Gramscian sense as *hegemonial* viewpoints. Although these are entwined with societal relationships of power, they are, in principle, contested and can be questioned or undermined. According to Hall, it is decisive that although ideologies cannot be derived from the economic basis or specific class positions, they cannot be understood independently of the material relations and powerful institutions which (re-)produce them. In addition, with his concept of articulation he developed a useful analysis tool for examining ideologies as the articulation or entwining of different elements to a specific chain of meaning (Hall 1981). It can thus be examined in what way concepts such as 'degraded areas' are positioned in different viewpoints, and ideas which are taken for granted, truths, and identifying or knowledge positions are produced and questioned by individual or collective subjects.

Proceeding from the above, *green grabbing* is when, in connection with strategies for dealing with the socio-ecological crisis, the control over land access and land use is concentrated on local elites or transnational enterprises, and peasant relations of ownership and labour are *restructured* in this process for expanded reproduction or the appropriation of surplus value. The restructuring processes, in the sense of continuous primitive accumulation, can manifest themselves in the integration of peasants into the palm oil sector, in their displacement from their land through either its purchase or new legal forms, or in their being expelled by force or the threat of force.

Analysing the palm oil expansion in Pará as a process of green grabbing therefore requires the contextualization of the regionally specific, historical relationships of land access and land use. Only in this way can we determine whether a restructuring of the relations of land access and land use, the criterion for green grabbing, has taken place. In addition, against the background of the above considerations, it includes three analytical dimensions: the material (relations of land ownership and land access), the political (power, class and other societal relations) and the discursive (ideologies, opinions, legitimation strategies and legitimation practices). These three analytical dimensions are inseparably entwined and can only be differentiated analytically. The following questions relate to them.

On the material analysis level we ask: how do palm oil investments change control over the relations of land access and land use? Who profits from this? On the political level we ask: how are power relations distributed? Who positions themselves, and how, to the changes? Is there resistance to the growing oil palm plantations? If so, from whom? And how is this articulated? On the discursive level we ask: how is green grabbing defined and justified? How is it connected with measures to deal with the socio-ecological crisis? Who has the right to define which areas are degraded? How is acceptance campaigned for in the various population groups? How are new truths created? How is contradiction silenced?

## Palm oil in Pará/Brazil

The African oil palm (*Elaeis guineensis*) was brought to Brazil by African slaves 400 years ago (Watkins 2011). It was introduced into the Amazon basin for the first time by researchers in 1942 and from the 1970s onward was promoted by the military dictatorship as part of its Amazon development policy (Silva *et al.* 2011; Furlan Junior *et al.* 2006). The target groups of the financial incentives under this development policy were the enterprises and banks which established the first plantations. As in numerous other large projects, land robbery and violent land conflicts took place (Acevedo Marin 2010). Already in 1980, the promotion of biodiesel on the basis of palm oil was planned in imitation of the Proalcool programme.[2] However, due to technical problems and falling oil prices, this plan ceased to be pursued at the end of the 1980s and the palm oil project in the Amazon basin was regarded (for the time being) as a failure (Homma and Furlan Junior 2001). In 2009, the plantation areas in Pará covered not quite 50,000 hectares, while, at roughly the same time, the soya fields grew to over eight million hectares (IBGE 2009).

Between 2002 and 2006 at the federal state level, pilot projects[3] were initiated in Pará to integrate peasants into the palm oil sector by means of farming contracts. Comprehensive state support for the Brazilian palm oil sector did not begin until 2010, however, with the state programme for sustainable palm oil production (*Programa de Produção Sustentável da Palma de Óleo*). In this programme the Brazilian government combined national energy and development policy goals with international climate policy targets and strategies for dealing with the socio-ecological crisis:

- *Economic policy:* in the short term, national requirements for foodstuffs and cosmetics are to be met, of which at present more than 50 per cent are met by imports.
- *Energy policy:* in the medium term, the basis for biodiesel is to be diversified. At present it is based to about 80 per cent on soya oil, a waste product of the animal feed export industry (USDA 2012).
- *Development policy:* the labour-intensive oil palm plantations are to create work and market access for small contract farmers and in this way contribute to the development of impoverished rural regions. Similarly to the national biodiesel programme,[4] a social seal is to guarantee the inclusion of peasant agriculture. If enterprises commit themselves to buying 15 per cent of their palm oil from the peasants, this seal will entitle them to preferential sales conditions on the national biodiesel market and free them from taxes.
- *Climate policy:* via the establishment of agro-ecological zones, it is to be

> ensured that only areas which were deforested before 2008 will be transformed into oil palm plantations and that no environmentally protected areas or territories of traditional communities will be endangered. This strategy of $CO_2$ avoidance through the prevention of deforestation and the agricultural use of already altered lands is intended to contribute to Brazil's voluntary climate protection goals (Brazilian Government 2010). Approximately 31.8 million hectares, which have already been transformed in the Amazon basin and the rainy coastal areas of northeast Brazil have, accordingly, been zoned as suitable for oil palm planting (EMBRAPA and MAPA 2010).

The main cultivation area is the northeast of the federal state of Pará. According to the Brazilian Agricultural Research Corporation EMBRAPA (*Empresa Brasileira de Pesquisa Agropecuária*) there are about 5.5 million hectares suitable for oil palm plantations in a contiguous region of 44 municipalities. So-called degraded grazing land has been given priority. In favour of the northeast region of Pará as the 'Brazilian palm oil centre' is also its excellent location near the state capital and port of Belém, as well as its greater distance to the biodiversity hot spots of the Amazon basin (Muller *et al.* 2006).

## Green grabbing in Pará

Since the adoption of the programme the area covered by oil palm plantations in Pará increased almost threefold to 140,000 hectares in 2013 (Glass 2013: 5). By mid-2012, a total of 649 peasant families had signed palm oil production contracts with the palm oil companies (MDA 2012b). Transnational enterprises such as the state energy company Petrobras, the mining company Vale and the US concern ADM (Archer Daniels Midland Company) invest in plantations and processing facilities and compete with local palm oil companies (e.g. Agropalma, Dendê-Tauá and Marborges) for land, plantation workers and potential contract farmers.

The current expansion of agroindustrial palm oil production is having an increasing effect on the landscape. Grid-patterned plantations are being established, with up to 10,000 hectares per unit – known as *pólos* in Portuguese – and with processing mills being constructed at their centres. The contract farmers' plots are scattered around the plantations within a radius of no more than 30 kilometres. Since the fruits must be processed within 24 hours of being harvested, a high logistical effort is required. This includes the tight organization of work on the plantations, a suitable transport infrastructure (roads, rivers, bridges, ports) and processing in the neighbourhood of the plantations (see Marín-Burgos, Chapter 9 and Dietz *et al.*, Chapter 3 in this volume).

The major driver of the current oil palm expansion in the region is Vale, the second-largest mining company in the world. This Brazilian enterprise explains its entry into the palm oil sector with its strategy of establishing itself on a global scale as a sustainable actor and of producing renewable energy for its own needs. By 2020, the fuel blend for its vehicle fleet is to contain 20 per cent biodiesel.

By 2013, Vale had established 40,000 hectares of oil palm plantations in Pará (USDA 2013). The production of agrodiesel is to begin in 2014 and will be expanded annually. Company employees estimate that the area farmed by the company will increase in the coming decades to four million hectares, with 15 per cent of production being obtained from small contract farmers.[5] This expansion is to be in line with sustainability. In a press release, Vale assures us that: 'All areas used to grow palm trees have been mapped and classified by the federal government as degraded areas' (Vale 2011).

### *Restructuring the relations of land access and land use*

The case-study region of the five contiguous municipalities of Moju, Acará, Bujaru, Tomé-Açu and Concórdia is by no means characterized simply by extensive grazing land, but also by heterogeneous peasant agriculture. It is one of the most densely-populated Amazon regions. According to census data, in 2010 the population of the case-study region was 234,016 in an area of 20,279 $km^2$, with somewhat more than half of the population living in the country and the rest in the towns (IBGE 2010a). At the same time, the region served an important function by supplying food to the metropolis of Belém. In the official statistics the region was classified as poor: according to the census poverty statistic, between 40 and 50 per cent of the rural population in the case- study region live below the Brazilian poverty line of 70 Reals monthly income per capita (approximately €35) (IBGE 2003, 2010b). According to the government, the palm oil sector therefore offers the peasants[6] a unique chance of lucrative access to the market and could at least slow down the marginalization of peasant agriculture, which has been taking place for decades.

This is not the case, however. Land speculation and contract farming inclusions have stimulated a comprehensive restructuring process of peasant land access and land use, which indicates that green grabbing is taking place.

#### *Land speculation*

The expansion of the oil palm plantations has caused the price of land to rise and has made land speculation a lucrative business for intermediaries – usually ex-mayors and large landowners. This contains a serious conflict potential because in Pará most of the land titles are invalid due to widespread land theft by means of document forgery and the lack of a superior land registry office (Treccani 2001). In the municipality of Moju alone, the claims to land in the land registry amount to more than five times the area of the municipality (Treccani 2013). There is a considerable risk that conflicts will be caused by differing claims to land.

The growing purchases of land from the peasants in the region have already started a process of displacement in the sense of green grabbing, without direct force being used. In the municipalities of Acará[7] and Bujaru (Nahum and Malcher 2012) the exodus of almost entire village communities could already be observed in 2011. Land purchase as such is a normal market process and not necessarily an indication of green grabbing. Rather, the extra-economic displacement effect of green grabbing arises in connection with the large asymmetries of power between buyers and peasants as well as the smouldering, sometimes decades-old land conflicts in the region. Families are increasingly isolated by the growing plantation areas due to the sales of their neighbours' land; they are no longer able to resist the pressure of the agricultural dealers, and thus they sell their own land. Pressure arises, for example, when agricultural dealers threaten peasants without land titles that they would lose their land in any case because of the missing documents, and that they should take the opportunity to sell their land before that happens.

### *Contract farming inclusion*

The contract farming inclusion of peasant farms sets off a restructuring process that results in the peasants largely losing control over the use of their land, even if they formally remain the owners.

The peasant forms of land use in the neighbourhood of the *pólos* are reorganized both spatially and temporally into agroindustrial suppliers for the agroindustrial palm oil complex: their land is converted into monocultural mini-plantations. In order to ensure efficient production, the day on which each individual stage in the work has to take place is specified, and this is monitored by the company's agro-technicians. If the peasants do not stick to the management regulations, the loan-issuing state bank suspends their account in consultation with the company concerned. The process of restructuring peasant land use is therefore accompanied by a disciplinary process.[8] With contract farming, the peasants surrender the control over their land and their labour and are at the same time directly subjected to the price trend of their products on the world market.

This dependence is strengthened by the fact that none of the traditional regional crops can be produced in between oil palms. Mixed cultivation is not provided for. The palm oil companies do not allow their contract farmers to deviate from monocultural planting on their palm oil plots for reasons of productivity and profit. Furthermore, the companies define the size of small-scale production as a uniform 10 hectares in order to keep organizational and transport costs as low as possible – although the state loan regulations for small-scale farmers (PRONAF-Éco)[9] would also allow the funding of smaller production units. This could lead in future to a reduction of the regional supply of food and strengthen the dependence of the families on the palm oil sector if the families do not possess adequate additional plots and labour to produce food.

The palm oil companies do not have to grab or buy land in order to obtain the peasants' plots. The planting of oil palms itself creates different long-term facts to those created, for example, by the grazing of a herd of animals. Palm oil producers are tied to the sector for at least 25 years by the life cycle of the oil palm, as the transformation of a palm oil plantation for another use is complex and expensive, and is therefore not an option for small farmers. The companies therefore do not shy away from planting palm oil plantations or contract farming plots, even without legal security concerning the land involved: somebody will (have to) farm them.

The contract farming inclusion of peasants into the agroindustry has been controversial for decades (Little 1994); it must not necessarily be the result of land grabbing (Borras and Franco 2012: 53). The way in which the peasants in Pará have been integrated into the palm oil sector, however, is an indication of a separation process in the sense of continuous primitive accumulation. With its objective of the more effective appropriation of surplus value, it is tied to the loss of control by the peasants over their land access and land use, although they remain the formal owners of their plots. The reorganization of the relations of production takes place almost exclusively in favour of the palm oil sector, which passes on the production risks (disease, bad harvests), compliance with national environmental and labour laws and fluctuating world market prices to the peasants without having to buy land or pay legal minimum wages.

It is uncertain how contract farming inclusion will continue to develop. The number of contract farmers in Pará has not, until now, fulfilled the companies' targets of having several thousand families under contract. According to one Vale employee, it is difficult to find 'suitable' families who own enough land, can supply enough labour and are creditworthy. Many peasants are also suspicious and have no desire to sign a contract.[10]

### *The political dimension of green grabbing*

The interviews with peasants and actors from civil society, private firms and the state in the region show that the restructuring processes of land access and land use cause criticism and controversy. Whereas some see in the palm oil programme a unique opportunity for the economic development of the region, others regard it as a renewed attempt by agribusiness to appropriate Amazonia. In the interviews, all the actor groups, apart from the company employees, regarded the increase in land purchases as entailing the risk of strengthening the migration of the rural population to the impoverished areas of the towns and cities, which is already taking place. In addition, the negative socio-ecological effects of plantation expansion, such as the contamination of river and drinking water by pesticides and the illegal deforestation of secondary forests, were problematized.

In contrast, positions with regard to the interpretation of small-scale contract farming vary: it is regarded either as a unique chance in the struggle against rural poverty or as part of an expropriation strategy. Opposing positions can also be found among the peasants: some of them wish unconditionally to take part in this 'project',[11] while others are hesitant or involve themselves in peasant associations and neighbourhood groups against contract farming production. Contradiction and resistance on the part of the peasants to the palm oil project can be implied from their distrustful hesitancy to accept Vale's contract farming offers.

However, in spite of the controversies and the criticism by various different actors in the interviews and group discussions, by the end of 2013 still no politically organized resistance to the state palm oil programme had been formed. No political alliances have been formed between critical factions of the rural population, the trade unions and social movements. It therefore remains largely without consequence that the public hearings which are required for major projects took place neither before the start of the state palm oil programme in 2010 nor during the first three years of its implementation.[12] The situation is similar with regard to the required environmental impact assessments and environmental licenses: in 2011, according to the federal environmental authority SEMA (*Secretaria do Estado de Meio Ambiente*),[13] not one single newly investing palm oil company had carried out the required environmental impact assessment, nor was any of them in possession of the required environmental license for the establishment of plantations. This, too, was problematized by state, trade union and civil society actors in the interviews, but it remained without political consequence until at least the time of writing (end of 2013).

In some places the conditions for political resistance simply do not exist: in some of the expansion areas, parts of the peasant population are socially so marginalized that they have neither places for political articulation nor state or civil society centres for dealing with their demands or complaints. Civil society actors hardly exist outside of the agricultural trade unions in the entire case-study region. Ironically, by the end of 2013, the latter still had no uniform position with regard to the expansion of the palm oil project: some of them, such as the trade union representatives in Moju and Bujaru, openly promoted small-scale contract farming in the palm oil sector, while others were hesitant and only a few individual trade unionists from Acará and Tomé-Açu have openly taken up a position which is critical of the government's palm oil programme.

### *The discursive dimension: the narrative of the degraded areas in the tropics*

In my opinion the lack of protest or politically audible resistance can be explained by the fact that the oil palm expansion has been established, even in the eyes of most of its critics, as without alternative and inevitable. For opponents of the state programme it is difficult to find allies against the widespread consensus in the region that the expansion of the oil palm plantations is a 'green' development in a 'degraded region'. According to one environmental secretary, it is meaningful to plant oil palms on 'grazing lands which produce nothing and are useful to nobody'.[14] An ex-trade union leader, now Agrarian Secretary in Concórdia, explains: 'Of course oil palm plantations have negative environmental effects. But in this way we create jobs for young people who have no education and no perspectives.'[15]

Resistance to the government programme is therefore forced to justify itself, as was shown in our interviews with critical actors. The latter often began their statements with a justification: 'Of course I am not opposed to planting grazing land with oil palm but…'[16] In their opinion, what basically spoke against the palm oil programme was the agroindustrial, monocultural model that was being supported by the state and was strengthening the displacement of peasant forms of land use. They seemed speechless regarding the definition of their region as largely degraded.

This speechlessness in public can be explained by the narrative of the degraded areas, which is by no means new but nevertheless powerful, according to which tropical forests are being linearly degraded by deforestation and by the growing population because of their sensitive ecosystem. In addition to the extensive grazing land, the traditional practice of shifting cultivation[17] in particular accelerates this process of degradation when the population is growing. Researchers from various disciplines have shown since the 1990s that the tropical ecosystems are much more complex and in part more resilient than classical tropic ecology has assumed (e.g. Nepstad *et al.* 1996). Neo-Malthusian explanations[18] and homogenizing assumptions concerning Amazonian shifting cultivation are also unsustainable against the background of diverse peasant strategies of adaptation and survival (Hurtienne 1999, 2005; Hecht 2005; Almeida 2008) and of new knowledge concerning complex, relatively intensive use systems in past civilizations (Heckenberger and Neves 2009).

Nevertheless, this naturalizing narrative continues to be influential and has been given a new meaning with the climate policy orientation of the palm oil programme: the palm oil plantations, according to a government information brochure, are to restore the degraded areas and thus contribute to Brazil's voluntary commitments to climate protection through the avoidance of deforestation and lower carbon emissions (Brazilian Government 2010). The oil palm plantations, according to the Agricultural Research Corporation EMBRAPA, were ideal for the 'reforestation' of degraded tropical soils because palms could grow in nutrient-poor soils and protect them from dehydration and erosion caused by wind and rain (Furlan Junior *et al.* 2006: 96). In addition, the palms sequestered $CO_2$, especially during their growth phase (ibid.: 102). Through the integration of peasant farmers into oil palm production, the peasants could also be offered an economically reliable alternative to the practice of degrading shifting cultivation (ibid.: 99). The homogenization of peasant forms of land use, and their classification as a cause of degradation, is not only maintained here, but worsened as also being damaging to the climate.

The decisive scientific instrument for the apparently objective identification of the areas to be restored is the spatial planning instrument of agro-ecological zoning (EMBRAPA and MAPA 2010). In the official zoning document, in contrast to political statements and interviews, the term 'degraded' is scarcely used (ibid.). Instead, the document only designates, on the basis of the deforestation data from the monitoring system PRODES (*Programa de Cálculo do Desflorestamento da Amazônia*), areas which were deforested before 2008, have soil characteristics and climate conditions (rainfall and solar radiation) that correspond to the requirements of the oil palm and fulfil the requirements of the agroindustrial palm oil complex (minimum amount of infrastructure and an adequate supply of labour). It is implied in the document, however, that about 80 per cent of the deforested areas are 'exhausted, degraded or abandoned grazing land' (EMBRAPA and MAPA 2010: 36). The characteristics of degraded grazing land are not defined. Decisions as to where the main production sites should be located are taken exclusively by a committee of experts, consisting of representatives of palm oil enterprises and state technical authorities. The rural population is included in the zoning plan only via economic and demographic data as potential workers or contract farmers. Their participation in decision-making is not provided for.

Agro-ecological zoning is not only an instrument of legitimation for the palm oil sector. In Pará it produces for the first time the natural resource 'degraded land' for renewed development by the agroindustrial palm oil complex (cf. Nalepa and Bauer 2012). The region can only be appropriated by the agroindustry through its definition, location and classification in combination with the corresponding credit lines of the support programme. That which is special and 'green' is the fact not only that the conditions for green grabbing are created by state institutions, but that these are legitimized by the political objectives of protecting primary forests and of contributing to national climate protection targets by reducing $CO_2$ emissions. This is possible because international climate policy is increasingly limited to the quantifiable (and therefore tradable) reduction of a single greenhouse gas ($CO_2$) (Brunnengräber and Dietz 2013).

This is an indication of a new legitimation narrative concerning what could become established as so-called climate-friendly Amazon policy: hot spots rich in biodiversity are to be protected by releasing so-called degraded areas to agroindustry. In order to achieve this, exceptions to the regulations must found at the level of the environmental laws, as seen in the debates on exceptions to the Forest Law (*código florestal*),[19] for example for reforestation with non-native oil palm and eucalyptus plantations. At the time of the passing of the palm oil programme in 2010, 'restoration' or 'reforestation' with oil palm plantations was not allowed by law. Nevertheless, it was already anticipated in a government brochure, as mentioned above (Brazilian Government 2010).

All of this has an effect on the way in which the palm oil programme can be justified, given a legal foundation and implemented as sustainable, desirable or without alternative in the case-study region, and explains why it is difficult, although not impossible, for critics to establish an counter-narrative. Engineers from Vale have explained in interviews that oil palms sequester much more $CO_2$ than natural forest and that they represent a unique chance of development for the degraded region.[20] The company had already created accomplished facts with the planting of its oil palm plantations in 2011, as if oil palm plantations for reforestation were already permitted by law. The signs around Vale's plantations emphasize this by describing them as a 'reforestation project'.

This climate policy upgrading of palm oil production is tied in the implementation process to the downgrading of peasant agriculture. One Vale manager characterized peasant manioc production as a vestige of a pre-modern era and linked it to negative attributes of the rural peasant population:

> Manioc is a culture which is already hundreds of years old, where you see no sign of development. The people do not change their life-style and increase their pressure on the forest. If you look at satellite photos then you will see that the peasant families have degraded the region. Together with the big landowners with the large deforested areas they have caused the greatest damage (*impacto*). [...] All [the peasant families] come from the manioc culture. The majority ... all of them plant only manioc. Some of them have planted maize or rice to generate income, but to feed their families they prefer to plant manioc. Because their families have grown considerably, the same areas are used more quickly and so the soil becomes poorer even faster. [...][21]

Here, he is repeating the clichés concerning Amazonian peasant shifting cultivation and linking the introduction of contract farming to a modernization project. The reference to the satellite pictures of the region reproduces the diagnosis of regional crisis from an apparently neutral macro-perspective: the region is 'exhausted' and Vale's 'palm oil project' offers the region a unique opportunity for 'sustainable development'. From this perspective, rural poverty is not the result of an unequal distribution of income and access opportunities in society, but the consequence of an allegedly natural process of degradation due to population growth and outdated, 'traditional' forms of land use.

By classifying both the peasant systems of land use and extensive grazing lands as degraded areas and degraded practices in connection with a state-promoted support programme, the lack of an alternative to this strategy is discursively confirmed far beyond the region itself. Allies in regional or international environment forums are hard to find for the largely isolated critics of the palm oil programme. Existing counter-narratives of the peasants and representatives of civil society who attach greater value to peasant manioc production as the production of a basic regional foodstuff, or who question the classification of the region as degraded, are therefore almost inaudible.

## Conclusion

Socio-ecological crises must not necessarily represent a limit to growth for capitalism, but can open up new fields of accumulation. The definition of green grabbing as an expression of continuous primitive accumulation enables us to differentiate new capitalist appropriation dynamics from already existing ones and to develop a differentiated analytical perspective for the political ecology of social change in connection with strategies for dealing with the socio-ecological crisis.

With green grabbing I have developed a flexible analytical tool for determining the extent to which control over relations of land access and land use in the sense of continuous primitive accumulation is restructured for expanded reproduction or the appropriation of surplus value. I have not defined the methods, dynamics and conflict constellations of green grabbing in advance; these must be worked out case by case. It is imperative that not only the material and political dimensions of green grabbing are examined, but also the discursive dimensions. As is clearly shown in the case study of the Brazilian Amazon region, these three analytic dimensions are inseparably linked to one another. None of them can be examined separately from the other two or derived from them.

State support for palm oil production in Pará is a continuation of the process of displacement of Amazonian peasant agriculture, which was established in the colonial era and which has been strengthened since the development of the Brazilian agroindustry in the 1970s. Land grabbing has therefore shown certain continuities in the decades since then. However, the extension of palm oil production is linked to *new* dynamics that indicate the existence of green grabbing. What is *new* about green grabbing is that it is initiated and legitimized by state activities in the fields of energy, climate and development policy (production of agrofuels in so-called degraded areas in an 'impoverished' region) for dealing with socio-ecological crises (scarcity of fossil energy and climate change). This launches a process of separation in the sense of continuous primitive accumulation because, in this restructuring process, the peasants lose control over their land and their labour to the palm oil companies – even if they partly remain the owners of their plots. The methods of separation used here are not direct force but state support measures, specific purchasing practices of intermediaries and the practice of the agroindustrial inclusion of peasant agriculture. As I have shown, the discursive dimension of the analysis is important in order to be able to understand why these restructuring processes, despite the criticism expressed in the interviews, have not led to politically organized protest. In the case under examination, the narrative of the degraded areas of Amazonia has proved to be particularly powerful. Only because it is widely accepted, from local to transnational political forums, that so-called degraded areas should be developed agroindustrially for the protection of the tropical forest, can the local population's participation rights, and environmental laws, be circumvented by the large-scale extensions of plantations, without causing transnational protest as in the case of other large projects in Amazonia. The bias of transnational environmental policy towards the protection of primary forests and the concentration of climate protection on the reduction of $CO_2$ closes political articulation spaces to peasant actors who live in these allegedly less valuable areas. The narrative of the degraded areas is not only a legitimation strategy with political implications, but it also has material consequences: agro-ecological zoning produces the new resource of the to-be-developed degraded areas, and thus makes possible the restructuring processes in the region.

The outcome of the green grabbing process in Pará is uncertain. It cannot yet be estimated whether the peasants will be either displaced or bound into the agroindustry via contract farming, as in the Brazilian sugar cane and soya sectors. The course of this process will be influenced by many factors, such as the price development of the *commodity*, the success or failure of involving the peasants in contract farming and the political will of the government. It also depends on the actors themselves, however; on their alliances and struggles for the control of land access and land use, and over the right to define allegedly degraded areas and their use, as well as the establishment of counter-narratives.

## Notes

1 This contribution is based on the results of my doctoral dissertation, prepared from 2009 to 2013 within the framework of the project 'Fair Fuels?' (www.fair-fuels.de/en, last access: 15 January 2014), which was supported by the German Federal Ministry for Education and Research (BMBF). In field studies of several months' length in 2010 and 2011, more than 80 interviews were conducted with actors from the private sector, the state and civil society in Brasília, Belém and five of the 44 municipalities designated for oil palm production (Moju, Acará, Tomé-Açu, Concórdia and Bujaru) in Pará, and supplemented with grey literature and secondary data.

2 Against the background of the oil crises in the 1970s, Brazil supported the production of ethanol for vehicles on the basis of sugar cane through the Proalcool programme, cf. Borges *et al.* (1984).

3 In three public-private-partnership projects between the federal state of Pará, the Brazilian palm oil company Agropalma and 150 families, an oil palm plantation with a total area of 1,500 hectares was established between 2002 and 2006. Each family received a plot of 10 hectares within this contiguous plantation.

4 With the national biodiesel programme (*Programa Nacional de Produção e Uso de Biodiesel – PNPB*) which was passed in 2004 a new market access for the impoverished peasant agriculture was to be created (www.mda.gov.br/portal/saf/programas/biodiesel, last access: 15 January 2014).

5 Interviews with engineers in 2010 and 2011, and the company's own homepage (www.vale.com/brasil/EN/aboutvale/initiatives/biodiesel/Pages/default.aspx, last access: 27 February 2013).

6 Various different rural classes and forms of land use are subsumed in the category of peasants. Following the definition of the Brazilian Agricultural Ministry MDA (*Ministério de Desenvolvimento Agrário*), I include in this category those who live on their land and who generate 90 per cent of their income from it, who do not employ more than two agricultural labourers and whose land area does not exceed the regionally defined maximum size (between 25 and 350 hectares); cf. Rodrigues (2009). In the case-study region, peasant families who are eligible to become contract farmers for palm oil production own between 25 and 100 hectares of land.

7 The trade union representative from Acará reported in 2011 on the village of Bucaia, where within two years the majority of the peasants had sold their land and moved to the poor quarters of the town of Acará (interview, March 2011, Acará).

8 Statement by an employee of Banco da Amazônia in an Interview in May 2011 in Belém.

9 PRONAF (*Programa Nacional de Fortalecimento da Agricultura Familiar*): national programme for strengthening peasant agriculture.

10 Interviews May 2011, Moju and Concórdia.

11 In the interviews in the region, talk was usually of *the* project, which underlines that no other projects of support for agriculture are being implemented except the palm oil programme.

12 In April 2011, for example, federal labour party PT (*Partido dos Trabalhadores*) member Ganzer publicly criticized the fact that public hearings had not taken place, but this remained without consequence until the time of writing (end of 2013), cf. Flexa (2011).

13 Interview with two SEMA employees in May 2011 in Belém and statement by a SEMA employee in the workshop 'Os impactos sociais e ambientais dos investimentos em dendê no Pará', 21 October 2013 in Belém.

14 Statement by the Environmental Secretary from Bujaru during a group discussion in May 2011 in Bujaru.
15 Interview with the Agricultural Secretary, June 2011, Concórdia.
16 For example, a representative of the MST (*Movimeto dos Trabalhadores Rurais Sem Terra*) in Belém, a trade unionist from Concórdia (both in June 2011), a trade unionist from Tomé-Açu in April 2011 and three employees of the state technical consultation authority for small farmers EMATER (Empresa de Assistência Técnica e Extensão Rural) (May, June, July 2011) all began their statements of their position with these words.
17 In the classical theories of tropical ecology and agrarian economics, shifting cultivation is described as the traditional, Amazonian form of land use: small areas of up to 5 hectares of tropical forest are burned down and the soil, which is fertilized by the ash and deacidified, is planted directly among the tree stumps with annual crops such as manioc. Following one or two years', use these areas are given over to the secondary forest. After a fallow period of between 10 and 15 years, the areas are cultivated again according to the same pattern (burning down and planting); Schmitz (2013: 330). It is criticized that this model of land use has been generally used to describe different Amazonian peasant forms of land use (ibid. and Hurtienne 1999, 2005).
18 Current approaches are described as neo-Malthusian which, like Thomas Robert Malthus, regard population growth as the main cause of the socio-ecological crisis and as the greatest risk for the continued existence of humanity.
19 See also the Brazilian Forest Code of 1965, Law No. 4.771/65, and the transitional regulation (medida provisória) No. 1.511 of 1996. The new Forest Law was passed at the end of 2012, following year-long controversies (Law No. 12.727 of 18 October 2012).
20 Interviews with three Vale managers (May and June 2011) in Moju and Concórdia.
21 Interview with a Vale manager (June 2011) in Concórdia.

## References

Acevedo Marin, Rosa E. (2010) 'Territórios Quilombolas Face à Expansão do Dendê no Pará', in Buenafuente, S.M.F. (ed.) *Amazônia. Dinâmica do Carbono e Impactos Socioeconômicos e Ambientais*, Boa Vista: Editora da UFRR, 165–184.

Almeida, Alfredo Wagner B. de (2008) *Antropologia dos Archivos da Amazônia*, Rio de Janeiro: casa 8/F.U.A.

Blaikie, Piers M. and Harold C. Brookfield (1987) *Land degradation and society*, London: Methuen.

Borges, Uta, Heiko Freitag, Thomas Hurtienne and Manfred Nitsch (1984) *PROALCOOL. Analyse und Evaluierung des brasilianischen Biotreibstoffprogramms*, Saarbrücken, Fort Lauderdale: Verlag Breitenbach.

Borras, Saturnino M. and Jennifer C. Franco (2012) 'Global land grabbing and trajectories of agrarian change: a preliminary analysis', *Journal of Agrarian Change* 12, 34–59.

Brazilian Government (2010) 'Palma de Óleo. Programa de produção sustentável', folder, Brasília.

Brunnengräber, Achim and Kristina Dietz (2013) 'Transformativ, politisch und normativ: für eine Re-Politisierung der Anpassungsforschung', *Gaia* 22, 224–227.

De Angelis, Massimo (2001) 'The continuous character of capital's "enclosures"', *The Commoner* 2, 1–22.

Deininger, Klaus, Derek Byerlee, Jonathan Lindsay, Andrew Norton, Harris Selod and Mercedes Stickler (2011) *Rising Global Interest in Farmland. Can it yield sustainable and equitable benefits?*, Washington, DC: World Bank.

Demirović, Alex, Julia Dück, Florian Becker and Pauline Bader (eds) (2011) *VielfachKrise. Im finanzdominierten Kapitalismus*, Hamburg: VSA.

EMBRAPA and MAPA (2010) *Zoneamento agroecológico do dendezeiro para as áreas desmatadas da Amazônia Legal*, Rio de Janeiro: Brasília.

Fairhead, James, Melissa Leach and Ian Scoones (2012) 'Green grabbing: a new appropriation of nature?', *The Journal of Peasant Studies* 39 (2), 237–261.

Flexa, Evandro (2011) 'Programa planta êxodo e colhe miséria. Em Tomé-Açu empresas compram terras do pequeno agricultor para cultivar palma de óleo', *O Liberal* 3.4.2011, 10.

Furlan Junior, José, Franz J. Kaltner, Gil F. P. Azevedo and Ivonice A. Campos (2006) *Biodiesel. Porque tem que ser Dendê*, Belém: EMBRAPA.

Glass, Verena (2013) *Expansão do dendê na Amazônia brasileira: elementos para uma análise dos impactos sobre a agricultura familiar no nordeste do Pará*, São Paulo: Repórter Brasil.

Hall, Stuart (1981) 'The whites of their eyes. Racist ideologies and the media', in Bridges, G. and R. Brunt (eds) *Silver linings. Some strategies for the eighties: 12th Annual meeting: Papers*, London: Lawrence and Wishart.

Hall, Stuart (1988) 'The toad in the garden: Thatcherism among the theoriests', in Nelson, Cary and Lawrence Grossberg (eds) *Marxism and the Interpretation of Culture*, Illionois: University of Illinois, 35–57.

Hecht, Susanna (2005) 'Soybeans, development and conservation on the Amazon frontier', *Development and Change* 36 (2), 375–404.

Heckenberger, Michael and Eduardo G. Neves (2009) 'Amazonian archaeology', *Annual Review of Anthropology* 38, 251–266.

Homma, Alfredo K.O. and José Furlan Junior (2001) 'Desenvolvimento da deindeicultura na Amazônia: Cronologia', in Muller, A.A. and José Furlan Junior (eds) *Agronegócio do dendê: uma alternativa social, econômica e ambiental para o desenvolvimento sustentável da Amazônia*, Belém: EMBRAPA, 193–207.

Hurtienne, Thomas (1999) 'Agricultura Familiar na Amazônia Oriental. Uma comparação dos resultados da pesquisa socioeconômica sobre fronteiras agrárias sob condições históricas e agroecológicas diversas', *Novos Cadernos NAEA* 2 (1), 75–94.

Hurtienne, Thomas (2005) Agricultura familiar e desenvolvimento rural sustentável na Amazônia, *Novos Cadernos NAEA* 8 (1), 19–71.

Hymer, Stephen (1971) 'Robinson Crusoe and the secret of primitive accumulation', *Monthly Review* 23, 11–36.

IBGE (2003) 'Mapa de Pobreza e Desigualdade – Municípios Brasileiros 2013', Rio de Janeiro (www.ibge.gov.br/estadosat/temas.php?sigla=pa&tema=mapapobreza2003, last access: 19 December 2013).

IBGE (2009) 'Pará. Lavoura Permanente 2009', Brasília: IBGE (www.ibge.gov.br/estadosat/temas.php?sigla=pa&tema=lavourapermanente2009, last access: 14 January 2014).

IBGE (2010a) 'Censo Demográfico', Brasília [www.ibge.gov.br/estadosat/perfil.php?sigla=pa, last access: 14 February 2014).

IBGE (2010b) 'Indicadores Sociais Municipais 2010: incidência de pobreza é maior nos municípios de porte médio', Brasília (www.ibge.gov.br/home/presidencia/noticias/noticia_visualiza.php?id_noticia=2019&id_pagina=1, last access: 19 February 2014).

Kalmring, Stefan (2013) 'Die Krise als Labor gesellschaftlicher Entwicklung. Fortgesetzte ursprüngliche Akkumulation und die große Krise der Kapitalakkumulation', in Backhouse, Maria, Olaf Gerlach, Stefan Kalmring and Andreas Nowas (eds) *Die globale Einhegung – Krise, ursprüngliche Akkumulation und Landnahmen im Kapitalismus*, Münster: Westfälisches Dampfboot, 70–109.

Kelly, Alice B. (2011) 'Conservation practice as primitive accumulation', *The Journal of Peasant Studies* 38 (4), 683–670.

Little, Peter D. (1994) 'Contract farming and the development question', in Little, Peter D. and Michael Watts (eds) *Living under contract. Contract farming and agrarian transformation in Sub-Saharan Africa*, Madison: University of Wisconsin Press, 216–247.

Marx, Karl (1962 [1867]) '*Das Kapital*. Kritik der Politischen Ökonomie. Band I', in *Marx Engels Gesamtausgabe (MEW) 23*, Berlin: Dietz.

MDA (2012a) 'PRONAF Eco Dendê. Reunião – Câmara Técnica da Palma de Óleo', powerpoint presentation (www.agricultura.gov.br/arq_editor/file/camaras_setoriais/Palma_de_oleo/7RO/App_Eco_Dende_Palma.pdf, last access: 19 December 2013).

MDA (2012b) 'Plantação de palma de óleo ganha força no Pará', press release, Brasília (www.biodieselbr.com/noticias/materia-prima/dende/plantacao-palma-forca-para-040612.htm, last access: 28. July 2013).

Muller, Antonio A., José Furlan Junior and Pedro Celestino Filho (2006) 'A Embrapa Amazônia Oriental e o Agronegócio do Dendê no Pará', factsheet, Belém: EMBRAPA.

Nahum, João and Tiago Malcher (2012) 'Dinâmicas Territoriais do Espaço Agrário na Amazônia: A Dendeicultura na Microregião de Tomé-Açu (PA)', *Confins* 16, 1–20.

Nalepa, A.R. and Dana M. Bauer (2012) 'Marginal lands: the role of remote sensing in constructing landscapes for agrofuel development', *The Journal of Peasant Studies* 39 (2), 403–422.

Negt, Oskar and Alexander Kluge (1983) *Geschichte und Eigensinn. Geschichtliche Organisation der Arbeitsvermögen. Deutschland als Produktions-Öffentlichkeit. Gewalt des Zusammenhangs*, Frankfurt am Main: Zweitausendeins.

Nepstad, D.C.C., C.A. Pereira and J.M.C. da Silva (1996) 'A comparative study of tree establishment in abandoned pasture and mature forest of eastern Amazonia', *Oikos* 76 (1), 25–39.

Perelman, Michael (1983) 'Classical political economy and primitive accumulation: the case of Smith and Steuart', *History of Political Economy* 15 (3), 451–494.

Rodrigues, José G.B. (2009) 'Identificação da Agricultura Familiar: Uma Análise dos Critérios da Declaração de Aptidão ao PRONAF – DAP no Contexto brasileiro e Internacional', Brasília. (www.iica.int/Esp/regiones/sur/brasil/Lists/DocumentosTecnicosAbertos/Attachments/504/Identifica%C3%A7%C3%A3o_da_Agricultura_Familiar_Jose_Germano.pdf, last access: 13 October 2012).

Schmitz, Heribert (2013) 'Bauern, Landnutzung und Entwicklung in Amazonien: Der Beitrag von Thomas Hurtienne seit 1994', in Backhouse, Maria, Olaf Gerlach, Stefan Kalmring and Andreas Nowak (eds) *Die globale Einhegung – Krise, ursprüngliche Akkumulation und Landnahmen im Kapitalismus*, Münster: Westfälisches Dampfboot, 324–337.

Silva, Felix L. de, Alfredo K.O. Homma and Heriberto W.A.P. Pena (2011) 'O cultivo de dendezeiro na Amazônia: Promessa de um novo ciclo econômico na região', in *Observatorio de la Economìa LatinoAmericana. Economía do Brasil* (www.eumed.net/cursecon/ecolat/br/11/shp.html, last access: 3 November 2013).

Treccani, Girolamo D. (2001) *Violência e grilagem: instrumentos de aquisição da propriedade da terra no Pará*, Belém: UFPA.

Treccani, Girolamo D. (2013) 'Questões juricidos', powerpoint presentation, workshop: Os impactos sociais e ambientais dos investimentos em dendê no Pará, 21.10.2013 in Belém.

USDA (2012) 'Gain report. Brazil. Biofuels annual. Annual report 2012', São Paulo.
USDA (2013) 'Brazil. Oilseeds and products annual, 2013–14 Record soybean production forecast at 85 mmt', GAIN Report Number: BR0908, São Paulo.
Vale (2011) 'Vale speeds up investments in biodiesel, press release', Rio de Janeiro. (saladeimprensa.vale.com/en/releases/interna.asp?id=20446, last access: 23 November 2013).
Vidal, John (2008) 'The great green land grab', *Guardian.* (www.guardian.co.uk/environment/2008/feb/13/conservation, last access: 13 October 2013).
Watkins, Case (2011) 'Dendezeiro: African oil palm agroecologies in Bahia, Brazil, and implications for development', *Journal of Latin American Geography* 10 (1), 9–33.
Zeller, Christian (2010) 'Die Natur als Anlagefeld des konzentrierten Finanzkapitals', in Schmieder, Falko (ed.) *Die Krise der Nachhaltigkeit. Zur Kritik der politischen Ökologie*, Frankfurt am Main: Lang, 103–135.

# 11 Transnational space and workers' struggles

## Reshaping the palm oil industry in Malaysia

*Oliver Pye*

### Introduction

In November 2013, hundreds of palm oil workers protested against the eleventh annual meeting of the corporate-led Roundtable on Sustainable Palm Oil (RSPO) in Medan, Sumatra. A statement by a new coalition of labour unions, SERBUNDO, claimed that RSPO member companies were contradicting RSPO 'Principles and Criteria' by 'employment of labors [*sic*] without transparent contracts, the repression of trade unions, arbitrary firing, violence against women and child labor' (Serbundo 2013). This is one of the first times that workers themselves have made their voice heard in the politicization of palm oil. Environmentalists and indigenous peoples have been much more vocal in their critique of the expansion of this sector. Transnational campaigns against the use of palm oil for agrofuels in the European Union have tended to pit indigenous peoples and small-scale farmers against transnational corporations, while labour issues have played only a minor role (Pye 2010). Recent publications that *do* discuss workers in the palm oil industry highlight the plight of only the most vulnerable victims: the child labourers, the trafficked and the modern slaves (Accenture for Humanity United 2013; International Labour Rights Watch and Sawit Watch 2013; World Vision 2013). In part, this is because agrofuels campaigning aims to change agrofuel policies in the Global North, so research often focuses on the negative impacts of these policies in the Global South. In lobbying work or consumer-oriented campaigns, workers like indigenous peoples tend to be subsumed under the category of 'victim' in association with the rainforest and endangered species.

Exploitation and repression in the palm oil industry is not specific for agrofuels and applies to the 'flex-crop palm oil' in general. For the plantation and mill workers, the final destination of the palm oil, be it chocolate bars or greenwashed diesel, is irrelevant in the first instance. But the crops used for agrofuels tend to be characterized by exploitative labour conditions that are part of large-scale monoculture production. The materiality of agrofuels production creates vast expanses of plantation agriculture (be it palm oil, sugar cane or rape seed) where the bulk of work is done by wage labourers. Given that agrofuels depend primarily on wage labour in their production, the relative silence on labour in the debates around agrofuels is strange. This silence corresponds to a more general problem identified by Herod (2001: 28), who argues that critical geography has 'marginalized conceptually the roles of workers in actively shaping the economic geography of capitalism'. With their labour power, agricultural workers create the agrofuel landscapes and industrial plants, while transport workers play a crucial role in the global production networks that characterize the corporate food regime in general and the new agrofuel commodity chains in particular. There would be no agrofuels without labour.

This chapter hopes to contribute to developing a 'labour geography' (Herod 2001) perspective on agrofuels by looking at the experience of migrant workers from Indonesia who work in the palm oil plantations and mills of Malaysia. Much of the material for this article comes from ongoing research with groups of workers from North Sumatra, Java, Lombok and Sulawesi, who worked in Peninsular Malaysia and Sabah and then returned home to their villages, and with groups of workers currently working in plantations and mills in Sabah.[1]

Malaysian corporations account for over half of the global production of palm oil and have invested heavily in Indonesia over the past decade. Part of this expansion is connected to an emerging and anticipated market for agrofuels in the European Union (Gerasimchuk and Koh 2013), that has been underscored by the European Union's acceptance of the RSPO as a standard for the Renewable Energy Directive sustainability criteria (European Commission 2012). Officially, half a million workers are employed in the Malaysian plantations and mills and nearly 80 per cent of the 'harvesters, fresh fruit bunch (FFB) collectors, loose fruit collectors and field workers' are migrants (Abdullah *et al.* 2011: 2). The real number is probably significantly higher because of the large number of undocumented 'irregular' workers. This chapter argues that these workers are part of a 'global precariat' (Standing 2011), as they enter a precarious labour regime established by a coalition between private corporations, state capital and the government in Malaysia. The social precaritization of the workers reflected in temporary employment, insecure and low wages, outsourcing and subcontracting, is an intended consequence of spatial control that is complemented by political precarity, i.e. a lack of political and social rights.

The politically repressive and precarious conditions of migrant workers in the Malaysian palm oil industry, sometimes characterized as 'bonded labour' or 'modern slavery' (Li 2011: 77), would suggest that resistance by the workers themselves is impossible or extremely difficult and risky. This is undoubtedly the case, and some of the workers from the research project tell horrific stories of debt bondage, violently enforced labour, political repression, caning and more. However, other workers tell different stories, work according to collective agreements and do indeed save money, which they invest in a house or a smallholding back home (biographical interviews with migrant workers during 2010 and 2011). These success stories explain the continuing lure of Malaysia for the 'surplus population' (i.e. the chronically underemployed, Li 2009) in Indonesia. They also show that the control of the state or the corporations over the plantation labour is far from total. The necessary understanding of the plantation migrant workers as victims of a ruthless labour regime needs to be qualified by a simultaneous understanding of them as resourceful agents of their own biography. In fact, workers are developing innovative resistance strategies that are located somewhere between acquiescence and open revolt.

This chapter starts by discussing the precarious labour regime imposed upon migrant workers in Malaysia. It then looks at the practices of political precarization and of spatial control that Kelly (2002), in his discussion of the estates of special export zones, argues prevent labour organization developing. It will then be argued that, despite these politics of control, workers are developing everyday resistance and that a key spatial dynamic in this is that workers are producing transnational spaces of social reproduction. Although the migration regime tries to impose a system by which workers enter Malaysia for three to five years and then return home, the social practice of workers creates networks that link places and people in Indonesia with the workplaces in Malaysia. Following in the footsteps of relatives and neighbours, workers establish communication and experience networks which offer them opportunities to overcome the spatial control of capital and the state. This 'scaling up' opens up space for resistance. Three key modes of resistance discussed here are: first, border negotiation strategies that undermine the territoriality of the plantation landscapes; second, the widespread practice of absconding (*lari*); and third, the occurrence of wildcat strikes (*mogok*).

A political ecology of agrofuels cannot be complete without taking into account the lives and struggles of the workers who produce them. In the politicization of agrofuels, the strategic question arises of whether workers' interests and those of environmental justice and land rights activists are fundamentally opposed or whether workers' movements could be potential allies. This is not an academic question, but depends on how activists from different movements relate to each other. The chapter concludes with some thoughts on how insights from the workers' experience could inform future campaigning around agrofuels.

## A precarious labour regime

Precarity as a concept emerged originally in a European context as a critique of the increase of flexibilized, temporary, insecure work at the expense of stable employment conditions (Castel 2009). But the precariat, as Guy Standing (2011: 27) argues, is rapidly 'becoming a global class'. One factor in the creation of a globalized flexible labour force is the migration of millions of workers into the industrializing centres of the South, who are paid 'remarkably low' wages and 'treated as a disposable, itinerant labour force' (ibid.: 28). Standing argues that migrants now 'make up a large share of the world's precariat' (ibid.: 90).

In contrast to previous debates about 'the informal sector' in the Global South, which implied the small-scale, petty traders' type of pre-modern employment, precarious labour conditions are at the very heart of the most modern, successful and globalized industries of industrializing countries. The palm oil industry is a case in point. It has become a billion dollar export industry for both Malaysia and Indonesia, based on transnational corporations with state-of-the-art industrial plants and global production networks. The owners of Malaysia's biggest private palm oil corporations are among the richest Malaysians. Robert Kuok – head of the Kuok group that controls most of the investment in the huge Wilmar corporation – is Malaysia's richest man on Forbes 'list of richest Malaysians'.[2] He is worth US$11.5 billion. His nephew, Kuok Khoon Hong, who runs Wilmar and owns 10 per cent of it, is worth $2.2 billion. Lee Shin Cheng, the owner of the IOI corporation, is worth $4.3 billion. Indonesian migrants who work in their plantations and mills earn between $60 and $200 a month. The millions of dollars of profits made in the sector are generated from low-paid, precarious, migrant labour.

The creation of a precarious labour regime in the palm oil industry was a *political* project, based on 're-regulation' (Standing 2011: 26) rather than deregulation, i.e the labour regime is highly controlled and not just left to the market. In both Malaysia and Indonesia, state and corporate interests are intimately connected in the palm oil industry. In Indonesia, state corporations and agro-industrial conglomerates formed an 'oligarchy' with the dictator Suharto (1965–1998) (Aditjondro 2001). In Malaysia, state capital (such as the huge Sime Darby or FELDA), private transnational capital (e.g. IOI, Kuala Lumpur Kepong, the Ganteng group) and state and industry institutions such as the Malaysian Palm Oil Board (MPOB), the Malaysian Palm Oil Council (MPOC) and the Malaysian Palm Oil Association (MPOA), work so closely together for the interests of the corporations that I have termed them elsewhere the 'palm oil industrial complex' (Pye 2008; 2014).

Crushing the labour unions was key to precaritizing the labour force. The plantation sector had been the site of militant workers' movements that had emerged in connection with the national liberation movements against colonial rule from the 1920s to the 1950s. In Malaysia, workers' struggles united Indian and Chinese labour in 1945–1948, with a general strike in 1947, in an effort to form a Pan Malayan General Labour Union. But the movement was crushed by the British colonial government, with police baton-charging strikers in 'strong concerted action to curb left wing unions' (Ramasamy 1994: 82). The British promoted ethnicity-based trade unionism in the form of the National Union of Plantation Workers, and the independent Malaysian state subsequently discouraged strikes and promoted 'harmonious industrial relations' (ibid.: 172). In Indonesia, the plantation workers union SARBUPRI had 100,000 estate worker members in Medan and nearly one million in the whole of the country in the 1950s (Stoler 1995: 125). After the nationalization of foreign companies, labour militancy was eclipsed by a mass squatter movement, with 130,000 ha occupied in North Sumatra by 1959 (ibid.: 129). But in 1965, the labour movement was crushed by Suharto's military regime, and tens of thousands of labour activists were killed in the Medan region in anti-communist massacres. Until the fall of Suharto in 1998, a state-controlled union federation prevented any independent workers' associations from legally forming. The rate of temporary workers in the estates increased dramatically during Suharto's New Order (Stoler 1995: 168). Today, the palm oil industry relies heavily on workers paid on a daily basis (*buruh harian lepas*) and on outsourced piece-rate payment, where the wives and children of the male workers have to contribute unpaid labour to meet the quota (Kelompok Pelita Sejahtera 2008).

In Malaysia, precarious labour has been at the heart of palm oil expansion from the onset. Malaysian state capital, such as the Federal Land Development Agency (FELDA), was instrumental in converting rubber plantations into oil palm in the late 1970s and early 1980s in Peninsular Malaysia. This shift was accompanied by the flexibilization of the labour regime. The lack of militancy on the part of the National Union of Plantation Workers (Ramasamy 1994) had been rewarded by low wages, but fairly stable employment. Now, FELDA started to subcontract labour to labour agencies, setting up a model that came to characterize the industry as a whole (Wong and Anwar 2003). Very soon, the labour contractors started to source their workers from Indonesia.

The use of migrant workers offers the palm oil industry 'flexible labour relations' (Standing 2011: 31). Since the 1990s, a 'formal guest worker system' has been in place in Malaysia, in which workers are recruited offshore and are then issued employer and location-specific work permits (Wong 2006: 221). The official system prescribes short-term contracts of three years (with the possibility of two one-year extensions), i.e. all contracts are temporary. Low wages are also an inbuilt part of the migration system and are usually a combination of piece-rate (*berongan*, payment per tonne of harvested palm oil, for example) and low daily wages (*kongsikong*). Whilst workers can earn more through piece-rate payments, the risk is passed on to the workers, who do not get paid if they cannot work because of weather conditions, or if there are no, or less abundant, harvests. At the same time, workers dislike working for a daily wage, as this is usually so low that they feel unmotivated. Nevertheless, those who work directly for employers in this official system are at the better end of the precarity scale.

The official precarious regime, in which workers work directly for an estate or plantation owner, is backed up by a vast subcontracting system of employment agencies and work gangs that rent out labour to different companies. This gives the industry greater 'numerical flexibility' (Savaranamuttu 2012) so that they can hire when needed and fire in slack times. The outsourcing of work to labour contractors also divides workers working in the same estate, making collective action more difficult. Often, workers at the same plantation will be employed by different labour agencies who pay different wages and deduct different amounts for permits, etc. If workers have an issue with the management or the foreman (*mandor*) of that particular company, these can deflect many of the issues because they are not the employer. It is more difficult for workers from one estate to unite because they work for different employers; and more difficult for workers working for one labour agency to unite because they are dispersed across the country in small groups, without connection to one another.

In another form of outsourcing, a 'gang' of workers, employed as 'quasi-freelancers' are given a contract to harvest a particular area, a common practice in the 57 different FELDA estates in the Tungku area of Sabah. Workers are attracted to this kind of outsourcing because the company offers more per fruit bunch than to workers employed directly at the company. For management, the benefit of outsourcing is that workers internalize the productivity logic (the quicker they harvest, the more they earn) and carry the risk. During a group discussion in one of these gangs in the FELDA estates, the workers articulated their dissatisfaction with the contract system after earnings had fallen during the dry season in which fewer fruit bunches can be harvested. At first, the gang leader stressed the work ethic, stressing that 'we are all working for a better future' and this was echoed by a gang member who agreed that 'the result depends on ourselves: if we work hard then we will be rewarded' (Group discussion, FELDA Gang, 1 July 2012). However, as the discussion progressed, gang members challenged this view, comparing wage labour favourably to the contract system (see below).

## Spatial control and the production of denizens

What Herod (2001: 45) terms the 'technology of spatial control during the exercise of power' is crucial in the production of a precariatized labour force. In his discussion of the Special Economic Zones of Southeast Asia's export industry, Kelly argues that space is used to control labour by:

> constructing the individual as an autonomous unit of negotiation; constructing the workplace as a container for dispute resolution; establishing the industrial estate as a denationalized and desocialized space; constructing spaces of national sovereignty and imagined national/ethnic community; and, the distancing of homeplace from workplace through the use of migrant workforces.
> (2002, 398)

These spatial dynamics are certainly at play in the palm oil industry.

Two scales are particularly relevant for spatial control: one that I term the 'mill-estate-scale', which is the basic unit of production, and the national scale, which is crucial for the othering of the migrant workers and their definition as 'denizens' (i.e. people with restricted civil, political and economic rights distinct from citizens (Standing 2011: 94)).

To begin with, the materiality of the oil palm, the harvested fruit of which must be processed within 24 hours to prevent it going rancid, leads to a concentration of plantations around the mill to minimize transportation time. The size of the mill and the profitability of running it 24 hours/seven days a week meanwhile expediate large-scale plantations, in turn leading to a landscape of huge estates clustered around mills. These mill-estate-spaces are usually far away from towns or villages and are under the territorial control of the corporation (i.e. what Kelly means by 'denationalized'). For example, the workers' housing is provided by the corporation in this space, and the roads leading to the mills and estates are closed by bars and guards. The social mobility of the workers is restricted by this spatial control.

At the national scale, social precarity is exacerbated by the *political precarity* inherent in the migration regime. Temporary migrant workers have 'no rights to socio-economic mobility and no political rights', making them the 'classic denizens' (Standing 2011: 94). The permit system chains workers to a particular employer and subjects them to his arbitrary control. As Kelly (2002: 408) points out, 'the contractual prevention of job-hopping impedes the mobility of workers and leaves them with little recourse in workplace disputes if they are to avoid returning home without the socially expected savings'. Contract brokers often upfront fees and some workers can never exit the cycle of debt and dependency this creates. Others are sold and resold along a chain of brokers linking Indonesia to different parts of Malaysia. The official migration system imposes biopolitical control in various ways: marriage, pregnancy, illness or injury can result in immediate deportation. Any kind of protest can lead to the employer terminating the contract or calling in the police. Workers who enter illegally because of the lengthy and costly process of the official channels or because they do not pass the health test etc., or who become illegal by overstaying or running away from their boss, face the constant threat of deportation, imprisonment and corporal punishment. Another extreme form of political precarity affects the family: migrant children are not meant to exist (no pregnancy allowed), are therefore illegal and consequently have no right to visit Malaysian government schools.

Illegal entry, or legal entry and illegal work, is often pursued by the migrant workers because of the high costs of the official system, or because they are rejected by the official system, or because they are recruited by informal agents working for subcontractors. The illegal status results in many forms of political precarity that are added to those of the legal system. Permit control in the legal system corresponds to the threat of snitching or of police intervention, leading to similar dependency on the employer. Added to this is the constant fear of police razzias and check-points, restricting the freedom of movement of the workers (no trips to the city) and often leading them to run from razzias, hide with friends or in the forest and even to live 'in the forest' to escape attention by police. The fear of imprisonment, caning and other physical abuse, and of deportation, is compounded either by their own experience (several of the respondents had been in prison, a couple had been caned, and several had been deported) or by deportation waves (such as occurred in 2002) and the arrest of friends and relatives. In turn, this means that they have no political security from which they can tackle their social precarity. Political precarity is increased when their freedom of movement is restricted still further by the subcontracting gangs who physically coerce them to remain on a particular plantation and prevent them from moving around or leaving the country.

This territorializing production of denizens at the national scale is highly gendered, connecting both to the scale of the body and producing a transnational scale of social reproduction at the same time. The official system is geared towards male migrant workers and the assumption is that they leave their wives and children at home. The biopolitical control over women's bodies and reproduction in the regular health checks mentioned above reinforce the division between workers and the rest of their family. In this way, the cost of social reproduction (bringing up children, housework, education, etc.) is outsourced to women in Indonesia, who are expected to do this work for free (Pye and Julia 2014). Women who do find their way into the palm oil sector are confronted by a gendered division of labour (Lai 2011; see also White *et al.*, Chapter 4 in this volume). The more lucrative harvesting work is given to men, while women are employed on a lower daily wage, particularly spreading fertilizers and spraying pesticides. Work by the NGO Tenaganita (Tenaganita and PAN-Asia-Pacific 2002) has shown that women's bodies are particularly vulnerable to the effects of the pesticides used in the palm oil industry and that pesticide sprayers experience a wide range of health problems, including impacts on the foetus during pregnancy.

## Producing a transnational scale of social reproduction

The combination of social and political precarity and the spatial politics of control and 'denizisation' divides and individualizes workers, encouraging the internalization of work ethics and impeding collective responses. Kelly (2002: 408) argues that, because of the spatial control over the estates, 'the potential for labour to organize is limited to particular sites, while employers and local government officials engage in networks that transcend these scales,' and he concludes that 'the processes through which labour politics are played out are constructed in such a way that any form of labour collectivity is precluded'. While these problems of different scales are undoubtedly serious and although the lack of political and social rights have prevented workers' organizations such as trade unions from developing in the palm oil industry, I would argue that such a conclusion represents a one-sided view from the perspective of capital and understates the active role of workers in the production of the palm oil landscape (Herod 2001).

As Herod (2001: 46) argues, workers 'actively produce economic spaces and scales in particular ways (both directly and indirectly, consciously and unconsciously) as they implement in the landscape their own spatial fixes in the process of ensuring their own self-reproduction'. In the palm oil industry, hundreds of thousands of migrant workers in the Malaysian plantations create a multitude of networks linking specific places to each other and producing transnational social spaces. Across time and space, the linkages between specific *kampung* in Indonesia to places and networks in Malaysia can empower workers, who develop various means of everyday resistance that reassert their own humanity over their alienation as labour power in the plantations. Everyday survival strategies and social practice of workers challenge the spatial logic or utopia of capital and contribute to the production of new social spaces.

These place–time contingencies link places and people over large distances and over longer periods of time than the prescribed three- or five-year work contract. The spatial distance has different impacts on the family experience, creating transnational families linked by connection and disconnection. At the one extreme, the men who migrate on their own leave their family behind and are separated from them for years at a time. Usually, there is no possibility of returning home during the work permit period, both because of passport reasons and because the wages would not cover a trip home. Although this model often applies for the shorter cycle of migration – and the separation can be one reason to return home early or not to migrate again – some workers extend this way of life for longer periods. One respondent, for example, had been working for 20 years in Malaysia, and had returned home only thrice during this time. Being somewhere else and being left at home becomes normal, the usual, with relationships being based on the phone and occasional stints of interim unemployment as family time in the home village.

At the other end of the family continuum, migrant workers – especially to Sabah – migrate as a couple, or meet up whilst in Malaysia, and can take children with them or have them there. This can have all sorts of consequences. During the first migration wave to Sabah, it was possible for some of the migrant workers to gain permanent residency status or even, eventually, Malaysian citizenship. This is the story of those young workers in the plantations who are second-generation Indonesians. Growing up in Sabah, they have only ancestral roots in a village in Indonesia, perhaps grandparents, without ever having been there. One respondent, a young man who grew up in a 'Bugis village' (in Sabah; the Bugis are the main ethnic group in southern Sulawesi, a main source of migrant workers in Sabah), was on his way to Sulawesi to meet his sister for the first time in his life. She had been left with her grandparents when her parents set out to work in the plantations. In more recent years, when the children reach school age, a decision has to be taken regarding their education. If the workers are lucky to have an NGO school[3] on or near their plantation, they can stay on in Sabah for a while longer. After that, enabling their children to go to school means that they either return to Indonesia or send the children back to live with relatives.

The result is disconnected and dislocated families that span the transnational space. One respondent's brother had worked in Malaysia since 1996; he returned three times to the village during that time. His wife later went to join him and left her one-and-a-half-year-old son in the village, who was then brought up by his grandmother. In another case, the distance created by economic and political precarity led to the separation of one woman worker from two husbands. The first time, the husband returned to Indonesia because of illness and died there without her being able to visit him because of the cost. Her second husband left her while they were in different countries and took two of her children with him, who she didn't see for several years. Another family transnationalized when a migrant worker pair sent their first born, six-year-old child home to Java to live with its grandmother, because the grandmother was feeling lonely. When their next child was four years old, the mother went back home. The father planned to return a year later.

A wide variety of transnational and transgenerational family networks emerge. Migrant workers who left their parents and siblings behind in Indonesia, migrant workers who originally travelled to Malaysia as children with their parents then stayed on and whose parents are now back in the village, workers who have left their children behind, or sent some of them back. In the transnational family, having some of the family in the other place becomes a normal and long-term state of affairs.

In the Sabahan spaces, where more workers are there with their families and where more permanent livelihoods have been established than in Peninsular Malaysia, new milieus are formed along extended family ties. Newer workers tread in the paths of uncles, aunts, brothers and parents who have gone before them and use this grapevine to find their way, look for a more decent job, etc. Intergenerational continuities and spatial continuities linking specific villages with certain agents, companies or areas in Malaysia create 'village clusters' (*bergerombol sesama kampong*). Extended family and village connections are plugged into embedded local structures that developed over decades, with naturalized migrants serving as the rock of stability on which solidarity networks can be based. The 'Bugis network' (one based on the Bugis ethnicity) is one that is firmly entrenched within Sabahan society – encompassing 'illegals', temporary migrants, recurring migrant workers, workers with permanent resident status or citizenship and second-generation Bugis-Sahabans. The 'Indon network' (more generally based on coming from Indonesia) also functions in the case of police razzias. If there is a razzia, the information is passed on, so other illegals in the area go and hide. There is also solidarity from the local Malaysian population.

However, there is also a more sinister side to the transnational networks. Class divisions and exploitative social relations also shape the new networks: the contractors sell 'their' workers on to other contractors of the same 'tribe', for example to pay off a debt of their own. The workers then have to work to pay off this debt, which is taken over by the new owner. Because they are also transnationally connected, they can reach into the home villages of the workers and so it is difficult to escape them and the thugs that ensure that the 'debt' is paid. Debt bondage also works via the village informal lending system and having to pay this back, even across the transnational social space. This is the darker side to the Bugis network: they can find you as a runaway worker and there is the threat of retaliation at home. At the extreme end of the debt bondage scale, this type of network dominates the experience of workers who are isolated from other kinds of social networks based on family members or friends. For them, Sabah is filled with hostile strangers and police from whom danger and not solidarity can be expected. The ties to their home village are characterized by exploitation and violence.

## Networks enable everyday resistance

From the perspective of migrant networks, the bounded territorial spaces of the estates become relative. For example, the guards controlling access to the estate suggest a tight control by management. However, these guards are also low-paid workers who are connected to other workers via family ties or through transnational networks to the Indonesian village, and they use discretion as to whom they let in and out. During research in one estate (19 September 2011), by way of illustration, the author stayed at the house of one respondent whose father was such a guard. One morning, the company owner had flown in from Kuala Lumpur and was making the rounds, checking up on the workers and management. This was communicated via the guards' walkie-talkies, so that we knew where the 'big boss' was and when it would be safe to leave the estate and which gate to use to travel onwards. This kind of 'everyday resistance' (Scott 1977) goes on all the time, but there are three general practices by which workers are reshaping the palm oil industry in Malaysia.

The first area of resistance is related to border negotiation. Hundreds of thousands of migrants actively challenge the Malaysian state's right to impose border controls and find many ways of circumventing the official prescriptions and restrictions regarding permits, passports, health checks, children etc. This can include illegal entry, but also more stealthy means, such as using multiple passports and entering legally under a different name, or forging birth certificates to gain resident rights (see also Idrus 2008). Evasive strategies are often conducted with the collusion of the palm oil companies, including the large corporations. For example, the health checks required by authorities are often ignored or postponed (in case of pregnancy), because companies want to keep their skilled workers. Tens of thousands of 'illegal' children are testimony to the success and widespread nature of this practice (Tenaganita 2006). Women play a decisive role in the production of a new transnational scale of social reproduction by asserting their right over their own reproduction and by finding ways in which to bring up and educate their children. Multiple entries and longer-term stays 'denationalise the territorial space' (Mazzucato 2004), transforming it from an 'in and out of temporary labour power' into one characterized by durable personal and social relations.

A second widespread tactic is absconding (*lari*) to work where the wages are higher (Pye *et al.* 2012). For this, the 'Indon network' or the 'Bugis network' comes into play. Not only do the subterranean networks offer information on current rates, experiences of 'good' and 'bad' employers etc., they also provide the informational infrastructure needed to move around and secure employment elsewhere. The illegal status this implies is preferred because the workers gain their freedom to choose and change employers (see also Idrus 2008). In this way, respondents were able in some cases to double their pay. The group discussion of the work gang in the FELDA estate in Sabah (see above) is insightful for how this can develop. The gang had been dissatisfied for some time with the pay resulting from the harvesting contract, which had dropped from around 1,000 Ringit (around €250) to 600 Ringit, but the gang leader invoked their work ethic to push them to higher results (thereby internalizing the outsourcing system). During the discussion, it was revealed that the gang members had already decided to leave that estate and look for a better deal elsewhere. They compared former wage labour favourably with the contract system, which had provided reasonable wages, plus sick leave and bonuses on public holidays. Their 'moral economy' (Scott 1977) of what was fair was related to a wage that left enough for some savings, but also to the style of the manager (the former one had been fair, the current one disrespectful). The gang members were also influenced by the fact that other gangs had already left the area. The discussion then revolved around the best tactic, with the gang leader (with wife and children in Sabah) pressing his workers to wait until they had a new contract in a different FELDA estate so that they could retain their work permit and legal status whilst the younger, single gang members were itching to leave and prepared to forsake legality if they could earn more sooner. Despite being a subcontracted and spatially-bounded work gang, the workers had produced their own networked space that went beyond the estate in which they worked and which gave them a bigger picture of what was going on and more options on how to respond (Group discussion, FELDA Gang, 1 July 2012).

We have seen how workers use their transnational networks to circumvent and adapt to the precarious labour regime they are confronted with. Surviving in illegality and *lari* are adaptation strategies that unconsciously or indirectly challenge the precarious regime. However, this kind of everyday resistance can empower workers to pursue more direct and collective action. A surprising number of respondents also talked about successful and unsuccessful collective actions, ranging from direct negotiations with bosses leading to increased pay, wildcat strikes, and petitioning the Indonesian Embassy to intervene on their behalf. This tendency has been noticed with concern by the plantation owners, who characterize the Indonesian workers as 'organized in a gang-like fashion', and who '*down-tools* readily and cause industrial unrest on plantations' (Daud 2006: 46, italics in original). In one of the group discussions, workers from a mill in Sabah discussed a wildcat strike (*mogok*) in which they had participated a few days before. Again, the decisive impetus for the strike was the moral economy of the workers and a sense of injustice over their treatment. The issue in this case was a wage raise for the staff members (i.e. office workers and management), who tend to be Malaysian citizens rather than Indonesian denizens. The wage increase was seen as unfair because the corresponding wage increase for the workers, which had been promised, was withheld. The experience of the strike itself radicalized the workers, giving them a feeling of their own power. After the strike, their self-confidence had increased and they were both enraged and amused at the subsequent actions of management, who had been trying to find out who the 'ringleaders' of the strike were. During the discussion, the workers connected the important wage issue to more principled questions regarding the status and rights of migrant workers in Sabah and the disrespect shown to them by Malaysian management (Group discussion, mill workers, 6 July 2012).

## Conclusion

It was the labour of Indonesian workers that converted the rubber plantations of Peninsular Malaysia to palm oil and that cleared the forests of Sarawak and

Sabah. They drained the land and planted the oil palm seedlings; they built the roads and the mills. Because this work is alienated wage labour, it is driven and determined by capital and management but, at the same time, work conditions and the political and social context of the work are contested by the workers themselves.

Everyday resistance by workers offers various opportunities for political interventions on labour issues in the palm oil sector. In particular, the production of a transnational scale of social reproduction by workers could be the basis for new kinds of organizing that reflect the migrant networks and the place–time contingencies connecting places in Indonesia with the plantations and mills in Malaysia. This could be used to transcend the estate scale that is still predominant in the practice of wildcat strikes and to scale up industrial action. Successful strategies would have to organize both 'legal' and 'illegal' workers, to acknowledge and accept the challenge of organising 'denizens', and also forge solidarity ties between environmental justice movements and labour struggles.

Taking the struggles of workers as the starting-point in this has two advantages. First, collective action by workers themselves helps them to develop the political consciousness necessary for collective organizing. Second, seeing the labour movement as a potentially proactive ally gives the environmental justice movement a different leverage than mobilizing consumers and pressuring the big brands, which can be accommodated by the sustainability and branding departments of the corporations (as in the RSPO, see Pye 2014). It would offer an alternative to current approaches focusing on 'slavery' that seek to integrate labour standards into the RSPO. One such report, for example, recommends 'interventions for key stakeholder groups, namely governments and corporations, to eliminate the industry's dependency on and exposure to slavery' (Accenture for Humanity United 2013: 3), i.e. the coalition that set up the precarious labour regime in the first place. Workers are not mentioned *at all* as a 'key stakeholder group' and are offered no course of action.

An organized workers' movement that spans the transnational social space between Malaysia and Indonesia could become a relevant force for organizing up along the transnational commodity chain. At the moment, consumer-oriented campaigning works by pressuring management along these chains. But linkages between groups of workers involved in the production of agrofuels could open up new possibilities. The social and political issues of agricultural production could be more actively linked to the urban political questions of climate change and to a transition towards a different kind of energy production and mobility system (Neale 2011).

## Notes

1 The main body of research was conducted in the context of the project 'The Making of Social Movements under Conditions of Precarisation and Transnationalism in Southeast Asia' funded by the German Research Foundation (2009–2013). This consisted of a longitudinal study with 10 groups of workers consisting of biographical interviews, subsequent in-depth interviews and finally group discussions. The interviews series

were conducted over a period of three years with five groups of workers who had returned to their villages in northern Sumatra, Central Java, Lombok and Makassar (Sulawesi) and with five groups of workers in plantations and mills in Sabah, Malaysia, in the area between Tawau and Sandakan. Biographical narratives offered insights into the life planning and biographical precarity of workers and into their transnational networks. In-depth interviews followed usually a year later, creating the possibility of catching up with what had happened in the meantime. The ranking and positioning and repositioning exercises were focussed on framing and political consciousness. The group discussions took place sometime later, involving 'natural groups' of workers who knew each other. The workers chose the topics they thought most relevant and these usually developed into controversial discussions. Unless otherwise stated, results presented here are from this project and will not be individually referenced.

2 See www.forbes.com/malaysia-billionaires/ (last access 22 March 2014).

3 Because migrant children are no longer allowed to visit Malaysian state schools, the Malaysian NGO Humana Child Aid has set up primary schools in some plantations in Sabah. See www.humanachildaid.org/projects/plantation-projects/ (last access 22 March 2014).

## References

Abdullah, Ramli, Azman Ismail and Ayatollah Khomeini A. Rahman (2011) 'Labour Requirements in the Malaysian Oil Palm Industry 2010', *Oil Palm Industry Economic Journal* 11(2), 1–12.

Accenture for Humanity United (2013) 'Exploitative Labor Practices in the Global Palm Oil Industry' (http://humanityunited.org/pdfs/Modern_Slavery_in_the_Palm_Oil_Industry.pdf, last access 26 February 2014).

Aditjondro, George J. (2001) 'Suharto's Fires. Suhartos Cronies Control an ASEAN-wide Palm Oil Industry with an Appalling Environmental Record', *Inside Indonesia 65* (www.insideindonesia.org/feature-editions/suhartos-fires, last access 26 February14).

Castel, Robert (2009) 'Die Wiederkehr der sozialen Unsicherheit', in Castel, Robert and Klaus Dörre (eds) (2009) *Prekarität, Abstieg, Ausgrenzung. Die soziale Frage am Beginn des 21. Jahrhunderts.* Frankfurt, New York: Campus Verlag, 21–34.

Daud, Amatzin (2006) 'Labour Constraints in the Plantation Industry', *Oil Palm Industry Economic Journal* 6(2), 37–48.

European Commission (2012) 'Commission Implementing Decision of 23 November 2012 on Recognition of the 'Roundtable on Sustainable Palm Oil RED' Scheme for Demonstrating Compliance with the Sustainability Criteria under Directives 98/70/EC and 2009/28/EC of the European Parliament and of the Council', *Official Journal of the European Union L* 326/53 (http://eur-lex.europa.eu/LexUriServ/LexUriServ.do?uri=OJ:L:2012:326:0053:0054:EN:PDF, last access 26 February 2014).

Gerasimchuk, Ivetta and Peng Yam Koh (2013) 'The EU Biofuel Policy and Palm Oil: Cutting Subsidies or Cutting Rainforest?' (www.iisd.org/gsi/sites/default/files/bf_eupalmoil.pdf, last access 26 February 2014).

Herod, Andrew (2001) *Labor Geographies. Workers and the Landscapes of Capitalism.* New York, London: The Guilford Press.

Idrus, Nurul Ilmi (2008) 'Makkunrai passimokolo'. Bugis Migrant Women Workers in Malaysia', in Ford, Michele and Lyn Parker (eds) *Women and Work in Indonesia.* Oxon, New York: Routledge, 155–194.

International Labour Rights Watch and Sawit Watch (2013) 'Empty Assurances' (www.laborrights.org/stop-child-forced-labor/resources/empty-assurances, last access 26 February 2014).

Kelly, Philip F. (2002) 'Spaces of Labour Control: Comparative Perspectives from Southeast Asia', *Transactions of the Institute of British Geographers, New Series*, 27(4), 395–411.

Kelompok Pelita Sejahtera (2008) *Buruh Harian Lepas. Studi Kajian Hubungan Kerja, Upah dan Kesejahteraan di Perkebunan Sumatera Utara.* Medan: KPS.

Lai Wan Teng (2011) 'Gender and Livelihood: A Case Study of the Mah Meri and the Oil Palm Plantations of Carey Island', *Asian Journal of Women's Studies* 17(2), 66–98.

Li, Tania Murray (2009) 'To make live or let die? Rural Dispossession and the Protection of Surplus Populations', *Antipode* 41(S1), 66–93.

Li, Tania Murray (2011) 'Centering Labor in the Land Grab Debate', *The Journal of Peasant Studies* 38(2), 281–298.

Mazzucato, Valentina (2004) 'Transcending the Nation. Explorations of Transnationalism as a Concept and Phenomenon', in Kalb, Don, Wil Pansters and Hans Siebers (eds) *Globalization and Development. Themes and Concepts in Current Research.* Dordrecht: Kluwer Academic Publishers, 131–162.

Neale, Jonathan (2011) 'Our Jobs, Our Planet. Transport Workers and Climate Change' (http://climateandcapitalism.com/wp-content/uploads/sites/2/2014/02/our-jobs-our-planet-feb-2014.pdf, last access 26 February 2014).

Peluso, Nancy Lee, Suraya Afiff and Noer Fauzi Rachman (2008) 'Claiming the Grounds for Reform: Agrarian and Environmental Movements in Indonsia', in Borras, Saturnino M., Marc Edelman and Cristobal Kay (eds) *Transnational Agrarian Movements Confronting Globalization.* Chichester: Wiley-Blackwell, 91–121.

Pye, Oliver (2008) 'Nachhaltige Profitmaximierung: Der Palmöl-Industrielle Komplex und die Debatte um 'nachhaltige Biotreibstoffe', *PERIPHERIE* 112, 429–455.

Pye, Oliver (2010) 'The Biofuel Connection: Transnational Activism and the Palm Oil Boom', *The Journal of Peasant Studies* 37(4), 851–874.

Pye, Oliver (2014) 'RSPO Governance Unravelling. The Round table on Sustainable Palm Oil and the Palm Oil Industrial Complex', in Cramb, Rob and John McCarthy (eds) *The Oil Palm Complex: Agrarian Transformation, State Policy, and Environmental Change in Indonesia and Malaysia.*

Pye, Oliver and Julia (2014) 'Climate Politics and the Gendered Palm Oil Landscapes of Southeast Asia', in Dannenberg, Petra and Birte Rodenberg (eds) *Klimaveränderung, Umwelt und Geschlechterverhältnisse im Wandel – neue interdisziplinäre Ansätze und Perspektiven.* Münster: Westfälisches Dampfboot.

Pye, Oliver, Ramlah Daud, Yuyun Harmono and Tatat (2012) 'Precarious Lives: Transnational Biographies of Migrant Oil Palm Workers,' *Asia Pacific Viewpoint* 53(3), 330–342.

Ramasamy, Palanisamy (1994) *Plantation Labour, Unions, Capital, and the State in Peninsular Malaysia.* Oxford, Kuala Lumpur: Oxford University Press.

Savaranamuttu, Johan (2012) 'The Political Economy of Migration and Flexible Migration Regimes. The Case of the Oil Palm Industry in Malaysia', in Pye, Oliver and Jayati Bhattacharya (eds) *The Palm Oil Controversy. A Transnational Perspective.* Singapore: ISEAS, 120–139.

Scott, James (1977) *The Moral Economy of the Peasant. Rebellion and Subsistence in Southeast Asia.* New York: Yale University Press.

Serbundo (2013) Indonesian Trade Unions Alliance (Serbundo). Stop Exploitation of Plantation and Industrial Labor. Stop a Large Scale Plantation Expansion' (http://understory.ran.org/wp-content/uploads/2013/11/Labor-Statement-RSPO34979D.pdf, last access 26 February 14).

Standing, Guy (2011) *The Precariat. The New Dangerous Class.* London, New York: Bloomsbury Academic.

Stoler, Ann Laura (1995) *Capitalism and Confrontation in Sumatra's Plantation Belt, 1870–1979.* Ann Arbor: University of Michigan Press.

Tenaganita (2006) 'Acting Today for Tomorrow's Generation', documentation of the Regional Conference on Stateless/Undocumented Children in Sabah, 16–18 November 2005, Kota Kinabalu.

Tenaganita and Pesticide Action Network Asia Pacific (2002) *Poisoned and Silenced. A Study of Pesticide Poisoning in the Plantations.* Kuala Lumpur: Tenaganita and Pesticide Action Network Asia Pacific.

Wong, Diana (2006) 'The Recruitment of Foreign Labour in Malaysia: From Migration System to Guest Worker Regime', in Kaur, Amarjit and Ian Metcalfe (eds) *Mobility, Labour Migration and Border Controls in Asia.* New York: Palgrave MacMillan.

Wong, Diana and T.A.T. Anwar (2003) 'Migran Gelap: Indonesian Migrants in Malaysia's Irregular Labor Economy', in Battistella, Graziano and Miluja M. Asis (eds) *Unauthorized Migration in Southeast Asia.* Quezon City: Scalabrini Migration Center.

World Vision (2013) 'Forced, Child and Trafficked. Labour in the Palm Oil Industry' (http://campaign.worldvision.com.au/wp-content/uploads/2013/04/Forced-child-and-trafficked-labour-in-the-palm-oil-industry-fact-sheet.pdf, last access 26 February 2014).

# 12 Agrofuel networks

## A case study of the regional agrodiesel network in Chhattisgarh (India)

*Shishusri Pradhan*

### Introduction

India is among the top five consumers of energy in the world (World Bank 2008) and 70 per cent of its crude oil demand is met through imports (Government of India 2011). Among the renewable sources of energy being explored, agrofuels have emerged as a key option, especially liquid agrofuels for transportation. The Government of India (GoI) in its Vision 2020 report stressed the importance of an agriculture-based energy policy and argued that the production of fuel oil and biomass would result in the creation of farm jobs and alternative markets, and reduce oil imports (GoI 2002). In 2003, the Planning Commission's *Report of the Committee on the Development of Bio-fuel* introduced the Ethanol Blending Programme (EBP) and the National Mission on Biodiesel (NMB)[1] as policy options that supported the production and blending of bioethanol (from sugarcane) and biodiesel (from *Jatropha curcas*) in petrol and diesel respectively (GoI 2003).

Among the two agrofuel options, the production and blending of agrodiesel was taken up actively because it was claimed that *Jatropha curcas* (hereafter Jatropha) is a hardy shrub that can grow on marginal lands, prevent erosion, reclaim lands and reduce competition with food crops, as it is a non-edible oil seed (Ariza *et al.* 2010). The biodiesel programme stressed the importance of not diverting arable land for the production of energy-crops and emphasised that Jatropha would be cultivated on land under the scheme of Joint Forestry Management (JFM), hedges around agricultural land, 'culturable fallow lands' and dry, marginal 'wastelands' (GoI 2003). The programme also claimed that agrofuels would 'generate massive employment for the poor belonging to the Scheduled Tribes, Scheduled castes and other under privileged categories living mostly in backward areas which have experienced the adverse impact of forest degradation, and loss of natural resources' (ibid.: 149). Hence the biodiesel programme was promoted as a pro-poor development initiative that would create jobs for rural people, and under which the non-edible feedstock would be grown on dry, marginal wastelands and in turn not lead to food security issues.

Once the biodiesel programme had been introduced as a policy mission at the national level, various state governments were keen to follow it; however, the

NMB was not adopted uniformly across various states in India. This is because the regional networks across the states were diverse in their composition, types of actors involved and the manner in which they supported and introduced the biodiesel mission. These regional networks comprised actors from the government (politicians, bureaucrats, policymakers), research centres, private companies, and non-governmental organizations (NGOs), who played a crucial role in popularizing the local biodiesel narratives and in turn ensuring the uptake of the mission. I contend that the NMB spread across India because of the creation of three types of networks: government-led networks, research centre-led networks, and private company/NGO-led networks. In this chapter, I focus on the creation and role of a government-led network in introducing and actively promoting the production of biodiesel in the state of Chhattisgarh. Chhattisgarh was one of the first states to actively adopt the biodiesel programme in India; it wanted to attain the status of a biofuel-reliant state by 2015, aiming to plant 160 million saplings of Jatropha, and the state government was proactive in setting up a special biodiesel agency to achieve these objectives of the biodiesel mission.

This chapter draws on a political ecology perspective to analyse the role of powerful actors, problems of expertise, the environmental subjects and local knowledge (Blaikie and Springate-Baginski 2007) involved in the development of the National Biofuel Mission in Chhattisgarh. I first sketch the political ecology of agrofuels networks in India and then take a closer look at the government-led network in Chhattisgarh. To do so, I draw on network terminology to delineate the creation of a regional agrodiesel network in Chhattisgarh: I argue that the state government was crucial in generating an extensive agrodiesel programme that hoped to achieve both energy security and rural development objectives. I then take a look at the implementation on the ground and how a secondary network of local state authorities and farmers embraced the agrofuel narrative. However, despite initial enthusiasm, empirical research conducted during this first implementation period unearthed contradictions within the programme, leading to the politicization of agrofuels around the issue of wastelands and common property resources.[2] I conclude that these contradictions led to a decoupling of the state-led network from the implementing networks on the ground.

## Political ecology of agrofuel networks in India

To understand socio-environmental issues and changes, political ecology draws on various theses (patterns of explanation), one of which deals with the role and 'uneven power of different players contending over the use and the management of natural systems' (Peet *et al.* 2011: 27). Specifically, political ecology emphasizes the manner in which 'power is exercised *within* – rather than over – individuals, communities and societies' (ibid.: 32). It addresses not only how power is exerted through control and political institutions, but also the role of narratives, discourses, expertise and networks. Political ecologists have drawn on network terminology to reveal the role of actors, how power is exercised within these networks and in turn how actors support and promote discourses.

Actor networks trace the multiple associations and relationships between actors, both human and non-human, within a network. 'Once actors/actants are enrolled, social interests are temporarily stabilised. These allegiances function to build specific forms of truth, through moments of translation and finally mobilisation' (Yearley 2005; cf. Pradhan and Ruysenaar 2014: 301). Proponents of actor-network theory argue that the actors' interests are themselves the 'outcomes of negotiations and interactions' (Yearley 2005: 55). Political ecology is increasingly concerned with the role of actors – how they construct knowledge and embrace particular narratives and ideologies.

Manuel Castells (2004) posited that flows and networks – instead of state and societies – are the new architects of global modernity. John Urry (2003) suggests that networks and flows function in three spatial patterns, namely regions, global integrated networks (GINs) and global fluids. Regions comprise actors, objects and networks that are clustered together in a geographical area, normally within a country. They are characterized by conditions within the country and are restrained by boundary restrictions; they dominated the pre-globalization area. GINs consist of complex, enduring and predictable networked connections between people, objects and technologies spanning different regions. GINs cross regional boundaries and deliver the same kind of outcome at all nodes, with limited adaptation to local circumstances. As the networks overcome regional boundaries, things are made close by the linkages within the network. In turn, 'such a network of technology, skills, texts and brands' (ibid.: 56–57) ensures that the 'products are predictable, calculable, routinised and standardized' (ibid.), and the same type of product is delivered across the entire network. There are many global enterprises organized through GINs, including McDonald's, American Express, Coca-Cola, Greenpeace and so on (Mol 2007). Global fluids are spatial patterns that are restricted neither by boundaries nor by stable relations; they are characterized by flexibility, gel-like movement and permeable boundaries. They do not have a marked starting-point, sequence or goal. Migrating people, the internet and social movements are examples of global fluids (ibid.).

Arthur Mol (2007) employs Urry's conceptualization of networks to investigate the emerging biofuel system. He says that initially agrofuel networks were strongly dominant in a region and were bounded by boundaries with limited global integration. However, with time, a global integrated biofuel network (GIBN) has emerged in which the actors, objects and relations are not bounded by region and there is an increase in the transboundary flow of biofuels. He cites the case of Brazil as an example of a regional biofuel network, where the government actively invested in bioethanol production from sugarcane and instituted an elaborate infrastructure supporting it. However, with the growing demand for biofuels in the world market, a GIBN has developed which has led to a proliferation of national biofuel regions, starting with Brazil and spreading to a number of developed and developing countries. He argues that these developments resulted in major changes in agrofuel networks and flows. Initially in local regions, farmers, cooperatives, and individual processors were the main actors, but large companies and government agencies are increasingly emerging as powerful actors in the global and local biofuel networks. Based on the success of Brazil and the USA in producing bioethanol from sugarcane and corn, the global agrofuel narratives gained prominence, with the GIBN actively supporting the development of agrofuels the world over. Subsequently, national agrofuel networks were created across many countries and the national networks created and supported national narratives promoting the development of agrofuels in the respective countries.

India's attempts to promote agrofuels were based on the assumptions of the global agrofuel narratives. The biodiesel programme was more popular than the bioethanol programme due to the choice of a non-edible oilseed, reducing issues of food security. The national biodiesel narratives promoted rural development, the creation of jobs at the farm level, the revitalization of the agriculture sector, energy security and environmental mitigation. According to scholars of actor-network theory (ANT), scientific activity, the pursuit of new knowledge and the stabilization of new ideas in a society is dependent on the ability of key actors either from the government, research organizations, industry or NGOs, or individuals, to build long chains of networks of association. Bruno Latour (1996) argues that reality results from the building of successful alliances and these alliances are composed of actors, institutions and technologies, and non-human actors.

The agrodiesel narratives and agrofuel ambitions of India were supported by the creation of strong networks comprising actors who patronized agrofuels. The key actors were government officials, policymakers, bureaucrats, scientists and representatives from industry and NGOs. Pradhan and Ruysenaar (2014) trace the creation of a policy network at the national level in India and describe how actors in this network recruited policy entrepreneurs who publicized the agrodiesel narratives and in turn enrolled other actors who supported and promoted the objectives of the NMB. Policy entrepreneurs were influenced by actors from the GIBN and in turn they introduced the NMB as a policy option in India. Despite the NMB being introduced as a policy mission at the central level, it did not spread uniformly across the country. I argue that the NMB spread across India based on the type and role of regional networks supporting it across the various states. A range of actors comprising government officials, policymakers, bureaucrats, scientists and representatives from private companies and NGOs formed alliances to promote the agrodiesel initiative across various states and these alliances were characterized by local conditions, ruling governments, political, social and cultural factors. Actors from the regional networks were influenced by the national agrofuel policy network and many of the officials from private companies were from international companies that were part of the private companies representing the GIBN.

## Chhattisgarh: a regional government-led network

Chhattisgarh is located in central India and was formed when 16 Chhattisgarhi speaking south-eastern districts of Madhya Pradesh gained statehood on 1 November 2000. More than 80 per cent of the state's population relies on agriculture for subsistence. The state government of Chhattisgarh (GoC) has been introducing a range of rural development schemes and projects and the pro-poor, pro-development objectives of the NMB gained purchase among the government officials. As 83 per cent of the population is engaged in agriculture and allied sectors, it was envisaged that the state could become self-sufficient[3] in biodiesel to be blended at 5 per cent by 2015 by planting Jatropha (GoC 2005).

To achieve this status many actors had to be enrolled and the negotiations between them resulted in a mandate to cultivate Jatropha in mass quantities to produce agrodiesel. On 26 January 2005, the state government established the Chhattisgarh Biofuel Development Authority (CBDA), which was designated as the primary organization responsible for promoting widespread plantations of Jatropha and the production of agrodiesel. In this government-led regional network, the primary network was created with the CBDA as the principal agency; members of the CBDA expanded this network by enrolling actors from related government institutions and research centres. The CBDA established linkages with the Forest Department, Department of Agriculture, Forest Corporation, Minor Forest Produce Federation, Chhattisgarh Renewable Energy Development Agency (CREDA) and the agricultural universities. Members of these organizations and the CBDA formed the primary network of the Chhattisgarh government-led agrodiesel network.

As the benefits of the NMB as a rural development initiative had already been promoted in the Planning Commission's Report on Biofuel (GoI 2003), it was easy for the actors in the primary network to interest other actors to join and promote this initiative. Initially the state government and related agencies were the key actors; however, they successfully enrolled actors from the private sector. The key strength of the biodiesel initiative was its degree of interpretative flexibility. Even though the final goal was to produce agrodiesel, the production of agrodiesel encompassed a gamut of activities ranging from plantation, research on oil seeds, management of wastelands, oil-processing and checking the standards of the agrodiesel produced before it was blended with diesel. Hence, the programme covered a range of interests of the numerous actors involved. For example, the agricultural universities were keen on experimenting, identifying and developing the appropriate genotype. Research scientists stated:

> We are hoping to identify the right type of genotype which can then be used by the state government to grow Jatropha.... We are conducting trials and have given seeds to the CBDA officials which they will distribute to the farmers. In this way we can check the viability of the seed under natural conditions.
>
> (Interview with Research Scientist, Anonymous, 5 May 2009)

CREDA, the official renewable energy department in Chhattisgarh, was keen to produce agrodiesel to offset the usage of non-renewable fuels and it actively invested in the agrodiesel programme. As the final goal of the biodiesel mission was to blend agrodiesel with diesel, private companies, especially from the oil sector, were eager to support the agrodiesel initiative. Many national and international companies were involved in Jatropha plantations in Chhattisgarh. Government officials enrolled actors from universities and private companies to promote the agrofuel ambitions of the state. The Indian Oil Corporation Limited (IOC), formed a joint venture with the state agencies and announced plans to employ 33,000 farmers; it expected to produce 100,000 tons of biodiesel (Indian Oil Corporation Limited 2006). CREDA-HPCL Biofuel Limited is a subsidiary company of Hindustan Petroleum Corporation Limited (HPCL) and Chhattisgarh State Renewable Energy Development Agency (CREDA) for the plantation of Jatropha in the State of Chhattisgarh (www.hindustanpetroleum.com). D1 Oils and Reliance Pvt Limited also invested in Jatropha plantations in Chhattisgarh.

Once the actors in the primary network were stabilized, they undertook a range of activities. Across sixteen districts, 210 million Jatropha saplings were planted on 8,400 hectares of land (http://cbdacg.com/about.htm, last access 13 August 2014). To ensure that the plantations were carried out successfully across the various districts, the actors from the primary network formed a secondary network comprising actors from the various districts. They set up a task force in each district, which was headed by the collector of that district and in turn included representatives from NGOs, regional universities, private firms and farming communities. The government officials from the primary network enrolled actors from various districts and they constituted the secondary network. Actors from the primary network provided financial, political and legal support, and Jatropha seeds. The actors in the secondary network had to ensure that Jatropha was planted across the various districts. In this way, actors from both the primary and secondary networks reached a consensus on where and how to cultivate Jatropha. Jatropha saplings were effectively planted on 8,400 hectares between 2005 and 2006, and an additional 6,800 hectares of land were devoted to the plantation of 170 million saplings from 2007 to 2009.

The progress of a project requires the constant work of 'translation (of policy goals into practical interests; practical interests back into policy goals), which is the task of skilled brokers (managers, consultants, fieldworkers, community leaders, etc.)' (Mosse 2004: 647), who translate the work of the project into the language of the stakeholders supporting it. One of the key proponents of the biodiesel mission in Chhattisgarh was the Chief Minister, Raman Singh. He was the first Chief Minister to use agrodiesel in his official vehicle and promised to switch all state-owned vehicles using diesel and petrol to biofuels (Jain 2006). Labelled the 'poster boy of biofuels' (ibid.), he was an ardent supporter of biodiesel production, saying:

> This is just the start of the bio-fuel energy revolution in Chhattisgarh. Jatropha diesel will power all government vehicles within three months....

> The state is tipped to cultivate 80 million Jatropha seedlings in rural areas to make Chhattisgarh a bio-fuel self-reliant state by 2015.
>
> (Ibid.)

The government of Chhattisgarh played a key role in promoting agrofuels by recruiting and constructing strong chains of allies. Actors from the CBDA set up demonstration plantations in various districts to foster investments from the private sector. The different actors involved in this network expressed varying interpretations about the viability and success of the programme. For instance, Shri S.K. Shukla, the executive director of CBDA, had an optimistic outlook towards the initiative of developing biodiesel. He was sceptical, however, about the pace at which biofuel blending targets were being set. In his speech at the fourth International Conference on Biofuel in Delhi (2008), he said, 'The experience of Jatropha plantation and its commercial production is still limited, even then the Indian Government is taking a large step into the unknown' (Shukla 2008). In contrast, the assistant project manager at CBDA, Mr J.L. Gupta, firmly believed in the 'pro-poor' narrative of this development initiative to improve the livelihoods of the poor. He was actively involved in promoting the cultivation of Jatropha by farmers and landless labourers in Chhattisgarh and was keen to impart information on cultivation and pruning techniques. The majority of interviewees from the industry said that they were cultivating Jatropha not 'only to profit from it' but also to improve the livelihoods of the rural users. Officials from Reliance Life Sciences said that their vision was to 'establish first end to end biodiesel initiative in India' (www.rellife.com/biofuels.html, last access 13 August 2014). Officials from this company said that, according to them, the most important thing was to 'establish acceptable Jatropha economics for marginal farmer on marginal lands' (ibid.) and once that was done, they were aiming to have a steady supply of Jatropha seeds from the farmers at a rate of 5 INR/kg (US$0.08). To achieve this mission, their objectives aimed at: acceptance of Jatropha cultivation by the farmers, multiple engagement points with the farmers, cultivation and productivity on marginal lands, creation of 'farmer stakes' through 'producer companies' and partnership with NGOs (ibid.).

Reflecting on Law and Callon's (1992) analysis of the development of a technology or initiative, it can be affirmed that appropriate actors had formed the necessary networks in Chhattisgarh to promote the cultivation of Jatropha under the NMB. However, on mapping the trajectory of this initiative certain nuances were spotted which hindered the progress of the NMB and led to a stasis for intermittent periods of time. It was observed that the entire programme, its planning, cultivation practices, production of oil and dissemination of information was contrived entirely among the key actors of the primary network. Once the objectives were set, the secondary network was created to take the initiative further and then the farmers were involved. The main actors in the networks were government officials, and their role was to ensure that the local people believed in the credibility of the biodiesel mission. After the CBDA was set up, it launched a mass awareness campaign to convince the rural people that this initiative was being promoted to augment their efforts in seeking better sources of revenue. The NMB was publicized through the media, across local community forums and conferences, and free seeds were distributed to the farmers. Additionally the support exhibited by the Chief Minister, who had already introduced other successful programmes,[4] was conducive to the propagation and acceptance of this programme.

## Implementing the programme

There were two categories of rural users[5] in Chhattisgarh: the farmers who cultivated Jatropha on their land, mostly holding marginal, small or medium-sized areas of land; and the labourers who cultivated Jatropha on land owned by the government, private companies, large farmers and the community. In Chhattisgarh, three types of plantations were carried out: block plantations, roadside and farmer plantations. Farmer plantations were carried out on land owned by farmers but under the guidance of CBDA members. In Chhattisgarh, Jatropha plantation was undertaken across many districts, and data presented in this section represents farmers cultivating Jatropha in the villages situated in the districts of Bilaspur, Janjgir, Korba, Mahasamund, Raigarh, Raipur and Surguja.

The majority of the farmers in these areas were marginal or small farmers and the average size of the plots was 0.5 hectares. The farmers spoke about low soil fertility and how parts of their land were not productive for the cultivation of crops. Hence, when they were approached by officials from CBDA and informed about Jatropha and its properties, it caught their attention. Despite their interest in this new crop, the farmers' views on it were divided and they debated whether to cultivate it or not. They were worried about the long gestation period of three years and about revenue for this period of time. Their concern was solved by the CBDA officials, as they were told not to cultivate only Jatropha but to inter-crop it with other crops. Despite various reservations, when a few farmers received free seeds and help from the CBDA members to plant Jatropha, others also did the same. The farmers were expecting a good yield and had been reassured of a buy-back policy by the government officials. However, the farmers had not been trained in the appropriate farming, pruning and spacing techniques, and hence, within the first year, the plants grew too tall,[6] were overcrowded or were neglected. Many farmers said: 'They told us that Jatropha is a hardy crop. We don't have to add fertilisers or irrigate it. Hence we were happy and did not tend it properly. Do you know what happened? All the plants died!'

The lack of awareness of cultivation practices resulted in low yield rates and this was a major hurdle in the production of agrofuels. CBDA officials had overlooked the need to train the farmers in certain cultivation practices related to Jatropha farming. The next hurdle was the varying yield rates, which were attributed to the unavailability of the appropriate genotype of Jatropha and to varying farming practices. Farmers across different regions were provided with seeds of varying genotypes, which had a direct effect on the output and led to major misunderstanding among them. Furthermore, certain farmers did not cultivate Jatropha only on non-fertile or less productive soil, and it was being grown in between rice fields and fertile areas. On questioning the farmers about this, some of the farmers claimed that since the government subsidized rice at 2 INR/kg (US$0.03), it was beneficial to grow some other crop that would pay more. Another group said in hushed tones that they were forced by the officials to grow Jatropha or else they would not receive subsidies and loans. People in this area were growing Jatropha on arable land, which contrasted with the narratives of agrodiesel feedstock being grown on dry lands and not leading to food security issues.

The rural development narrative of the NMB also stressed the creation of jobs for farmers and labourers cultivating Jatropha on government and corporate land. The labourers were involved in block and roadside plantations; block plantations were near the hilly regions and forests. The majority of the labourers were employed under the Mahatma Gandhi National Rural Employment Guarantee Scheme (MNREGS). In the block plantation sites, more than 50 per cent of the labourers were women. The users in this area were displeased when the government appropriated their common land for the cultivation of Jatropha. The block plantations were mainly carried out under rain-fed areas and were not irrigated. The plantations at the roadside were also done in those areas which fell under the category of rain-fed areas and did not require irrigation. The users involved in roadside plantations perceived Jatropha cultivation differently from the ones who undertook block plantations. Most of the users had their own land or were employed by larger farmers. When they were approached by the officials from the forest department, they saw it as a source of additional income and were willing to take it up as they were planting on tracts of land near the highways and were not responsible for the output.

When these labourers went home, they spread the word about Jatropha and said the government would not be investing in it if was not profitable. They supported the narrative of Jatropha being an ideal feedstock and influenced the farmers in their region to grow Jatropha on their own land. Hence the actors involved in block plantations on common lands were reluctant grow Jatropha as it prevented them from using their common property resources (CPRs) for food, wood and fodder, whereas the actors involved in roadside plantations felt that the cultivation of Jatropha fitted in with their practices and they would not have to go to the cities to search for extra work.

The private companies involved in cultivating Jatropha and producing biodiesel followed two routes: they bought or leased land and employed farmers and labourers to work on the land, or they entered into contract agreements with the farmers, who would cultivate Jatropha on their own land. Companies like IOC and CREDA-HPCL formed linkages with the government and were allocated wastelands to use as Jatropha plantations. Reliance Pvt Ltd bought land and also leased large tracts of land from the government and from large farmers. Most of the rural actors were willing to work on land owned by the large companies, as they would be employed as day labourers and would be paid for their work irrespective of the yield. In Chhattisgarh, most of them were paid 50 INR/day

(US$0.82). Despite the rate of pay being low, many people wanted to work for these companies. On being asked why, a labourer named Kumar said, 'These are private companies; they are not as corrupt as the government-owned ones. Here we do not have to bribe the officers to get a job. Sometimes there is a delay in getting paid but we get paid' (Interview with labourer Kumar, 15 June 2009).

Officials from private companies supervised and checked on the progress of Jatropha cultivation and stressed that it was being cultivated on marginal land. However, in the absence of company officials, most of the interviewed labourers said that the land they were working on was good arable land. Hence, despite the narratives of agrodiesel feedstock being cultivated on marginal wastelands, it was being grown on good arable land.

## Wasteland or common property resources?

While successful Jatropha plantations were taking place on arable land, the objective of using 'wasteland' for agrodiesel production quickly became contested, leading to the politicization of agrofuels on the ground. According to Scott (1998), the process of classifying land as wasteland is an example of state simplification. Certain state processes, such as establishing land and population surveys, are undertaken to decipher the actions of populations, which in turn augment the state's ability to monitor its citizens. These processes often simplify 'complex, illegible, local social practices' (ibid.: 2), but such processes may be easy to write on paper yet difficult to implement in practice (Li 1999) and often they fail in their goal of improving the human condition (Scott 1998). Such initiatives often modify or change the landscapes of the communities in which they operate and subsequently alter the relationship between the state and its citizens. As Peet *et al.* argue, powerful actors tend to seize or misrepresent common land resources and 'the flow of values that issues from these' (Peet *et al.* 2011: 27), The expansion of agrofuels altered the relationship between the people and the government when their Common Property Resources (CPRs) were termed as wastelands and diverted for the cultivation of Jatropha.

According to the Wasteland Atlas of India (2008), the area classified as 'cultivable wastelands' in the year 2000 added up to forty-five million hectares. Twenty-six million hectares from this classification come under the category of forest land and provide sustenance to large numbers of forest communities. The interviewed rural users said these so-called 'wastelands' are used as pasture lands, sources of fuel wood and revenue, and are intrinsically linked to communities' livelihoods. The block plantations in the villages of Birkoni, Basna and Sinodha were created on wastelands identified by the government. The people living in these villages claimed 'it was their land and they had been using it for years and all of a sudden the government decided to use the land for something else'. The first batch of Jatropha was cultivated on the common lands in 2007–2008 and, within a year, the crop grew very tall as the villagers had not been told to prune it. They were asked to uproot those plants and were coerced into planting Jatropha again. A majority of them did not want to cultivate

Jatropha and when they tried to grow other crops, these were uprooted by *sahaabs* and their people.

Rural poverty is high among individuals and households who do not own their own land or only have access to marginal land, including agricultural labourers (Census of India 2011). In India, CPRs contribute between 12 and 25 per cent of poor households' incomes – the poorer the household the higher the contribution (Gundimeda 2005). The NMB exacerbated rural poverty by preventing the rural population of landless labourers and poor farmers from using their CPRs for subsistence farming, wood or fodder. Rural poverty and the access and control over land varies and affects men and women differently in India (Sen 2008). Women are particularly vulnerable when common lands are diverted for agrofuel feedstock production, because they have limited control over private resources and draw heavily on CPRs to meet the majority of their needs. In northern India, nearly half of the income of poor women is generated by resources from common land, compared to only one-eighth of poor men's incomes (Reddy *et al.* 1997). A majority of the women interviewed in Chhattisgarh said they used the CPRs for fodder and wood. They complained that they now had to plant Jatropha on their community land and hence had to walk for hours into the forests to collect fodder and wood.

Souparna Lahiri (2008) stressed the alarming rate at which common lands and pastures were being diverted for the production of Jatropha and the large number of people who had succumbed to this development. Many rural people in Chhattisgarh felt that they were being targeted under the 'Politics of Jatropha' (Friends of the Earth Europe 2009). Shiva (2009) argues that 'tribals farms are being forcefully appropriated for Jatropha plantations, aggravating the food and livelihood crisis in Chhattisgarh'. From being a state-led rural development programme, agrofuels became a political issue. There were protests and consultations against the cultivation of Jatropha on CPRs. When the former Indian President A.P.J. Abdul Kalam, an active supporter of agrofuels, visited and praised Chhattisgarh for actively promoting the production of agrofuels, his enthusiasm was not shared by everyone. While the state actively welcomed him to the 'land of Jatropha', social groups and people's organizations opposed the move and penned a letter to him, stating that 'any reference made to Chhattisgarh as the "land of Jatropha" undermines the significance of "rice" as the foundation of people's economy, cultural identity, and dignity and is an insult and open attack on their rights to life and livelihoods' (Lahiri 2008: 14).

Similarly, a group of NGOs, people's organizations and individuals from different parts of India initiated a national consultation on 3–4 December 2007 entitled *Biofuels in India: Will they deliver, or destroy?*; this was held in Pataspur village of the Medak district in Andhra Pradesh. The consultation stressed that indigenous people, pastoralists, small farmers and tribal communities all across India have a holistic view of life that is reflected in their interaction with the living world, which, in turn, provides for all their needs of food, fuel, fodder and energy. In a statement issued at the end of the consultation, the participants declared:

> We believe that the promotion of large-scale corporate-sponsored biofuels (agro-fuels) in the garb of improving energy security is yet another form of not only physically destroying the above, but also a psychological assault perpetrating the idea that farming as our people have done it is no longer good or tenable.... Rural and forest communities ... say that there is no such thing as wastelands. Most of these lands are grazing lands, common pastures, degraded forests, and also lands of small and marginal communities. They not only support a multitude of livelihoods but also have a critical ecological role. This is where the government and corporations are pushing for their fuels, displacing thousands of peoples.
>
> (Ibid.: 17)

Friends of the Earth Europe (2009) claims there is a clear-cut conspiracy to uproot and displace hundreds of thousands of Adivasis and Dalit[7] farmers from their CPRs by planting Jatropha on public land. In contrast I argue that it is not a conspiracy, but that the policymakers and bureaucrats did not foresee the problems that diverting CPRs for Jatropha production would cause. High-level technocrats and policymakers assumed that these CPRs were marginal, unused wastelands and could be used for Jatropha cultivation. However, once the NMB was drafted they were keen to ensure that the CPRs were used only for Jatropha production, as this was government land and they did not back away from the utilization of wasteland narrative.

The mission brought in its wake varying interventions, which mainly affected the poor labourers and farmers who were displaced from their land and were forced to migrate to the cities or to cultivate Jatropha. Others cultivated Jatropha because they were being paid (or had been promised they would be paid) for it. The problem of CPRs and the refusal and dissent expressed by the people captures the gap between expert (policymakers, bureaucrats) and indigenous knowledge and the invalidity of the wasteland narrative. The policymakers and technocrats did not anticipate the problems that would arise by using CPRs for Jatropha cultivation, but they continued to support the biodiesel initiative, while the rural users said that what had aimed to be a pro-poor initiative had evolved into an anti-poor initiative.

## Concluding discussion

The initial consensus on the viability and benefits of agrofuels resulted in the global agrofuel discourse gaining momentum and spreading actively across various countries. This chapter has highlighted the importance of networks and how narratives and their legitimacy are constructed and promoted by key actors in a network. The chapter delineated the indisputable significance of experts – the government officials – in giving shape to the objectives of the biodiesel mission and classifying Jatropha as an ideal feedstock and certain lands as wastelands. The discussion on the impact of the Jatropha boom in Chhattisgarh demonstrated that the efforts of the state government to create better conditions for the rural poor emerged as a mechanism for officials from the government and private sector to actually seize control of resources used by the rural poor. The rural actors contested the 'pro-poor' claim of the initiative and argued that this mission was exploiting the rights of poor people.

The NMB had been introduced as a test policy option in a mission mode and was supposed to be implemented in two phases. However, in 2008, prior to the start of the second phase, the NMB was scrapped and replaced by a draft biofuel policy, which was passed as the National Biofuel Policy in December 2009 (Business Standard 2009). Policymakers feared corporate land-grabbing, and were also doubtful that Jatropha was an ideal feedstock (Dey and Jayaswal 2008). However, despite the contestations of the use of the term wasteland, the new biofuel policy also incorporates the cultivation of feedstock on degraded wastelands. The national biofuel policy no longer specifically promotes the growth of Jatropha; instead, it focuses on the cultivation of non-edible oil seeds but does not name any particular feedstock. Various actors have said that the new policy has taken into account some of the drawbacks of the NMB, but has not reviewed the major challenges and problems of the NMB. Hence the policy has immediately come under scrutiny, with suggestions that it fails to address the competing renewable energy agendas of the Energy and Agricultural Ministries (Lane 2008). Despite the issues raised regarding the use of so-called wastelands, the government of India is still encouraging the cultivation of biofuel feedstock on wastelands. This proves that, more often than not, development policies are not important for what they say, but for who is supporting them and how they exercise their power. The networks of ministers supporting the biofuel policy are exercising their power over CPRs through complex forms of hegemony and governmentality.

This chapter has drawn from ANT and argued that the success of a programme depends upon the active enrolment of supporters and the importance of collaboration between actors. It stressed the importance of networks, citing the creation and role of the national biofuel policy network and in turn how regional networks shaped the manner in which the biodiesel initiative progressed across India. Hence it implies that networks, the actors and their continual support are imperative for the existence of a development initiative. ANT and the framework used by Mosse (2004) refer to the collaboration, compromise and translation of interests of actors in a network. The actors are able to enrol others by negotiating with them, so that they tie in their interests with the policy and development initiative. However, the NMB did pursue negotiations that were different to those seen under ANT. The rural actors (farmer and landless labourers) were also important actors in the biodiesel network, especially when the NMB was introduced as a pro-poor development initiative. It was thus necessary for the other actors, i.e. government officials, bureaucrats, policymakers and officials from private companies and NGOs, to enrol the rural actors into the biodiesel network. It was noticed that the enrolment of rural actors was not done by tying or representing their interests in the NMB, but that they were 'forced', 'manipulated' or 'lured' into cultivating Jatropha, rather than aligning their interests with the NMB.

I therefore argue that the NMB was successful as an initiative, not only because of the collaboration, comprise and negotiations between key actors. Rather, it progressed across various states, in this case Chhattisgarh, because of the power held by key actors to force or manipulate rural actors to cultivate Jatropha. This demonstrates that the regional network did not afford the same status to all actors, that rural actors' needs were misrepresented and that they were forced to grow Jatropha on their common land. Although initially successful in expanding the area devoted to Jatropha cultivation, the government-led programme contributed to the politicization of agrofuels, increasing dissatisfaction and disillusionment with agrofuels as a means of rural development. On this basis, the secondary network of farmers promoting Jatropha will wither and turn against the state's expansion policy, eventually undermining the grandiose agrofuel plans of the state.

## Notes

1 This chapter will focus solely on the National Mission on Biodiesel.
2 Unless otherwise stated, the data presented in this chapter is based on field research done in seven districts of Chhattisgarh, namely Bilaspur, Janjgir, Korba, Mahasamud, Raigarh, Raipur and Surguja. The research methods comprised interviews, focus group discussions and participatory methods. The research participants encompassed government officials, agricultural researchers, scientists, officials from private companies and NGOs, farmers and labourers. This research was carried out from April 2009 to June 2009 and March 2010 to May 2010. It involved nine group discussions and forty-three interviews with rural farmers and labourers, thirteen interviews with government officials and scientists, nine interviews with officials from the private sector and NGOs, and email surveys with various informants.
3 The state government of Chhattisgarh wanted to be able to cultivate and harvest an adequate amount of Jatropha to produce sufficient biodiesel to be blended at 5 per cent by 2015.
4 Providing rice at the rate of 2 INR/kg (US$0.03), has made the Chief Minister popular among the rural population.
5 In the case of NMB, 'user' technically refers to the people who buy and use agrodiesel in their cars. In this chapter 'user' does not refer to the end-user of agrodiesel, but focuses on the users of the NMB as a rural development initiative, namely the farmers and landless labourers who cultivated Jatropha.
6 If the plant is not pruned, it grows too tall and reduces the output of the seeds.
7 Dalits and Adivasis are groups belonging to scheduled tribes and scheduled caste communities in India.

## References

Ariza-Montobbio, Pere, Sharachchandra Lele, Giorgos Kallisand and Joan Martinez-Alier (2010) 'The Political Ecology of Jatropha Plantations for Biodiesel in Tamil Nadu, India', *The Journal of Peasant Studies* 37(4), 875–897.

Blaikie, Piers and Oliver Springate-Baginski (2007) (eds) *Forests, People and Power: The Political Ecology of Reform in South Asia*, London: Earthscan.

Business Standard (2009) 'Biofuel policy gets govt approval', *Business Standard* (www.business-standard.com/india/news/biofuel-policy-gets-govt-approval/380689/, last access 19 February 2013).

Castells, Manuel (2004) *The Network Society: A Cross-Cultural Perspective*, Cheltenham: Edward Elgar.

Census of India (2011) (http://censusindia.gov.in/, last access 18 March 2014).

Chhattisgarh Biofuel Development Authority (www.cbdacg.org, last access 1 September 2013).

Dey, Sushmi and Jayaswal, Rajeev (2008) 'Biodiesel Mission Set to Pull Down Shutters', *The Economic Times* (http://articles.economictimes.indiatimes.com/2008004/news/27723860_1_jatropha-plantation-farm-land-biodiesel, last access 10 March 2012).

Friends of the Earth Europe (2009) *Losing the Plot: The Threats to Community Land and the Rural Poor through the Spread of the Biofuel Jatropha in India*, Belgium: Friends of the Earth Europe.

Government of Chhattisgarh (2005) *Human Development Report Chhattisgarh*, Raipur: Government of Chhattisgarh.

Government of India (2002) *India Vision 2020*, New Delhi: Planning Commission of India.

Government of India (2003) *Report of the Committee on Development of Biofuel*. New Delhi: Planning Commission of India.

Government of India (2011) (www.barc.ernet.in/, last access 15 January 2014).

Gundimeda, Haripriya (2005) 'Can CPRs Generate Carbon Credits without Hurting the Poor?', *Economic and Political Weekly* 40(10), 973–980.

Hindustan Petroleum Corporation Limited (www.hindustanpetroleum.com, last access 17 March 2014).

Indian Oil Corporation Limited (2006) (www.iocl.com/Aboutus/sustainbility/Indian_oil_the_energy_of_india.pdf, last access 22 February 2014).

Jain, Sonu (2006) 'Chhattisgarh Jatropha Order to Rev Up Biofuel Production', *The Indian Express* (www.indianexpress.com/news/chhattisgarh-jatropha-order-to-rev-up-biofuel-production/11716/, last access 12 June 2013).

Lahiri, Souparna (2008) "Colonizing the Commons, It Is Jatropha Now!', *Mausam* 1(1), 14–18.

Latour, Bruno (1996) *Aramis: Or the Love of Technology*, Cambridge, MA: Harvard University Press.

Law, John and Michel Callon (1992) 'The Life and Death of an Aircraft: A Network Analysis of Technical Change', in Bijker, Wiebe E. and John Law (eds) *Shaping Technology/Building Society: Studies in Sociotechnical Change*, Cambridge, MA: MIT Press.

Lane, Jim (2008) 'India Ministerial Group Approves National Bio-fuel policy, Avoids key decisions on Bio-diesel, jatropha Support', *BioFuel Digest* (www.biofuelsdigest.com/blog2/2008/07/25/india-ministerial-group-approves-national-bio-fuel-policy-avoids-key-decison-onbiodiesel-jatropha-support/, last access 15 March 2012).

Li, M. Tania (1999) 'Compromising Power: Development, Culture and Rule in Indonesia', *Cultural Anthropology* 14(3), 295–322.

Mol, P.J. Arthur (2007) 'Boundless Biofuels? Between Environmental Sustainability and Vulnerability', *Sociologia Ruralis* 47(4), 297–315.

Mosse, David (2004) 'Is Good Policy Unimplementable? Reflections on the Ethnography of Aid Policy and Practice', *Development and change* 35(4), 639–671.

Peet, Richard, Paul Robbins and Michael Watts (eds) (2011) *Global Political Ecology*, Abingdon, New York: Routledge.

Pradhan, Shishusri and Shaun Ruysenaar (2014) 'Burning Desires: Untangling and Interpreting 'Pro-poor' Biofuel Policy Processes in India and South Africa', *Environment and Planning A* 46(2), 299–317.

Reddy, Amulya K.N., Robert H. Williams and Thomas B. Johansson (1997) *Energy after Rio: Prospects and Challenges*, New York: UNDP.

Reliance Life Sciences (www.rellife.com/biofuels.html, last access 10 March 2014).

Scott, James (1998) *Seeing Like a State: How Certain Schemes to Improve the Human Condition Have Failed*, New Haven, CT: Yale University Press.

Sen, Gita (2008) 'Poverty as a Gendered Experience: The Policy Implications', *Poverty in Focus* 13, 6–7.

Shiva, Vandana (2009) 'From Seeds of Suicide to Seeds of Hope: Why Are Indian Farmers Committing Suicide and How Can We Stop This Tragedy?', *Huffington Post* (www.huffingtonpost.com/vandana-shiva/from-seeds-of-suicide-to_b_192419.html, last access 3 March 2014).

Shukla, Shailendra K. (2008) 'Biofuel Development Programme of Chattisgarh', in *Fifth International Development Conference on Biofuels*, February 2008, New Delhi: Winrock International India.

Urry, John (2003) *Global Complexity*, Cambridge: Blackwell Publishing.

Wasteland Atlas of India (2008) (http://dolr.nic.in/wasteland.html, last access 14 June 2013).

World Bank (2008) 'India Country Overview' (http://web.worldbank.org/WBSITE/EXTERNAL/COUNTRIES/SOUTHASIAEXT/INDIAEXTN/0,,menuPK:295593~pagePK:141132~piPK:141107~theSitePK:295584,00.html, last access 17 September 2013).

Yearley, Steve (2005) *Making Sense of Science: Understanding the Social Study of Science*, London: Sage.

# 13 US agrofuels in times of crisis

## Food, fuel, finance

*Aaron Leopold*

### Introduction

This chapter wrestles with the question of whose interests are being served by agrofuels promotion policies in the United States. In trying to unravel the answer to this question, it considers the roles of key political economic actors involved in the agrofuels project in the United States from the 1970s to 2014, and finds that, to date, political support for agrofuels has not supported the broader economic, environmental or social benefits claimed by proponents, but rather has mostly benefited deeply entrenched, hegemonic business interests.

The analysis presented here is framed by a neo-Gramscian analytical perspective that is used to explore the dominant role of the neoliberally focused agricultural and financial industries[1] in shaping and controlling political economic discourses on agrofuels, redefining them not as environmental products, but as representing energy security and, more recently, as financial assets. While hegemonic control of discourses on agrofuels has shifted over time due to such factors as the global financial crisis and critiques from global civil society of their social and ecological credentials, the neo-Gramscian perspective helps better explain how and why the commodity price crises of 2007–2008 and 2010–2011 barely affected the agrofuels industry, despite important calls for reform. As a case study, this chapter considers the dominant role of agricultural multinational Archer Daniels Midland (ADM) in shaping agrofuels discourses in the US over the twentieth and twenty-first centuries, first as a producer, and then as a key financial trader in agricultural commodities, and the financialization of agrofuels feedstocks as fungible commodities. The chapter closes by questioning whether agrofuels' ever-increasing physical, political and financial likeness to fungible commodities has made their sustainability ungovernable.

### A neo-Gramscian analytical framework

A neo-Gramscian perspective towards the political ecology of agrofuels offers a critical analytical framework that encompasses a broader range of actors and structures than more realist-based political science and international relations approaches (Ekers *et al.* 2009). The basis of Gramscian thought, and the more nuanced neo-Gramscian perspective (in which scholars have brought Gramsci's discussion out of its original national context and onto the international and transnational stage), is its distinct conceptualization of the term 'hegemony'.

Rather than understanding hegemony as the dominance or supremacy of one actor or group over another, as in realist theories of international relations, hegemony in the Gramscian sense is equated with the establishment and perpetuation of a dominant group/class or socio-political idea by embedding the interests this agenda represents into the fabric of a society. Understood in this way, it is not enough for a class or ideology to simply dominate others, rather it must foster widespread societal appeal, consensus and acceptance of its world view, handling resistance through compromise and co-opting naysayers onto its side. A hegemonic project is then a strategic and compromise-filled process of marginalization and/or de-politicization of dissent that requires mass mobilization of political, economic and social forces (Bieler and Morton 2001; van der Pijl 2009).

In neo-Gramscian terms, a hegemonic political economic system is termed an *historic bloc*, represented by alliances between societal actors who together work to align discursive, organizational and material conditions to produce and maintain the specific modes of production and meaning that support the historic bloc. Hegemony is exercised via bureaucratic and coercive elements of the state, as well as through the economy and civil society (Levy and Newell 2002). The media, religious organizations, NGOs, academia etc. are seen as key to the ideological reproduction of core hegemonic ideals. Only by accessing and eventually co-opting these actors, can a hegemonic project influence news media, moral debates, science etc. to promote and reinforce the project's socioeconomic and socio-political worldview, goals and ideals. In this sense, civil society can serve to legitimize the very system it seeks to contest and can act dually as a source of resistance and reinforcement of hegemonic projects (ibid.).

Those who form the ideas behind hegemonic and counter-hegemonic ideologies are labelled *organic intellectuals* and are responsible for 'developing, and sustaining the mental images, technologies and organizations which bind together the members of a class and of an historic bloc into a common identity' (Cox 1983: 168). Intellectuals' ability to frame debates and how people think about issues, or their *discursive power*, is at the heart of a neo-Gramsican analysis of power. Hence, power does not only imply the ability to pursue interests via overt structural or direct forms of power, but entails the capacity to implant these interests in the minds of others (Fuchs 2005). Additionally, by moulding discourses in their favour, actors exercising discursive power create legitimacy for their position, representing the point of co-optation within Gramsci's hegemony. In the context of agrofuels, competing discourses on the 'reality' of their economic, ecological and social benefits have been very publicly battled over by a raft of academic, corporate, political and social actors. Together, these actors form the actor networks of hegemonic and counter-hegemonic intellectuals vying for discursive dominance over the 'truth' behind agrofuels – which are at the same time simply processed plants, but also complex and controversial tools of a broad political-economic and political-ecological movement.

The neo-Gramscian school of thought was born out of attempts to understand what or who were the main drivers shaping the development of global capitalism in the late twentieth century, which neo-Gramscian scholars determined as being squarely in the hands of transnational class networks (Cox 1983; Rupert 1995; van der Pijl 1998). The *transnational capitalist class* (TCC) as a hegemonic actor network, or historic bloc of neoliberally-oriented capital interests fuelled by the discourse-shaping organic intellectuals, is one such conceptualization developed by Kees van der Pijl and elaborated on by Bastiaan van Apeldoorn. It considers how small groups of similarly interested drivers and owners of capital have gained and maintained privileged access to decision-makers, or have been the decision-makers themselves, during the rise of globalized capitalism, and more particularly of the dominant neoliberal style of capitalism we see today that encourages privatization, deregulation of markets and free movement of goods and capital – all of which have played a key role in agrofuels crises (van Apeldoorn 2000).

The TCC is not a single group of individuals and cannot even be considered as a unitary actor. Rather it is an informal collection of (often temporarily) ideologically motivated actors whose primary goals are to promote global, business-friendly neoliberal policies. The next section focuses on the importance of one TCC key actor in the context of the rise of global agrofuels, the US-based multinational agricultural producer, processer and trading company, Archer Daniels Midland (ADM). In particular, Dwayne Andreas, ADM's CEO from 1971–1997, is presented as a key organic intellectual within the TCC, who played a pivotal role in the development of the agrofuels discourse in the US as one of rural economic, geopolitical and energy security. In considering the latest iteration of the ongoing crisis of agrofuels, the penultimate section of this chapter considers the broader role of the highly globalized financial trading industry, in which ADM became an important player during the neoliberal deregulatory era, as a TCC network promoting the commodification and financialization of agrofuels.

## A brief history of modern agrofuels in the United States

In response to the Organization of the Petroleum Exporting Countries (OPEC) oil embargo of 1973, which caused the international price of oil to quadruple over six months, the hitherto niche product of bio-ethanol was placed at the forefront of US energy politics (Keeney 2009). As alternatives were sought to decrease dependency on OPEC countries for American energy needs, one company in particular, ADM, lobbied the government to promote a considerable increase in ethanol production to combat both the fuel crisis and the global agricultural commodities glut.

As one of the largest agricultural commodity traders and processors in the world, ADM, led by the extremely politically influential CEO Dwayne Andreas, was already on its way to becoming the most influential single actor in US agricultural history (Bovard 1995). In 1973 Nixon, had declared that the US would achieve energy independence by 1980. With this declaration and the desperate situation of American agriculture during the global grain glut as ammunition, Dwayne Andreas is personally credited with playing a major role in getting numerous Midwestern states to institute tax and other incentives to promote ethanol production in the mid-1970s, and Congress to pass legislation in 1977 to invest US$60 million in guaranteed loans for ethanol producers (Bovard 1995: 7).

Ethanol's real corporate welfare windfall began in 1978 however, when support efforts to increase ethanol subsidization at the outset of the second OPEC oil embargo began – something that had also been suggested by Andreas (ibid.). Carter saw the oil embargo as a war-like act, and enacted a fuel ration in the US that exacerbated the effect of the crisis on the country's economy, making the issue highly visible to citizens. Andreas and Carter pushed Congress hard to make moves to ensure the US could produce more of its own energy, an effort which resulted in the Energy Tax Act of 1978. This tax both created a US$0.40 per gallon tax credit for ethanol use (this rate fluctuated over time, reaching US$0.60 in 1984) and also exempted gasoline with at least a 10 per cent ethanol blend from the US$0.04 per gallon federal excise tax (Keeney 2009). Finally, as a last ditch effort to gain votes in the Midwest before the 1980 Presidential election, Carter instituted a 0.54 cents per gallon tariff on Brazilian ethanol, essentially barring it from entry into the US. This tariff was also attributed to Andreas, who suggested it to Carter, according to a former ADM lobbyist (Lilley 2006). Further illustrating the close bonds between politics and ADM at the time, not long after the tariff on Brazilian ethanol was enacted, former Democratic National Committee chairman Robert Strauss became a member of ADM's board of directors (Bovard 1995).

These events further underline the co-optive hegemony of capital as well as its historical need for oil, or an oil substitute in this case, as an engine of growth. Indeed, Andreas' role as an architect behind the scenes set the stage for the broad promotion and embedding of a key component of the then emergent neoliberal ideology, namely its implicit dependence on cheap and abundant fuels to power economies and radically scaled-up international trade. Also illustrated here is how, while some actors within the TCC may have agendas that partially contradict their broader ideologies – such as the anti-neoliberal protectionist policies urged by Andreas and US agricultural lobbies more generally – these policies indeed served to further fuel capitalism's insatiable need for oil, or an oil substitute in this case, at any cost to broader society, as an engine of growth.

With broad subsidies and trade protectionist policies in place by 1980 and the turbulent commodity price history of the 1970s in hindsight, the early 1980s saw the first critical analyses of how widespread ethanol use might further complicate the increasingly challenging and unpredictable globalized food marketplace. Authors such as Göricke and Reimann (1982) and Bernton *et al.* (2010 [1982]) warned of many issues debated in the twenty-first century, including a possible food versus fuel crisis, issues of land-use change, and climate change, which had just begun to emerge on the political and civil society radar. Indeed, in the run-up to the 1979 US Presidential election, two government reports of particular importance were published on climate change. The first found that increasing concentration of atmospheric carbon dioxide was 'the most important environmental issue facing mankind' (ibid.: 195). The second concluded that global warming would 'threaten the stability of food supplies, and would present a further set of intractable problems to organized societies…' and that 'Enlightened policy in the management of fossil fuels and forests can delay or avoid these changes.' (ibid.: 197–8) Although these dire, prophetic words had already earned significant scientific backing and could not have been more well-informed, they were roundly ignored by policymakers and the media.

*Despite these causes for concern,* discursive control of the agrofuels narrative remained steadfastly in the hands of corporate interests, and ethanol support was radically ramped up in the 1980s, a decade in which ADM controlled more than 70 per cent of the US ethanol production (Bovard 1995). Even when oil prices plummeted, annihilating any hope that the ethanol industry could compete economically, subsidies were not removed but increased on the grounds that US farmers were in need of support and that ethanol was helping to clean urban air by lowering emissions, a position in fundamental contradiction to US governmental and independent scientific evidence on the air pollution effects of ethanol use in cities. By 1992 production nearly quadrupled to 4.16 billion litres, despite continued record low oil prices (Bernton *et al.* 2010 [1982]: xii; Bovard 1995).

In 1994, President Bill Clinton declared that the US Environmental Protection Agency (EPA) should ignore years of accumulated science-based complaints about ethanol's effects on air pollution levels, and summarily mandated its use in the most polluted US cities. At the time, US Assistant Secretary of Energy Sue Tierney explicitly told Congress that 'the primary beneficiary will be ADM' since they continued to control the vast majority of US ethanol production (Bovard 1995: 13). Tierney's move serves here to help underline the role of the TCC in shaping the development of global capitalism: ADM's success in bringing about policy shifts that not only benefited their industry, but that were publicly admitted to essentially be gifts to the company demonstrates the structural, instrumental, and discursive power over what is good or bad for the country: fuel is essential to growth, and growth is what is needed for a bright future; hence, fuel is required at any economic, environmental or social cost.

This new mandate did spark a major political backlash for environmental and economic reasons however, and resulted in a court case which ruled the Clinton administration lacked administrative authority to impose such a mandate, with the US Justice Department claiming that the mandate was part of the government's strategy to reduce dependence on imported oil (Herbert 1995). Despite this, or perhaps because of it, attention was drawn away from the variable per-gallon tax credit that remained at over 50 cents per gallon throughout the 1990s, seeing that the federal government still spent US$3.5 billion directly subsidizing ethanol from 1990–1995, and ADM received over US$2.1 billion of that amount (Bovard 1995).

These windfall subsidies were doled out despite the aforementioned social, environmental and economic critiques, and the fact that, as early as 1980, analysts had shown that ethanol production required nearly as much, if not more, energy than was embodied in the ethanol itself (i.e. it had a neutral or negative energy balance) (ibid.). US government scientists, along with those at the International Energy Agency (IEA), then the world's only international authority on energy, either did not address most of these issues, or considered them to be technocratic problems that could be overcome via sound planning and governance (IEA 1994). Through their reproduction of these core ideals of the TCC, national and international scientists who failed to caution against the immediate promotion of agrofuels production allowed TCC interests to further control the discussion over ethanol production, keeping it non-critical for decades despite serious initial concerns having been raised. Indeed, over its entire history, the IEA has consistently felt that agrofuels can be environmentally friendly, financially viable, geopolitically secure and energy positive alternatives to gasoline and diesel, lending support to agrofuel advocates in the US, as elsewhere (IEA 1994; IEA 2004; IEA Bioenergy 2010).

*Although* other powerful individuals and interest groups, such as the American Farm Bureau and Renewable Fuels Association, have had not insignificant roles in the rise of agrofuels in the US (Taxpayers For Common Sense 2013), this section has aimed to illustrate the discursive and structural influence that powerful individual facets of the TCC can wield when broadly promoting the interests of an historic bloc of hegemonic political economic interests. In attempts to understand the structural, instrumental and discursive power of actors such as Dwayne Andreas and his TCC peers, Stephen Gill coined their efforts the *new constitutionalism* of the hegemonic neoliberal globalization project, which they spurred into existence, claiming:

> the politico-legal dimension of the wider discourse of disciplinary neoliberalism [serves to] separate economic policies from broad political accountability in order to make governments more responsive to the discipline of market forces and correspondingly less responsive to popular-democratic forces and processes.
>
> (van Appeldoorn 2000: 160)

The entrenchment of neoliberal ideology into national law around the globe and via deregulation-enforcing, free market-inspired institutions such as the WTO, IMF and World Bank, has enabled the neoliberal globalization project to weave capital interests neatly not only into the daily workings of politics and economics, but also into societies in themselves, reducing opportunities for resisting hegemonic neoliberal premises as they become more and more engrained in society. In the context of agrofuels, the environmental and social concerns raised by a few actors in the 1980s and 1990s would become major geopolitical concerns in the early 2000s. By this time, however, the vested interests of the historic bloc of neoliberally-oriented capital interests, led by the actors within the TCC, had already entrenched themselves deep within the national political economic fabric of the US, and increasingly, globally.

## Agrofuels in the twenty-first century and the food versus fuel controversy

For many reasons, the beginning of the twenty-first century saw renewed support for agrofuel production by governments around the world, civil society organizations and industry: with early critical science left unheeded, countries and industry took interest in agrofuels as an avenue to reduce carbon dioxide emissions in line with Kyoto Protocol commitments; the attacks of September 11, 2001 created a fear-based motivation for the US to search for alternatives to Middle Eastern oil; and as the decade continued, global oil prices spiked to record levels. Additionally, and perhaps most importantly, in the US, key Midwestern agrofuel-producing states are national election battlegrounds that politicians would do almost anything to please. The power of these states and their lobbies is evidenced by this passage from a 2011 *New York Times* op-ed piece by Steven Rattner, President Barack Obama's former 'auto czar', who stated that in national and state-level elections, politicians have been known more for their ethanol appeasement strategies than their consistency:

> John McCain and John Kerry were against ethanol subsidies, then as candidates were for them. Having lost the presidency, Mr McCain is now against them again. Al Gore was for ethanol before he was against it … Tim Pawlenty pushed for subsidies before he embraced a 'straight talk' strategy [against them].
>
> (Rattner 2011)

It has been shown that both President George W. Bush and President Barack Obama have also been pro-agrofuels for electoral reasons (Barrionuevo 2007; Rohter 2008). In the early twenty-first century, in addition to dozens of state-level agrofuel promotion policies passed in part due to this pressure, national legislation passed in the US in 2004, 2005, 2007 and 2008 was also largely the result of intense lobbying pressure from industry and industry associations. These included enormous production/consumption mandates, extensions of the tariff on Brazilian ethanol, research and project funding and loan guarantees. As a result, US ethanol production capacity leapt from approximately 10.2 billion litres in 2003 to 40.1 billion litres in 2009, placing the US at number one in global ethanol production. During this time, the market had expanded from the near monopolistic dominance of ADM in the 1990s to a more oligopolistic grouping of ADM, Aventine Renewable Energy, Cargill, POET Biorefining and Valero, who until the financial crisis began in 2008, still dominated the market.[2]

By mid-decade, however, rapidly increasing rates of agrofuel production led critical civil society, governmental and scientific observers to raise the same questions concerning the ecological, economic and social sustainability that had been first brought up, but summarily ignored, in the 1980s (Congressional Budget Office 2009; Doornbosche and Steenblik 2007; Fargione *et al.* 2008; Pfuderer *et al.* 2008; Royal Society 2008; Searchinger *et al.* 2008; Shattuck

2009). That they were paid attention to at this time was largely due to the dramatic increase in food commodity prices that took place over the course of 2005–2008, with most of this radical jump coming over five months in 2008. Between January 2005 and June 2008, food commodity prices increased by 83 per cent, with corn prices nearly tripling, and wheat and rice prices jumping by 127 and 170 percent respectively (de Shutter 2010: 2).

Consumers in the United States rely heavily on processed foods, the prices of which are much more closely associated with the costs of transport, packaging, marketing and processing than with the actual food commodity price. As such, American consumers felt little direct effect on their food prices compared to those in countries more reliant on raw commodity purchases. Indeed, food riots broke out in 31 developing countries during the 2008 commodities price spike. These 'food vs. fuel' riots have been attributed with toppling Haitian Prime Minister Jacques-Édouard Alexis in April of 2008 (AFP 2008). These events grabbed the attention of the world and rushed the previously monochromatic agricultural topic of crop production further into the public limelight. Over the course of 2007–2008, the previously little-debated topic of 'biofuels' (connected with the earth and traditional farming) became a crosscutting debate about 'agrofuels' (connected with industry and science) over energy security, agricultural policy, environmental protection, development, global food security, transportation and mobility and bioengineering, among other things.

To proponents such as the newly booming agrofuels processing industry and the farmers who had not seen profits this high in their lifetimes (Tetreau 2013), agrofuels represented a way forward in terms of a new energy-secure future, driven by a new industry that only needed time to work through its problems and prove its worth. To the critics noted above, agrofuels came to represent the rapid and massive commodification of nature driven by actors who had thus far not adequately taken the consequences of the industry into account.

The emergence of critical science simultaneously to the beginning of the food crisis played a significant role in the re-evaluation of agrofuels, providing what Colin Hay (2001) calls an 'exceptional moment' in agrofuels history. During this exceptional moment, control of the growing contradictions and crises presented by agrofuels was wrestled from the organic intellectuals within the hegemonic bloc who previously controlled them, and was put into the hands of a more critical group of counter-hegemonic organic intellectuals, such as the vocal Tim Searchinger of Princeton, who argued that US corn-based agrofuels, on which the industry relied, could likely never be sustainable (Searchinger *et al.* 2008). Using commodity price increases, food shortages and the ensuing riots as ammunition to refocus the agrofuels discourse, the previously unheeded criticisms of agrofuels were taken up by most major news outlets in the US and civil society globally. For instance, many of the 200 NGOs that signed an EcoNexus call for a moratorium on agrofuels incentives and trade are US-based.[3] Industry organizations also joined the fray, such as the powerful automotive industry, which saw agrofuels as possibly damaging to the engines of their cars,[4] as did equally powerful food industry associations such as the Grocery Manufacturers Association, the American Meat

Institute and the National Restaurant Association (Bloomberg 2008), which saw agrofuels as a threat to their food businesses. The involvement of this new group of actors began to remould power structures within the agrofuels discourse as the food vs. fuel crisis played out. F&C Investors' analysis of 30 industry leaders went so far as to conclude that despite the massive political and financial capital already invested in its success, the agrofuel 'industry's future is highly uncertain because of political risks…' largely because 'Biofuels producers … misjudged government sensitivity to food price inflation and environmental concerns; and failed to find industry solutions to sustainability challenges…' (F&C Investors 2008: 3)

Despite this dire outlook, in April 2008, at the height of the food commodity crisis, US President George Bush remained steadfast in his position that corn ethanol was an issue of national energy security and that environmental problems were only growing pains on the way to more efficient, second- generation agrofuel technology. More critically, while admitting that ethanol likely played an important role in increasing food prices, he put US energy and economic interests over those of the starving when he stated that:

> …the truth of the matter is it's in our national interests that our farmers grow energy, as opposed to us purchasing energy from parts of the world that are unstable or may not like us. In terms of the international situation, we are deeply concerned about food prices here at home and we're deeply concerned about people who don't have food abroad. In other words, scarcity is of concern to us. Last year we were very generous in our food donations, and this year we'll be generous as well…
>
> (Bush 2008)

Bush's statement came two months before prices began falling in June of 2008, although they have never returned to the levels before the crisis. Proponents of agrofuels nonetheless were able to assuage fears of an ongoing global crisis by pointing towards falling global commodity prices and agreeing to actively engage with critics in designing sustainability criteria. This move, exemplified by the participation of WWF, Oxfam, Conservation International and Both ENDS on the Board of Governors[5] of the controversial Roundtable on Sustainable Palm Oil (in favour of biodiesel production) could be seen as representing the co-optation of some of the worlds most important civil society voices into the very system they were struggling against.

When many of the hundreds of new producers that rushed to the scene as agrofuel policies emerged went bankrupt or were bought out during the financial crisis that began in 2007, F&C Investors' predictions came true to some extent. It appears that this was due more to increasingly tough competition and a flooded marketplace than to concern over environmental and social issues as F&C had predicted, however. Despite, or perhaps because of, these challenges, policy support measures in the US remain essentially unaltered up through 2014. The only major policy change came in the form of US legislative inaction in letting the tariff on Brazilian ethanol and US ethanol tax credits expire in 2011, which for decades had protected US producers from international competition.

On the surface, the massive level of political support for the sector that key players such as ADM had asked for appeared to weaken their agenda-setting power in relation to others. At the end of 2013, only about 33 per cent of ethanol production was in the hands of the four biggest firms: POET Biorefining; Valero Renewable Fuels; Aventine Renewable Energy; and ADM.[6] From its peak of being a near monopolist up to the late 1990s, ADM's market share dropped to below 12 per cent in early 2014. As is explained below, however, this does not represent a waning of ADM's power in the agricultural marketplace but is, rather, representative of its strategic realignment away from increasingly competitive and minimally profitable ethanol processing and production, and onto the highly profitable business of ethanol feedstock, i.e. commodities, trading. Indeed, while its ethanol production market share radically decreased in the 2000s, ADM's stock price quintupled from 2004 to 2014.[7]

## Agrofuels and the financialization of nature

In 2012, ADM, Bunge, Cargill and Louis Dreyfus, together known as the ABCD group, controlled approximately 90 per cent of the global grain trade, up from 73 per cent in 2003 (Murphy *et al.* 2012). These companies are highly diversified vertically and horizontally, but are extremely secretive and challenging to gather information on. There are clear indicators that they represent a key new iteration of the discursive control of the TCC.

Shortly after the beginning of broad support and deep subsidies for US agrofuels in the late 1970s, the US, led by President Reagan, began pushing a broad set of ideologically-driven changes in the global political economy that saw much of the US and international financial sector deregulated (ibid.). Regulatory politics similarly became dominated by the neoliberal precepts of the emergent neoliberal-oriented historic bloc of capital interests, which hailed free markets and self-regulation as the hallmarks of a new, modern age of an efficiency-focused, market-driven world. The deregulation of the financial sector, which led to a radical consolidation of US banking including 8,000 mergers between 1980 and 1998, also eroded financial support for small-scale, locally- owned agriculture. This, in combination with falling commodity prices and increasingly expensive farm inputs, fostered the radical consolidation of US agriculture as well (Holt-Giménez 2008).

These changes allowed the ABCDs, the more global face of the TCC in the agricultural realm, to diversify away from their already broad services of seed, fertilizer and agrochemicals provision, agricultural and livestock production and storage, landownership, food processing, transportation and agrofuel production, to include investment and other financial services in commodity markets. In particular, this included investments and other financial services in commodity derivatives, which connect agricultural commodities to other non-food commodities such as oil in financial markets (Murphy *et al.* 2012; Holt-Giménez 2008).

Equally, these regulatory changes also saw traditional financial services companies move more deeply into commodity trading. This led to their involvement in the creation of the now infamous over-the-counter (OTC) derivatives that were also largely responsible for the financial crisis, and into which commodity investments were combined with mortgages and other investments that were bundled together as a way to supposedly mitigate investment risk.[8]

Deregulating financial markets increased the unpredictability, confusion and trepidation seen in financial markets, with investors increasingly unsure about what represented a 'safe' investment, and escalating both the financial and commodities markets' vulnerability to shock. These conditions, which are broadly attributed to causing the 2007 financial crisis, were, however, not so much a threat as an opportunity to the oligopoly of commodities traders at the ABCDs. Indeed, while their individual levels of profitability varied, all of the ABCDs grew and increased profits during the financial and commodity crises. Subsequently, this raised the question of whether they were performing the helpful market role of hedging agents or the role of speculators, which could be illegal.[9] There is an extremely fine line between the two, however, and in practical terms differentiating between hedging and speculation is often impossible, as traders are not responsible for reporting on each business segment of their activities (Murphy *et al.* 2012). At the height of the commodity price rise in 2012, caused in part by a sustained drought in the US, Glencore's (a large commodities speculator) director of agriculture trading, Chris Mahoney, nevertheless summarized the situation as being 'Good for Glencore.... The environment is a good one. High prices, lots of volatility, a lot of dislocation, tightness, a lot of arbitrage opportunities' (Neate 2012).

In a 2010 report by the United Nations Special Rapporteur on the Right to Food, it was noted that between 2003 and 2008 the volume of global index fund speculation had increased over 1.900%; outstanding maize futures contracts increased from 500,000 to nearly 2.5 million; and commodity index fund holdings expanded from US$13 billion to US$317 billion. (de Schutter 2010: 3) The report is a summation of a raft of evidence that has been produced since the crisis, which indicates that agrofuels, as noted previously, were not responsible for the price spike in terms of a supply or demand shock. Rather, agrofuel policies, production and consumption in real terms and in terms of expectations were major catalysts fuelling the food vs. fuel crises, which at the end of the day were speculative bubbles (BIS 2011; Lee *et al.* 2012; IATP 2011; Lagi *et al.* 2011; Murphy *et al.* 2012; UNCTAD 2012).

Views such as those of Glencore's Chris Mahoney, and the actions that stem from them, have led to this dangerous 'financialization' of nature. Today, it is often the case that instead of trading simple corn or wheat as individual commodities, we now see financial products such as housing loans bundled together and bought or sold as part of derivatives together with grain and oil futures, or other combinations of completely unrelated products. These derivates now link previously unrelated sectors financially, making the financial viability of one dependant on the other in ways that they had never previously been (Murphy *et al.* 2012).

Seeing how well the ABCDs were faring during and after the financial crisis, dozens of other important financial institutions began to enter into the commodities game. Many saw this as the reason for the second speculative spike in 2010–2011, which brought prices up to theretofore unheard of levels, but, mysteriously, without the vocal criticism from civil society that had been seen in 2007–2008. After the fact however, in 2012 and 2013, some companies both came under fire by, and responded to, campaigns and publications by the World Development Movement (WDM 2011), Oxfam (Murphy *et al.* 2012) and Friends of the Earth Europe (FoE Europe 2012) on the role their commodities speculation is likely having on global hunger and impoverishment. These companies included Deutsche Bank, Allianz, BNP Paribas, ING Bank, Axa, Crédit Agricole, Commerzbank, Austria's Volksbanken and Barclays. Most of them, apart from Deutsche Bank and Allianz, Germany's largest agricultural traders, recanted at least partially as civil society critics became more vocal and halted or reduced their agricultural commodities speculation (Kelleher 2013).

This development is interesting because it represents the ebb and flow of discursive power in the battle for hegemony. Some of these institutions represent neoliberal ideologies in their purest form and have financed agrofuels businesses. Moreover they have financed land deals in developing countries that have been labelled environmentally harmful and even at times represented land grabs that have displaced people and taken their livelihoods (FoE Europe 2012; Leopold and Dietz 2012). Hiding behind impact assessments and agrofuels promotion policies as justification, however, has deflected such attacks. Nevertheless, in this case, the counter-hegemonic organic intellectuals of only a few civil society organizations were able fundamentally to affect the business practices of major global players within the neoliberal hegemonic bloc by directly attacking practices for which only they were responsible.

Interestingly, Deutsche Bank, an institution that has benefited greatly from the rise of neoliberal globalization and which has often been criticized for the lack of social responsibility in its investment practices, responded to the campaign by stating that the:

> majority of studies agree that the fundamental cause of rising food prices is sharply rising demand that is not yet matched by supply. Demand is surging because of population and income growth in developing countries while production is limited by water scarcity, climate change, lack of infrastructure and harvest waste.
>
> (Deutsche Bank 2013)

While these latter points are indeed noted in many of the studies considered for this chapter, they have been found to be minute in comparison to the influence of speculators. Allianz reached similar conclusions, but conceded that speculation likely reinforced upward price trends. The discursive battle continued as the International Swaps and Derivatives Association (ISDA), a lobby group made up of over 800 financial institutions, launched www.commodityFACT.org. The website employs quotations from a limited selection of literature that place doubt on the role of speculation in increasing market volatility and food prices. It further argues that re-regulation of the financial industry would be a mistake. This organization is the newest incarnation of the TCC on the agrofuels scene, and while the 'facts' it presents run directly in the face of peer-reviewed literature (Lagi *et al.* 2011), the main lesson presented by the neo-Gramsian perspective utilized in this chapter is that, while knowledge may be power, it is how one uses that knowledge strategically towards winning the battle of hearts and minds that makes or breaks a hegemonic project.

## Discussion and conclusion

> Demand from financial investors in the commodity markets has become overwhelming during the last decade … with the volumes of exchange-traded derivatives on commodity markets now being 20 to 30 times larger than physical production, the influence of financial markets has systematically transformed these real markets into financial markets.
>
> (UNCTAD 2012: 1)

Discourses surrounding ethanol in the US were framed and developed essentially unchallenged by agribusiness, and by ADM in particular, over the twentieth century. ADM, its CEO Dwayne Andreas and others like them were able to use their economic might, political connections and ideas on how to deal with geopolitics to create a series of storylines that reinforced the idea that agrofuels were not only good for business, but that they were imperative for an energy-secure, economically healthy and environmentally friendly United States of America. Framing agrofuels as the solution to multiple national goals early on, these hegemonic intellectuals placed the burden of proof on latecomer critics rather than on the companies making these claims – claims which have now all been essentially proven misguided or completely false, especially concerning the particularly environmentally and economically unsustainable first-generation corn ethanol produced in the US.

It was only the dramatic rises in food commodity prices over the course of 2007–2008 that afforded counter-hegemonic forces discursive space to open a global critical discussion of agrofuels. In the US, however, this chapter has illustrated how human rights and environmental issues were deflected using traditional neoliberal hegemonic arguments around economic development and energy security. As a result, agrofuel production is currently being pushed more than ever before in these contexts, and to a certain extent in the EU as well, despite the numerous unresolved social, ecological and economic questions that remain (see Brunnengräber, Chapter 5 and Vogelpohl, Chapter 14 in this volume).

As a relatively new category of commodified nature, industrial agrofuels have re-opened discussions on what exactly 'environmental issues' are. Their complexity as energetic, agricultural, environmental, industrial and now financial products, has transformed these simple plants into geostrategic-environmental products, the production of which can be seen as mandatory or deplorable depending on one's point of view. To this analyst, agrofuels' ever-increasing physical, political and financial likeness to fungible commodities likely means that truly meaningful governance of their sustainability may have become impossible. At this point it will take not only a major unified, global effort to determine agricultural sustainability criteria, but another, probably completely separate, effort to stem harmful speculation on the commodities from which they are produced. The prospects of either seem highly unlikely in the short to medium term.

This pessimistic outlook is based on a few points, all of which can be attributed to the structural (political-systemic), direct (economic) and discursive power wielded by the TCC and those who represent their different interests in the plethora of sectors relevant to agrofuel governance. First, governing agrofuels has not been undertaken in line with the precautionary approach, but remains a game of 'catch-up' by decision-makers reacting to science carried out after the harm has already begun. With this 'act first, ask questions later' style of governance, entrenched interests have been allowed to establish themselves in advance of proposed governance shifts, and have often been able to prevent changes to the system. Second, the futures markets that are used to speculate on commodities have existed for over 150 years – they play a legitimate role in minimizing investment risk, meaning proponents such as Deutsche Bank and Allianz will always have this legitimate argumentation in their discursive toolbox. Additionally, hard-fought changes to the financial system in the US and EU have already been carried out in response to the recent financial crisis, making additional changes to the system unlikely in the short term. Third, even if a global agreement were to be reached, to date such agreements have never truly represented the interests of the poor and vulnerable, and those which attempt to do so have stagnated, with two of the most important global agreements being prime examples: the United Nations Framework Convention on Climate Change (UNFCCC) and the World Trade Organization (WTO).

From the neo-Gramscian perspective presented here, truly addressing the sustainability issues surrounding agrofuels will only be possible when counter-hegemonic actors are able to significantly reclaim discursive, direct and structural ground. This process has begun with slow progress on making economic actors more responsible for their actions on the ground and also now for the first time ever, in terms of investment and financial activities. That said, it is completely unclear at the moment whether these concessions from financial actors and others should be seen as a counter-hegemonic 'success', or simply as the co-optation of dissenting voices by the TCC and the hegemonic ideas and practices they represent.

## Notes

1 In the US, auto manufacturers throughout the 2000s were (and still are) against agrofuels for fear of damaging engines, although there is a lack of evidence on their actual

influence in this regard. With a few exceptions, historically up through the early 2000s, both the oil and automobile industries in the US were neither very enthusiastic nor adverse to agrofuels and do not appear to have made this a political issue until very recently (Bernton *et al.* 2010 [1982]).

2 Up-to-date information on industrial processing capacities as well as national production statistics can be found at: www.ethanolrfa.org/ (last access: 24 March 2014).
3 www.econexus.info/agrofuel-moratorium-call (last access: 24 March 2014).
4 Groups of automobile manufacturers sued the US EPA in 2010 and again in 2013 over blending mandates (Murray 2010; Schroeder 2013).
5 See, www.rspo.org/en/board_of_governors (last access: 24 March 2014].
6 See, www.ethanolrfa.org/pages/statistics (last access: 24 March 2014).
7 See, www.google.co.uk/finance?client=ob&q=LON:ADM (last access: 24 March 2014).
8 A detailed treatment of deregulation and the resultant structural changes and emergent financial products as they relate to agriculture is beyond the scope of this chapter, but can be found in Murphy *et al.* (2012), pp. 26–38.
9 Where hedging is a type of speculation that uses futures markets to improve markets and reduce risk by buying and selling when others are not. Speculation in its negative connotation is when buying low and selling high at high turnover rates is profit-oriented and can create instability unrelated to supply and demand, especially when speculative trade begins to outstrip normal trading, leading to the creation of bubbles and crashes (Albino *et al.* 2012).

## References

AFP (2008) 'Haiti PM ousted over soaring food prices', AFP (www.google.com/hosted-news/afp/article/ALeqM5hL0HvIfNZQ2nMgFdy9dSKLZ7t2Gw, last access: 24 March 2014).

Albino, Dominic, Karla Bertrand and Yaneer Bar-Yam (2012) *Food FAQ*. New England: Complex Systems Institute.

Barrionuevo, Alexei (2007) 'Springtime for ethanol', *New York Times*, 23 January 2007.

Bernton, Hal, William Kovarik and Scott Sklar (2010 [1982]) *The Forbidden Fuel: A History of Power Alcohol*. Nebraska: Bison Books, 2nd ed.

Bieler, Andreas and Adam David Morton (2001) 'The Gordian knot of agency-structure in international relations: a neo-Gramscian perspective', *European Journal of International Relations* 7(1), 5–35.

BIS (2011) 'BIS 81st Annual Report', Bank of International Settlements.

Bloomberg (2008) 'Food-related industries launch anti-biofuel campaign', *Bloomberg News* (www.chron.com/business/energy/article/Food-related-industries-launch-anti-biofuel-1783563.php, last access: 24 March 2014).

Bovard, James (1995) 'Archer Daniels Midland: a case study in corporate welfare', Cato Institute.

Bush, George (2008) 'Press conference by the President' (http://georgewbush-whitehouse.archives.gov/news/releases/2008/04/20080429–1.html, last access: 24 October 2013).

Congressional Budget Office (2009) 'The impact of ethanol use on food prices and greenhouse-gas emissions', Congressional Budget Office.

Cox, Robert (1983) 'Gramsci, hegemony and international relations: an essay in method', *Millennium – Journal of International Studies* 12(2), 162–75.

de Schutter, Olivier (2010) 'Food commodities speculation and food price crises: regulation to reduce risks of price volatility', United Nations Special Rapporteur on the Right to Food. Briefing Note 02.

Deutsche Bank (2013) 'Questions and Answers on investments in agricultural commodities' Deutsche Bank (www.db.com/cr/en/concrete-questions-answers.htm, last access: 26 March 2014).

Doornbosche, Richard and Ronald Steenblik (2007) *Biofuels: Is the Cure Worse than the Disease?*, Paris: Organisation for Economic Co-operation and Development (OECD).

Ekers, Michael, Alex Loftus and Geoff Mann (2009) 'Gramsci lives!', *Geoforum* 40(3), 287–291.

Ethanol History (2011) 'Ethanol history – from alcohol to car fuel' (www.ethanolhistory.com/, last access: 10 March 2014).

F&C Investors (2008) 'Biofuels and sustainability: an investor perspective', F&C Investors.

Fargione, Joseph, Jason Hill, David Tilman, Stephen Polasky and Peter Hawthorn (2008) 'Land clearing and the biofuel carbon debt', *Science* 319(5867), 1235–8.

FoE Europe (2012) *Farming money: How European banks and private finance profit from food speculation and land grabs*. Friends of the Earth Europe.

Fuchs, Doris (2005) *Understanding Business Power in Global Governance*. Baden-Baden: Nomos.

Göricke, Fred and Monika Reiman (1982) *Treibstoff statt Nahrungsmittel: Wie eine falsche energiepolitische Alternative den Hunger vermehrt*. Reinbek bei Hamburg: Rowohlt GmbH.

Hay, Colin (2001) 'The crisis of Keynesianism and the rise of neoliberalism in Britain: an ideational institutionalist approach', in Campbell, John and Ove Kaj Pedersen (2001) *The Rise of Neoliberalism and Institutional Analysis*. Princeton, New Jersey: Princeton University Press, 193–215.

Herbert, Josef (1995) 'Appeals Court Throws Out EPA's Ethanol Mandate', *Associated Press*, 5 August.

Holt-Giménez, Eric (2008) 'The world food crisis: what is behind it and what we can do' (www.worldhunger.org/articles/09/editorials/holt-gimenez.htm, last access: 24 March 2014).

IATP (2011) 'Excessive Speculation in Agriculture Commodities: Selected Writings From 2008–2011', IATP.

IEA (1994) *Biofuels*. Paris: Organisation for Economic Co-operation and Development (OECD) and International Energy Agency (IEA).

IEA (2004) *Biofuels for Transport. An International Perspective*. Paris: International Energy Agency.

IEA Bioenergy (2010) *Strategic plan 2010–2016*. Paris: International Energy Agency.

Keeney, Dennis (2009) 'Ethanol USA', *Environmental science and technology* 43(1), 8–11.

Kelleher, Ellen (2013) 'Food price speculation taken off the menu,' *Financial Times*, 3 March 2013.

Kovarik, William (1998) 'Henry Ford, Charles Kettering and the fuel of the future', *Automotive History Review* 32, 7–27.

Lagi, Marco, Yanvi Bar-Yam, Carla Bertrand and Yaneer Bar-Yam (2011) *The Food Crises: A Quantitative Model of Food Prices Including Speculators and Ethanol Conversion*. Cambridge, MA: New England Complex Systems Institute.

Lee, Bernice, Felix Preston, Jaakko Kooroshy, Rob Bailey and Glada Lahn (2012) *Resource Futures*. London: Chatham House.

Leopold, Aaron and Kristina Dietz (2012) 'Transnational contradictions and effects of Europe's bioenergy policy: evidence from Sub-Saharan Africa', Fair Fuels Working Paper 4.

Levy, David L. and Ans Kolk (2002) 'Strategic responses to global climate change: conflicting pressures on multinationals in the oil industry', *Business and Politics* 4(3), 275–300.

Levy, David and Peter Newell (2002) 'Business strategy and international environmental governance: toward a neo-Gramscian synthesis', *Global Environmental Politics* 2(4), 84–101.

Lilley, Sasha (2006) *Green Fuel's Dirty Secret*. San Fransisco: CorpWatch.

Meyer, C. (2010) 'Biofuels – Alcohol, the fuel of the future? – Part 4', *Energize* (May 2010), 19–21.

Pfuderer, Simone, Grant Davies and Ian Mitchell (2008) 'A note on rising food prices', World Bank Policy Research Working Paper No. 4682.

Murphy, Sophia, David Burch and Jennifer Clap (2012) *Cereal Secrets: The World's Largest Grain Traders and Global Agriculture*. Oxford: Oxfam.

Murray, James (2010) 'Auto industry sues EPA over biofuel blend ruling', *Green Business* (www.businessgreen.com/bg/news/1933888/auto-industry-sues-epa-biofuel-blend-ruling, last access: 24 October 2013).

Neate, Rupert (2012) 'Glencore food chief says US drought is "good for business"', *Guardian*, 21 August 2012.

NREL (no date) *Ethanol-Blended Fuels*. Colorado: National Renewable Energy Laboratory.

Philpott, Tom (2006) 'How cash and corporate pressure pushed ethanol to the fore', *Grist*, 7 December 2006.

Rattner, Steven (2011) 'The great corn con', *New York Times*, 24 June 2011.

Rohter, Larry (2008) 'Obama camp closely linked with ethanol', *New York Times*, 23 June 2008.

Royal Society, The (2008) *Sustainable biofuels: prospects and challenges*. London: The Royal Society.

Rupert, Mark (1995) *Producing Hegemony: The Politics of Mass Production and American Global Power*. Cambridge: Cambridge University Press.

Schroeder, Joanna (2013) 'Engine products group appeals E15 partial waiver', *Domestic Fuel* (http://domesticfuel.com/2013/03/26/engine-products-group-appeals-e15-partial-waiver/, last access: 24 October 2013).

Searchinger, Timothy, Ralph Heimlich, R.A. Houghton, Fengxia Dong, Amani Elobeid, Jacinto Fabiosa, Simla Tokgoz, Dermot Hayes and Tun-Hsiang Yu (2008) 'Use of US croplands for biofuels increases greenhouse gases through emissions from land-use change', *Science* 319(5867), 1238–40.

Shattuck, Annie (2009) 'Will sustainability certifications work? A look at the roundtable on sustainable biofuels', in Jonasse, R. (ed.) *Agrofuels in the Americas*. Oakland, California: Food First.

Taxpayers for Common Sense (2013) *Political Footprint of the Corn Ethanol Lobby*. Washington, DC: Taxpayers for Common Sense.

Tetreau, Matt (2013) 'Rebels with a cause: US corn farmers' storage revolution', J.P. Morgan Asset Management (http://insights.jpmorgan.co.uk/adviser/commentary-and-analysis/rebels-with-a-cause-us-corn-farmers-storage-revolution-insight-blog/, last access: 24 October 2013).

UNCTAD (2012) *Don't Blame the Physical Markets: Financialization is the Root Cause of oil and Commodity Price Volatility*. Geneva: UNCTAD.

van Appeldoorn, Bastiaan (2000) 'Transnational class agency and European governance: the case of the European round table of industrialists', *New political economy* 5(2), 157–81.

van der Pijl, Kees (1998) *Transnational Classes and International Relations*. London: Routledge.

van der Pijl, Kees (2009) *A Survey of Global Political Economy*. Brighton: University of Sussex, Centre for Global Political Economy.

WDM (2011) *Broken Markets: How Financial Market Regulation Can Help Prevent Another Global Food Crisis*. London: World Development Movement.

# 14 Immunization by neoliberalization

## The strange non-death of the win–win narrative in European agrofuel policy

*Thomas Vogelpohl*[1]

### Introduction

The narrative of the multiple benefits that a more aggressive promotion of agrofuel development would yield has been a staple of the agrofuels debates since the early 1980s. This so-called win–win narrative emphasizes the (putative) advantages of an agrofuel policy for the protection of the environment, energy security, rural development in both the southern and the northern hemisphere and various socio-economic benefits in general (e.g. Borras *et al.* 2010; Franco *et al.* 2010).

In Europe, however, the narrative only began to manifest itself politically in the early 2000s. In the wake of the rise of the environmental policy paradigm of ecological modernization, which centres on the emphasis of the ecological and economic double benefit of environmental policy – i.e. win–win solutions – the European agrofuels directive was adopted; it called for increasing the agrofuel share of the total fuel market to at least 2 per cent by 2005 and 5.75 per cent by 2010 (European Union 2003). At the time, these goals were non-binding; however, after missing the 2005 target and with the outlook for 2010 not looking any better, the EU stepped up the political pressure behind its agrofuel policy. This resulted in the 2009 Renewable Energy Directive (RED), which mandates the attainment of a 10 per cent renewable energy share of the total transport fuel market in the member states – a binding target that is largely to be achieved by means of agrofuels (European Union 2009).

The adoption of the RED came during a period of growing criticism of agrofuels, i.e., at a time when the validity of the win–win narrative was being strongly called into question. This criticism is reflected in the RED to the extent that the 10 per cent target is linked to specific sustainability criteria, and that all renewable forms of energy, not just agrofuels, count toward attainment. However, the many criticisms of the directive did not stop after 2009. And yet, the directive still remains in force, in its original, unmodified form. The agrofuels win–win narrative thus has proven itself, at least in a formal sense, to be quite robust.

This chapter aims to show how it became possible for the narrative to take hold in the European Union and how it was able to overcome the wave of ongoing criticism that began in 2007. Closely related is the question of how the narrative became entrenched in the corresponding policy decisions. By answering this, the chapter also aims to place this development in the context of overall processes of neoliberalization within environmental policy, which, as a rather recent policy field, has been increasingly characterized by the neoliberal *zeitgeist* of the last four decades.

The chapter is structured as follows. First, I briefly sketch my analytical framework and methodological approach. Following that, the various components of the agrofuels win–win narrative are outlined and its development and, especially, its modifications in the context of European agrofuel policy are illustrated. Next, the extent to which these modifications occur under the aegis of neoliberalism, and how they manifest themselves in the RED – thus representing the neoliberalization of European agrofuel policy – is analysed. Finally, the results of the analysis are summed up and a conclusion is drawn.

## Analytical framework: the neoliberalization of nature from a discourse perspective

The influence of neoliberalism on environmental policy can be seen in the coupling of environmental policy goals to free market mechanisms, a development frequently referred to as the 'neoliberalization of nature', particularly in the field of political ecology (e.g. Heynen and Robbins 2005). Neoliberalization here means that nature is not only valuated and exploited, but that this happens under the aegis of neoliberalism, i.e., by the privatization and commodification of nature, for the implementation of which, at least, a selectively strong government role is necessary. In marked contrast to the neoliberal narrative, however, this process of conforming nature to fit neoliberal market logic has tended to accelerate rather than lessen the degradation of natural resources and has intensified the unequal distribution of access to these resources (e.g. Castree 2010).

To place the development and survival of the agrofuels win–win narrative and its role in European agrofuel policy in the context of the neoliberalization of nature means that it will be analysed how the agrofuels win–win narrative is given a specific neoliberal twist, and how this then manifests itself in the respective policy decisions, which in turn give rise to specific material effects.

This intention calls for a discourse-analytical approach. Discourse analysis makes it possible to show how such processes of neoliberalization proceed at the narrative level and how they manifest themselves in political decisions. Neoliberalism can be understood here as an interpretative scheme. Interpretative schemes can be described as 'fundamental meaning and action-generating schemata, which are circulated through discourses and make it possible to understand what a phenomenon is all about' (Keller 2011: 57). Various interpretative schemes interact to lend phenomena a specific structure. These phenomenal structures do 'not describe any essential qualities of a discourse topic, but rather the corresponding discursive attributions' (ibid.: 58). Narratives in this context set the individual dimensions of these phenomenal structures – the so-called phenomenal components – into specific relationships with one another. A narrative serves, so to speak, as the common thread through a phenomenal structure and thus plays a key role in the process of ascribing meaning to such phenomena.

Particularly important for the political effectiveness of a narrative are the storylines and their interrelations (Hajer 1995). Storylines are shorthand representations of an overall narrative that are characterized above all by their ability to support multiple interpretations, in effect concealing or obscuring differences and contradictions, thus facilitating the adoption of political decisions.

In the context of this chapter, this means that it is necessary to analyse the extent to which the components of the agrofuels phenomenon are woven together into a win–win narrative by storylines drawing upon a neoliberal interpretive scheme and how this is reflected in EU agrofuel policy.

Methodologically, the chapter builds upon the analysis of around 50 policy documents published in the relevant time frame (from 2003 to 2009).[2] The analysis is guided by Grounded Theory and aims to identify the phenomenal structure of agrofuels in these documents. In order to categorize the phenomenal components, a set of categories is used that is based on typical structures of discursive phenomena referred to by Keller (2008) and Strauss and Corbin (1996: 75–93) that are adjusted to the purposes of this chapter. In a second step, the textual analysis will be related to the interpretive scheme of neoliberalism that is re-invoked by the storylines that weave the identified phenomenal components of agrofuels together into a win–win narrative.

## Establishment of the win–win narrative and its manifestation in the EU agrofuels directive

The agrofuels phenomenon as identified in the documents analysed consists of six core components:

1 scientific-technical potentials of agrofuels: potentials for the technological development of agrofuels; quality of agrofuels and compatibility with conventional combustion engines; land consumption by agrofuels and land availability;
2 agrofuels as an instrument of environmental protection: overall climate impact of agrofuels; impacts of agrofuels on biodiversity, local water supplies and soil conditions;
3 agrofuels as an agricultural product: contribution of agrofuels to rural development (regional and global value and job creation); role of farmers in the context of agrofuels production;
4 agrofuels as a source of energy: contribution of agrofuels to the security of the energy supply and to a reduced dependence on imports of crude oil; energy efficiency of agrofuels (ratio of energy input to energy yield);
5 socio-economic aspects of agrofuels: competitiveness of agrofuels (compared to fossil fuels); overall economic effects of an expansion of agrofuels production;

6 policy options: eligibility of agrofuels for subvention or other forms of promotion; policy instruments to be utilized for the promotion of agrofuels.

As indicated above, this phenomenal structure should not be confused with the essential or 'real' nature of the phenomenon 'agrofuels', which – according to the interpretivist approach applied here – is not directly accessible. Rather, it reflects the way agrofuels are construed as a policy object in the documents analysed, the components it consists of and the aspects that are debated regarding these components.

The introduction of an agrofuel policy in Europe reflects the link between those specifications of the first five phenomenal components that rate agrofuels positively and the aspect of policy options. This means that support for agrofuels promotion is typically associated with attributions suggesting a certain technical and quantitative potential, as well as positive impacts with respect to the environment, the energy supply, rural development and socio-economic concerns: the agrofuels win–win narrative (e.g. Borras *et al.* 2010; Franco *et al.* 2010).

At first, this is hardly surprising. But then one must ask why this narrative only first found expression in a policy decision in the early 2000s, as it has existed in this form since the 1980s. Thus we find, for example, in the 1992 effort by the European Commission to put forward an EU-wide 10 per cent limit on excise duties for agrofuels (the Scrivener proposal), that the proposed measure, along with micro- and macro-economic benefits, would offer 'global advantages with respect to the common agricultural policy, the energy policy, and the Community policy on the protection of the environment' (CEC 1992: 4; own translation). In the same year, however, the upper house of the German legislature recommended defeat of this very EU proposal 'on macroeconomic, budgetary, environmental and energy policy grounds' (Bundesrat 1992: 1, own translation) – a 'lose-lose' narrative, so to speak. This citation shows that the Scrivener proposal was not rejected on the basis of agricultural concerns, as the benefits to agriculture of such a policy were (and still are) relatively undisputed. But to be politically effective, it lacked a well-established positive association linking the agricultural component to the other components of the agrofuels phenomenon.

The ecological modernization discourse offers an important overarching frame for such an association. The concept of ecological modernization essentially describes the utilization of the forces of modernization inherent in the capitalist development model for the benefit of the environment (Jänicke 2000). In this chapter, however, ecological modernization is not assessed according to its chances of success but is viewed, rather, as an environmental discourse that systematically 'produces certain meaningful connections between phenomena, events, and actors of environmentally relevant action' (Hajer 1997: 109, own translation). The emphasis on economic-environmental win–win solutions, i.e., the idea of environmental protection and economic growth as a positive-sum game, takes on a central role within the concept. It thus systematically produces economic-ecological win–win narratives that in the course of the rise of ecological modernization as the dominant environmental policy paradigm during the 1990s acquired a substantial measure of societal significance (Jänicke 2000; Hajer 1997).

But if the overriding win–win narrative of ecological modernization is to be applied to agrofuels, then agrofuels must first be perceived as an ecological innovation – a positive conceptual connection, to characterize it in terms of discourse theory, needs to be established between the environmental and the agriculture-related phenomenal components.

The discursive framework for this consists of several closely linked discourses. The multifunctionality discourse emphasizes the positive externalities of agricultural production beyond food production – above all with respect to environmental as well as social impacts (e.g. Potter and Tilzey 2005: 590–591). It systematically construes positive conceptual connections between agricultural production and environmental protection on a general level. The renewable raw materials discourse can be viewed as part of a broadly-defined multifunctionality discourse, as it emphasizes the applicability of agricultural products outside of the classic utilization scheme (food production) and their function as a natural resource for industry and energy production. A key component of the renewable resources discourse is thus the bioenergy discourse, which links agricultural production not only to energy production, but specifically to 'green' or renewable energy (e.g. McHenry 1996: 382–384). The renewable energy discourse initially linked renewable sources of energy to the reduction of dependence on foreign oil. Later, however, another aspect became at least as important: namely, the establishment of a conceptual connection between renewable energy and climate protection (e.g. Rowlands 2005), which established itself over the course of the 1990s as the main environmental policy issue (e.g. Mert 2013).

This discourse complex, which arose in Europe in the 1980s – and by the 1990s, at the latest, was also institutionally established – provided the necessary framework for the agrofuels win–win narrative assuming a greater importance and effectiveness during the 1990s. Agrofuels were no longer only or primarily perceived as agricultural products, but instead took on a broad range of positive attributions. In terms of discourse analysis, positive conceptual connections were established between the various components of the phenomenon 'agrofuels', particularly among its environmental, energy, agricultural and socio-economic dimensions.

Although agrofuels, until 2000, were nothing more than a niche product, they became the screen onto which a number of hopes and expectations from many areas were projected (see Brunnengräber, Chapter 5 in this volume). Thus it became possible for a policy promoting agrofuels that 10 years previously had been rejected using a 'lose-lose' narrative, to be accepted in 2003 as a win–win solution:

> This Directive aims at promoting the use of biofuels ... with a view to contributing to objectives such as meeting climate change commitments,

environmentally friendly security of supply and promoting renewable energy sources.

(European Union 2003: 44)

The agrofuels directive that was finally adopted in 2003 – although non-bindingly – called for increasing the agrofuels share of the total fuel market in the EU to at least 2 per cent by 2005 and 5.75 per cent by 2010 (ibid.).

## Neoliberalization of the win–win narrative and its manifestation in the RED

The focus of the following analysis is the period between 2003 and 2009, i.e., the period in between the adoption of the agrofuels directive and the RED. Although the number of incentive measures adopted since 2003 has shown a strong increase, the declared goal of achieving a 1 per cent share of the overall fuel market in the transport sector of the EU in 2005 was (with only 1 per cent) missed by a large margin (CEC 2007: 6). In January 2008, in a comprehensive climate and energy package, the Commission introduced its proposal for a renewable energy directive that included a mandatory 10 per cent share of renewable energies in the transport sector by 2020, linked to specific sustainability criteria (CEC 2008).

The development of this proposal, as well as the revisions leading to the final directive, thus took place during the same period of time when criticism of agrofuels and doubts about their advantageousness were growing louder. As a result, the draft was revised and modified in 2008. However, the core demands of many non-governmental organizations (NGOs), specifically for expansion of the sustainability criteria to include social and stronger environmental criteria as well as elimination of the bindingness of the 10 per cent target, were left out.[3] A compromise was finally arrived at that modified the 10 per cent target to encompass not only agrofuels, but renewable energy in general (European Union 2009). Despite the ongoing debates about EU agrofuel policy and the 10 per cent target, as of early 2014, the RED is still the operative regulatory policy for agrofuels at the EU level.

In the end, the win–win narrative proved to be so robust that it made possible a continuation – or even intensification – of EU agrofuel policy despite the emerging criticism. From a discourse analysis perspective, it is desirable to consider the modification that the narrative went through in this period to make it resilient to criticism.

### *Re-alignment of the win–win narrative under the aegis of neoliberalism: technological innovation, competitiveness and climate protection*

In particular, the increased attention given to the phenomenal component of the scientific-technical potentials of agrofuels marks a significant new conceptual connection within the win–win narrative in this context. 'Technological innovation' is now a primary component of the re-aligned narrative, as can be seen in the following vision for a European agrofuels sector in the year 2030:

> By 2030, the European Union covers as much as one quarter of its road transport fuel needs by clean and $CO_2$-efficient biofuels. A substantial part is provided by a competitive European industry. This significantly decreases the EU fossil fuel import dependence. Biofuels are produced using sustainable and innovative technologies; these create opportunities for biomass providers, biofuel producers and the automotive industry.
>
> (BIOFRAC 2006: 3)[4]

In publications such as this, socio-technological and economic imaginaries of a future European agrofuels sector are woven together, resulting in a conditionalized win–win narrative that links the positive-sum aspect of agrofuels to technological innovation and competition in future markets (Levidow *et al.* 2012). This bears the neoliberal stamp in that it links the future promise of technological innovation to a neoliberal interpretation of competitiveness in which competitiveness is defined as 'survival of the fittest in a fully open environment of a global free market' (van Apeldoorn 2002: 172), rather than as the economic assertiveness of state-supported 'European champions'. It is made clear within the framework of the re-aligned win–win narrative that these envisioned markets, in which the symbiotic interplay of competition and technological innovation is meant to take place, can only be global.

Marie Widengard speaks in this context of the 'imperative to go global', i.e., of the construction of a need to make of agrofuels a global commodity in order to maintain its positive-sum character (Widengard 2011: 45). Thus, in the course of the re-aligned win–win narrative, the need for an 'expansion to vast areas of the Global South and the establishment of international trade … to reap maximum efficiency' is construed (Kuchler and Linnér 2012: 585). A key storyline in this context is the availability of 'marginal land', i.e. the 'vast areas of the Global South' that – in contrast to the situation in Europe – are available and can be readily exploited for agrofuels production (Levidow *et al.* 2012: 174–175; see also Backhouse, Chapter 10 in this volume):

> There is plenty of undeveloped arable land that could be used to increase the supply of both food and fuel, not to mention the often dry, marginal land that is currently a curse for poorer nations, but which could become home to fields of jatropha or similar biofuel-producing crops.
>
> (Watkin 2008)

In the context of the re-aligned agrofuels win–win narrative, this (ostensibly) available marginal land is to be commodified by means of technological innovation, i.e. by the development of 'second-generation biofuels'.

> We have all seen the maps showing the vast tracts of land that would be required to replace petrol to any significant degree. That is why research and

> development into second generation biofuels that are cleaner, more versatile, and can be used on more marginal land is so important.
>
> (Mandelson 2007: 1–2)

The neoliberal agrofuels discourse is thus linked to projections of global competition in future markets and technological innovation that assume not only a central but also a self-reinforcing role in the re-aligned win–win narrative. The neoliberal competitiveness discourse thus enters into a sort of symbiotic relationship with a technology-centred innovation discourse. 'Market competition stimulates innovation, which enables growing eco-efficiency, which in turn triggers further market competition' (Pellizoni and Ylönen 2012: 6). With respect to agrofuels, this close and self-reinforcing relationship between technological innovation and competition can, for example, be seen in the EU Strategy for Biofuels:

> While existing technologies do not at present offer cost-competitive solutions for the EU, the benefits of encouraging the development of biofuels should outweigh the cost. In this context, the development of second generation biofuels ... could further contribute to their cost-effectiveness.
>
> (CEC 2006: 4)

Technological innovation and competition in global markets thus become necessary prerequisites for environmentally-friendly agrofuels, with the singular aspect of climate protection playing the pre-eminent role in the environmental component of the phenomenon of agrofuels. Only under these conditions is it possible for the components of the agrofuels phenomenon to be assembled into a (then neoliberally conformed) win–win narrative.

Not only in the case of agrofuels are environmental problems such as climate change viewed in the light of the neoliberal discursive frame as market inefficiencies and technical problems to be solved through the implementation of a market-friendly regulatory framework (Parr 2012). Once this framework is in place, competition in the free and open markets will lead to technical advances (and vice versa) and thus to more efficient climate protection (Levidow *et al.* 2012).

This neoliberal-technical climate protection discourse, which is firmly established, particularly at the international level (e.g. in the market-based mechanisms of the Kyoto Protocol), also becomes apparent in the re-aligned agrofuels win–win narrative. 'Second-generation biofuels' are represented here as a future technological fix, an envisioned technical solution that in accordance with the internal logic of this re-aligned win–win narrative will someday help to solve the technical problem of climate change – if properly established in an appropriate competitive environment (i.e. global and with minimized trade restrictions) (ibid.):

> Linking greenhouse gas benefits to encouraging the provisions of biofuels would help to increase their benefits and send a clear signal to the industry

> of the importance of further improving production pathways in this respect. It would also allow market-based signals to be sent to fuel and feedstock producers, to further reduce carbon emissions in the transport sector.
>
> (CEC 2006: 10)

> That means that Europe should be open to accepting that we will import a large part of our biofuel resources.... We should certainly not contemplate favouring EU production of biofuels with a weak carbon performance if we can import cheaper, cleaner, biofuels. Resource nationalism doesn't serve us particularly well in other areas of energy policy – biofuels are no different.
>
> (Mandelson 2007: 2–3)

The re-aligned agrofuels win–win narrative is thus characterized particularly by the newly established conceptual connections linking together the phenomenal components of socio-economic effects, technological development potential and climate protection, for which the primary storylines are technological innovation, competitiveness, global future markets and climate protection. The discursive framework for these connections is set by a neoliberal competition and technology discourse in which the positive-sum aspect of the narrative is linked to neoliberal terms.

### *The win–win narrative at work: the 10 per cent target and sustainability certification*

The reflection of this re-aligned agrofuels win–win narrative in EU policy can be seen in the 2006 EU Strategy for Biofuels (if not before), which aims at '[preparing] for the large-scale use of biofuels by improving their cost competitiveness through the optimized cultivation of dedicated feedstocks, research into "second generation" biofuels, and support for market penetration by scaling up demonstration projects and removing non-technical barriers' (CEC 2006: 4).

It is possible to find numerous other examples of the infiltration of agrofuel policy communications by the re-aligned win–win narrative. As mentioned above, the (imagined) positive-sum game of technological innovation, competition in global markets and climate protection – which is central to this narrative – functions only when global markets for innovative agrofuels are available. The primary policy demand thus deriving from this win–win narrative is the creation of such a global market, more specifically:

> [the creation of] a framework which gives investors the confidence they require to invest in better, capital-intensive forms of biofuel production, and informs vehicle manufacturers of the fuels for which vehicles should be designed (it is therefore necessary to set minimum biofuel targets for 2015 and 2020).
>
> (CEC 2007, 8)

Amelia Sharman and John Holmes note in this context that 'imperatives located in the political space dominated scientific evidence and led to a process of "policy-based evidence gathering" to justify the policy choice of a ten per cent renewable energy/biofuels target' (Sharman and Holmes 2010: 309). Such imperatives are an effect of the re-aligned win–win narrative, which construes the adoption of a preferably ambitious and binding target and, therefore, the establishment of a preferably big and secure global market as an inherent necessity. This is to be seen, for example, in the preamble to the RED, which affirms the 10 per cent target as an 'appropriate and achievable objective', and adds that:

> a framework that includes mandatory targets should provide the business community with the long-term stability it needs to make rational, sustainable investments in the renewable energy sector which are capable of reducing dependence on imported fossil fuels and boosting the use of new energy technologies.
>
> (European Union 2009: 17)

Although this neoliberal story of the symbiosis of global competition, technological innovation and climate protection is set sometime in the future and the assumed causalities are yet to be proven, it nonetheless already has effects today.

On the one hand, it serves as the basis for far-ranging policy decisions. The 10 per cent target has already led to a global market for agrofuels – a development with consequences: agrofuel imports in the EU are increasing, as are investments in agrofuel production and the production of raw materials in other parts of the world (Lamers *et al.* 2014), resulting in a spatial and material expansion of the raw materials base for the European fuel industry. Thus we find in the re-aligned win–win narrative and associated 10 per cent target a trend toward a 'neoliberalization of nature', which comprises the integration of extensively or non-utilized natural resources into neoliberal structures of capital valorization.

On the other hand, the neoliberal narrative of the future 'techno-fix' to be secured by global markets for 'second-generation biofuels' largely immunizes the re-aligned win–win narrative from an effective articulation of criticism (Levidow *et al.* 2012: 165). The reference to the future as well as the circular nature of the argumentation within the re-aligned win–win narrative means that many criticisms of the impact of agrofuel support are effectively neutralized or even appropriated.

The most frequent criticisms by civil society actors concern the climate impact of agrofuels (indirect land-use change; ILUC), their effect on food prices ('food vs. fuel') and the social and environmental impacts of raw materials production in the southern hemisphere ('land-grabbing' and rain forest destruction). These criticisms, however, are – at least partly – deflected by the re-aligned win–win narrative that tells the story of highly efficient high-tech agrofuels of the future, that will (ostensibly) be the solution to such problems. This can only happen, however, if competitive global markets are first established, in which the above-mentioned imagined symbiosis of global competition, technological innovation and climate protection can take place, as the following quotation demonstrates:

> The frameworks which apply today were originally introduced to have an initial positive effect on climate protection and supply security. However, they do not yet reflect the many new developments – quite the reverse: they are currently a hindrance and delay true competition between biofuels. They therefore need to be revised.... Rewriting the framework will enable the market penetration of specific biofuels to be controlled specially in favor of sustainability factors, and will therefore precisely maximize their efficiency with respect to climate protection aspects. This will only be possible if the existing fuel options are evaluated differentially according to the aforementioned sustainability criteria. This will enable biofuels to be assessed reliably, and thus clearly and unequivocally reveal the options with the highest environmental, economic and social compatibility. This must form the only basis for the legislative promotion of products and processes.
>
> (BASF 2007: 1)

Thus, a large part of the criticism of today's agrofuels – especially the climate-related one – is 'turned into an extra argument for pursuing a techno-fix: that is, high targets [are] necessary to stimulate biofuel innovation that would minimize environmental destruction' (Levidow *et al.* 2012: 174–175). Rather than a concession to the critics of agrofuels, the coupling of the 10 per cent target to sustainability criteria (which had already been planned before the wave of criticism from 2007 onward) and the certification of agrofuels constitute the 'market mechanism [that] will be required to ensure that $CO_2$ efficiency of biofuels is acknowledged and rewarded', called for by BIOFRAC (BIOFRAC 2006: 4).

This absorption of criticism into the neoliberal win–win narrative also manifests itself in the RED, as the 10 per cent target was retained, and its coupling to sustainability criteria does not in any way contradict the narrative. Instead, certification is easily integrated into a neoliberal logic of environmental management, since as a market-based instrument it 'reinforce[s] the profit-making logic of capital and (re)inscribe[s] neoliberal regulatory practices by working with the market (rather than against it)' (Higgins *et al.* 2008: 1777–1778).

Eventually, a version of the sustainability certification was included in the RED, which – contrary to the demand of many environmental and developmental NGOs – contains only a few environmental and no social criteria. The environmental criteria are limited – in accordance with a 'carbonification of environmental discourses' (Mert 2013) – to $CO_2$ reductions, along with biodiversity, while other environmental aspects are disregarded. Furthermore, 'advanced biofuels' count twice toward attainment of the 10 per cent target. This represents a more or less ideal transfer of the neoliberally re-aligned win–win narrative into policy practice, as the sustainability criteria will lead to the creation of market incentives that should drive future technological innovations, which in turn will make possible an expansion of this market. Here, however, we also see the contradictions inherent in the RED, which is supposed to encourage the development of technological innovations that someday will mitigate certain undesirable dynamics such as land-grabbing and deforestation – the same dynamics that, if not a direct result of the binding 10 per cent target in the directive, were at least co-induced by it. Furthermore, the example of the omission of social criteria with reference to WTO rules makes clear that the EU sustainability certification for agrofuels is not intended to function as a protectionist instrument and is meant to compromise the desired global trade in agrofuels and the corresponding natural resources as little as possible (Levidow *et al.* 2012: 177).

Altogether, it becomes clear that the interpretation of agrofuels to be found in the binding 10 per cent target for agrofuels, as well as in the associated sustainability certification, is one that has been strongly influenced by a neoliberally re-aligned win–win narrative. The 10 per cent target is supposed to lead to the establishment of a market primarily for mass-produced, globally-traded and climate-efficient, high-tech agrofuels. The sustainability certification is subordinate to these aspects and will be implemented under the premise that it does not undermine the neoliberal character of the RED but, rather, strengthens it.

## Conclusion

The process by which the agrofuels win–win narrative in the EU became established and modified can be described as a process of neoliberalization. In the framework of a general ecological modernization discourse, positive conceptual connections were established among the agrofuel phenomenal components of environmental protection, agriculture, energy security and socio-economics. In this version of the win–win narrative, agrofuels acquire a fundamentally positive meaning that is virtually unconditional. This discursive transformation finds expression in the 2003 agrofuels directive, which did call upon the member states to adopt incentive measures and set goals, but granted them relatively wide scope in their implementation.

Between 2003 and 2009, however – a period during which agrofuels were being most sharply criticized – the win–win narrative was modified, linking the multiple benefits of agrofuels to certain conditions. This conditionalization of the win–win narrative mainly consisted of a re-alignment of the components of environmental protection, socio-economics and scientific-technological potentials. It drew on neoliberally informed socio-technical and economic visions of the future in which technological innovation ('second-generation/advanced biofuels') and competition in global markets enter into a symbiotic relationship from which the climate is also to profit. In terms of policy options, this neoliberally re-aligned win–win narrative leads to the imperative to establish these global markets and to put in place mechanisms (such as sustainability certification) to make possible the envisioned symbiosis of technological innovation and competition in global markets.

These policy options are reflected in the 2009 Renewable Energies Directive. The RED represents a neoliberalization of European agrofuel policy, which is expressed in the binding 10 per cent target (which establishes a reliable global market for agrofuels) as well as in the linking of the goal to sustainability criteria that, by means of incentivizing technological innovations, are intended to make possible an expansion of the global market and international trade. The neoliberal re-alignment of the win–win narrative thus rendered possible the continuation – and even intensification – of the EU agrofuels policy. Thereby, it also gave rise to specific material effects, because – although the nvisioned future symbiotic relationship of technological innovation, competition and climate protection in the global agrofuels markets is by no means certain to materialize – it already puts the market into place, on the basis of which investments in land and raw materials are being transacted and international trade in the sector is being expanded (Lamers *et al.* 2014).

The European agrofuels sector is thus being fitted into an 'emergent global biofuels complex' (Borras *et al.* 2010: 577). This implies a specific 'neoliberalization of nature', by which previously extensively or non-utilized natural resources become commoditized and adapted to a global, more strongly integrated agro-energy sector.

Borras *et al.*, however, note that 'a *global* biofuel complex is still incipient, as neomercantilist practices (protected subsidized national biofuel sectors, with off-shore complements managed through tariff structures) continue alongside emergent globalizing recombinant corporate/state arrangements' (ibid). The European agrofuel policy – particularly the RED – is a prime example of this, since we are not dealing here with a 'pure, unadulterated' form of neoliberalism, but rather with the contours of a specific 'actually existing neoliberalism' (Brenner and Theodore 2002), in which neomercantilist institutions such as import restrictions are still ever present.[5] Thus EU publications always emphasize that while international trade is certainly welcomed, at the same time it is important to 'pursue a balanced approach' regarding the relation between imports and exports 'in order meet the interests of domestic producers and EU trading partners alike' (CEC 2006: 14).

Neoliberalization is to be understood here as a discursive process of mediation, in which neoliberal interpretive schemes manifest themselves in actual political measures but are mediated and 'watered down' by context-specific, historically-evolved institutions and practices (like the agricultural protectionism of the EU in this case). The 'actually existing neoliberalism' of European agrofuel policy is thus not the result of a uniform process of discourse institutionalization in which one discourse is 'victorious' over another. Instead it is much more what Maarten Hajer refers to as the dominance of a discourse coalition – a coalition 'formed among actors (that might perceive their position and interest according to widely different discourses) that, for various reasons (!) are attracted to a specific (set of) story-lines' (Hajer 1995: 65).

In the case of EU agrofuel policy, a comprehensive discourse coalition arises that is kept together by a broad set of storylines, including international trade, competitiveness, efficiency, technological innovation, climate protection and sustainability, as well as rural development, value creation and employment. This discourse coalition consists of, and is reproduced by, a number of diverse actors, since the political object, agrofuels, as construed by these very storylines, is attractive for many stakeholders. These include various industry factions, ranging from the automobile industry, which sees an option in agrofuels for the continuance of private transport based on the internal combustion engine, to the mineral oil, agricultural and biotech industries, which see new fields of business in a global agrofuels market, and the single farmer, who views this market as a place to expand sales for his products. At the same time, the more compliant NGOs see in agrofuels a possibility to gain influence and establish and promote their ideas and initiatives in the area of sustainability standards and certification.

The various criticisms of agrofuels that began to emerge in the mid-2000s have, as a result of this broad discourse coalition – neoliberalized (but not made wholly neoliberal) by the neoliberally re-aligned win–win narrative – been either appropriated or largely marginalized in European agrofuel policy. Thus the less radical, more strongly conciliatory criticism, which admits to a fundamentally positive aspect in agrofuels if they can be sustainably produced and utilized, has already, with the support of neoliberal environmental management tools such as sustainability criteria and certification, been made part of this dominant discourse coalition.[6] At the same time, the future-oriented imaginary of symbiotic interaction between technological innovation and competition in global markets – being based on a circular line of reasoning and thus irrefutable – serves to immunize the neoliberally re-aligned win–win narrative against criticism of social and environmental problems, since – so the story goes – this symbiotic interplay will finally solve the social and environmental problems associated with agrofuels. Furthermore, the multi-interpretability of the underlying storylines makes politically effective articulation of more radical criticism of, for example, patterns of agro-industrial production or Western consumption patterns, more difficult or even impossible.

## Notes

1 This chapter draws on analyses carried out in the course of the project 'Fair Fuels?', which was funded between 2009 and 2014 by the German Federal Ministry of Education and Research in the context of its Social-Ecological Research programme (research grant number: FKZ 01UU0905). I would like to thank the editors of this volume for their detailed and constructive comments on an earlier draft of this chapter.

2 These documents include Commission communications and reports, proposals and drafts of EU directives, and regulations, minutes and summaries of parliamentary plenary and Council meetings, reports and recommendations of parliamentary committees, Council positions, reports and minutes of public hearings and conferences, as well as statements and communications by the civil society and industry actors involved (such as press releases, position papers and contributions to the EU public consultation on 'Biofuel issues in the new legislation on the promotion of renewable energy', in May and June 2007).

3 The set of sustainability criteria ultimately adopted include a minimum $CO_2$ reduction value (to rise from 35 per cent to 60 per cent by 2018), a ban on the extraction of raw materials from carbon- and biodiversity-rich areas, and the obligation to comply with 'Cross Compliance' regulations of the EU Common Agricultural Policy (European Union 2009).

4 The Biofuels Research Advisory Council (BIOFRAC) was constituted in 2004 by leading companies of the oil and automotive industry and the biotechnology sector, the European farmers' union, agrofuel associations and research institutions. In 2006, it was transformed into the European Biofuels Technology Platform (EBTP).

5 The term 'neomercantilism' is used to describe national economic strategies that, in effect, strive to achieve export surpluses in the trade of goods and services and often go along with a protectionist trade policy.

6 This becomes clear, for instance, in the ongoing political debate about ILUC, which is ultimately a struggle within this discourse coalition, in which a neoliberal competitiveness and technoscience discourse is reproduced by environmental NGOs, and questions about the general pros and cons of agrofuels are no longer discussed (e.g. Palmer 2012; Levidow 2013).

## References

BASF (2007) 'Contribution to the CEC's public consultation on biofuel issues in the new legislation on the promotion of renewable energy (May-June 2007)' (http://ec.europa.eu/energy/res/consultation/doc/2007_06_04_biofuels/industry_private/basf_en.pdf, last access: 19 February 2014).

BIOFRAC (2006) *Biofuels in the European Union. A vision for 2030 and beyond. Final report of the Biofuels Research Advisory Council*, Brussels: Commission of the European Communities (http://ec.europa.eu/research/energy/pdf/biofuels_vision_2030_en.pdf, last access: 11 February 2014).

Borras, Saturnino M., Philip McMichael and Ian Scoones (2010) 'The politics of biofuels, land and agrarian change: editors' introduction', *The Journal of Peasant Studies* 37(4), 575–592.

Bowyer, Catherine (2011) *Anticipated indirect land use change associated with expanded use of biofuels and bioliquids in the EU – an analysis of the National Renewable Energy Action Plans*, Brussels: IEEP (www.ieep.eu/assets/786/Analysis_of_ILUC_Based_on_the_National_Renewable_Energy_Action_Plans.pdf, last access: 19 November 2013).

Brenner, Neil and Nik Theodore (2002) 'Cities and the geographies of "actually existing neoliberalism"', *Antipode* 34(3), 349–379.

Bundesrat (1992) *Empfehlungen der Ausschüsse zum Vorschlag für eine Richtlinie des Rates über den Verbrauchsteuersatz auf Kraftstoffe aus landwirtschaftlichen Rohstoffen. Drucksache 193/1/92 vom 15.06.92*, Bonn: Bundesrat. (http://dipbt.bundestag.de/doc/brd/1992/D193+1+92.pdf, last access: 19 February 2014).

Castree, Noel (2010) 'Neoliberalism and the biophysical environment 2: theorising the neoliberalisation of nature', *Geography Compass* 4(12), 1734–1746.

CEC (1992) *Vorschlag für eine Richtlinie des Rates über den Verbrauchsteuersatz auf Kraftstoffe aus landwirtschaftlichen Rohstoffen. KOM(92) 36 endg.*, Brussels: Commission of the European Communities (http://dipbt.bundestag.de/doc/brd/1992/D193+92.pdf, last access: 16 March 2014).

CEC (2006) *An EU Strategy for Biofuels. COM(2006) 34 final*, Brussels: Commission of the European Communities (http://eur-lex.europa.eu/LexUriServ/site/en/com/2006/com2006_0034en01.pdf, last access: 5 February 2014).

CEC (2007) *Biofuels progress report. Report on the progress made in the use of biofuels and other renewable fuels in the member states of the European Union. COM(2006) 845 final*, Brussels: Commission of the European Communities (http://eur-lex.europa.eu/LexUriServ/site/en/com/2006/com2006_0845en01.pdf, last access: 13 February 2014).

CEC (2008): 'Proposal for a directive of the European Parliament and of the Council on the promotion of the use of energy from renewable sources. COM(2008) 19 final' (http://eur-lex.europa.eu/LexUriServ/LexUriServ.do?uri=COM:2008:0019:FIN:EN:PDF, last access: 11 November 2013).

European Union (2003) 'Directive 2003/30/EC of the European Parliament and of the Council of 8 May 2003 on the promotion of the use of biofuels or other renewable fuels for transport', *OJ L* 123, 42–46.

European Union (2009) 'Directive 2009/28/EC of the European Parliament and of the Council of 23 April 2009 on the promotion of the use of energy from renewable sources and amending and subsequently repealing Directives 2001/77/EC and 2003/30/EC (Text with EEA relevance)', *OJ L* 140, 16–62.

Franco, Jennifer, Les Levidow, David Fig, Lucia Goldfarb, Mireille Hönicke and Maria Luisa Mendonça (2010) 'Assumptions in the European Union biofuels policy: frictions with experiences in Germany, Brazil and Mozambique', *The Journal of Peasant Studies* 37(4), 661–698.

Hajer, Maarten A. (1995) *The Politics of Environmental Discourse. Ecological Modernization and the Policy Process*. Oxford: Oxford University Press.

Hajer, Maarten A. (1997) "Ökologische Modernisierung als Sprachspiel. Eine institutionell-konstruktivistische Perspektive zum Umweltdiskurs und zum institutionellen Wandel, *Soziale Welt* 48(2), 107–132.

Heynen, Nik and Paul Robbins (2005) 'The neoliberalization of nature: governance, privatization, enclosure and valuation', *Capitalism Nature Socialism* 16(1), 5–8.

Higgins, Vaughan, Jacqui Dibden and Chris Cocklin (2008) 'Neoliberalism and natural resource management: agri-environmental standards and the governing of farming practices', *Geoforum* 39(5), 1776–1785.

Jänicke, Martin (2000) *Ecological modernization: innovation and diffusion of policy and technology. FFU-report 00–08.* Berlin: Freie Universität (http://userpage.fu-berlin.de/ffu/download/rep_00–08.PDF, last access: 16 March 2014).

Keller, Reiner (2008) 'Diskurse und Dispositive analysieren. Die Wissenssoziologische Diskursanalyse als Beitrag zu einer wissensanalytischen Profilierung der Diskursforschung', *Historical Social Research/Historische Sozialforschung* 33(1), 73–107.

Keller, Reiner (2011) 'The sociology of knowledge approach to discourse (SKAD)', *Human Studies* 34(1), 43–65.

Kuchler, Magdalena and Björn-Ola Linnér (2012) 'Challenging the food vs. fuel dilemma: genealogical analysis of the biofuel discourse pursued by international organizations', *Food Policy* 37(5), 581–588.

Lamers, Patrick, Frank Rosillo-Calle, Luc Pelkmans and Carlo Hamelinck (2014) 'Developments in international liquid biofuel trade', in Junginger, Martin, Chun Sheng Goh and André Faaij (eds) *International Bioenergy Trade. History, Status & Outlook on Securing Sustainable Bioenergy Supply, Demand and Markets*. Dordrecht: Springer, 17–40.

Levidow, Les (2013) 'EU criteria for sustainable biofuels: Accounting for carbon, depoliticising plunder', *Geoforum* 44, 211–223.

Levidow, Les, Theo Papaioannou and Kean Birch (2012) 'Neoliberalising technoscience and environment: EU policy for competitive, sustainable biofuels', in Pellizoni, Luigi and Marja Ylönen (eds) *Neoliberalism and Technoscience: Critical Assessments.* Farnham: Ashgate, 159–186.

Mandelson, Peter (2007) 'The Biofuel Challenge', speech given at the International Conference on Biofuels, 5–6 July 2007, Brussels (http://eeas.europa.eu/energy/events/biofuels/sessions/s2_04_mandelson_biofuel.pdf, last access: 16 March 2014).

McHenry, Helen (1996) 'Farming and environmental discourses: a study of the depiction of environmental issues in a German farming newspaper', *Journal of Rural Studies* 12(4), 375–386.

Mert, Aysem (2013) 'Discursive interplay and co-constitution: carbonification of environmental discourses', in Methmann, Chris, Delf Rothe and Benjamin Stephan (eds) *Interpretive Approaches to Global Climate Governance: (De)constructing the Greenhouse.* Milton Park: Routledge, 23–41.

Palmer, James (2012) 'Risk governance in an age of wicked problems: lessons from the European approach to indirect land-use change', *Journal of Risk Research* 15(5), 495–513.

Parr, Adrian (2012) *The Wrath of Capital. Neoliberalism and Climate Change Politics.* New York: Columbia University Press.

Pellizoni, Luigi and Maria Ylönen (2012) 'Introduction', in Pellizoni, Luigi and Marja Ylönen (eds) *Neoliberalism and Technoscience: Critical Assessments.* Farnham: Ashgate, 1–24.

Potter, Clive and Mark Tilzey (2005) 'Agricultural policy discourses in the European post-Fordist transition: neoliberalism, neomercantilism and multifunctionality', *Progress in Human Geography* 29(5), 581–600.

Rowlands, Ian H. (2005) 'Global climate change and renewable energy: exploring the links', in Lauber, Volkmar (ed.) *Switching to Renewable Power. A Framework For the 21st Century.* London: Earthscan, 62–82.

Sharman, Amelia and John Holmes (2010) 'Evidence-based policy or policy-based evidence gathering? Biofuels, the EU and the 10% target', *Environmental Policy and Governance* 20(5), 309–321.

Strauss, Anselm and Juliet Corbin (1996) *Grounded Theory: Grundlagen Qualitativer Sozialforschung.* Weinheim: Beltz.

van Apeldoorn, Bastiaan (2002) *Transnational Capitalism and the Struggle over European Integration.* London/New York: Routledge.

Watkin, Tim (2008) 'The biofuel solution', *Guardian*, 4 June 2008 (www.theguardian.com/commentisfree/2008/jun/04/thebiofuelsolution, last access: 24 March 2014).

Widengard, Marie (2011) 'Biofuels governance: a matter of discursive and actor intermesh', in: Matondi, Prosper B., Kjell Havnevik and Atakilte Beyene (eds) *Biofuels, Land Grabbing and Food Security in Africa.* London, New York: Zed Books, 44–59.

# Index

Page numbers in *italics* denote tables.